W9-AFS-143

TEACHER THINKING TWENTY YEARS ON

This book is dedicated to the founders of ISATT – The International Study Association on Teachers and Teaching. These are the forward thinkers on whose shoulders we stand as we think, write, discuss and together look beyond national boundaries. This book is also dedicated to the teachers and learners worldwide who have allowed ISATT members access to their professional worlds and their private thoughts and actions in it over the last twenty years.

Teacher Thinking Twenty Years on: Revisiting Persisting Problems and Advances in Education

EDITED BY

Michael Kompf
Brock University, Canada

and

Pam M. Denicolo
University of Reading, UK

LISSE ABINGDON EXTON (PA) TOKYO

LB2840.T433 2003+ c.3

Library of Congress Cataloging-in-Publication Data

<hier CIP gegevens>

Printed in the Netherlands by GPB Gorter b.v., Steenwijk, The Netherlands

Copyright © 2003 Swets & Zeitlinger B.V., Lisse, The Netherlands

All rights reserved. No part of this publication or the information contained herein may be reproduced, stored in a retrieval system, or transmitted in any form or by any means, electronic, mechanical, by photocopying, recording or otherwise, without written prior permission from the publishers.

Although all care is taken to ensure the integrity and quality of this publication and the information herein, no responsibility is assumed by the publishers nor the author for any damage to property or persons as a result of operation or use of this publication and/or the information contained herein.

Published by: Swets & Zeitlinger Publishers
www.szp.swets.nl

ISBN 90 265 1954 0

Contents

Acknowledgements

From the outset of this project, we have relied on the services and good nature of those around us. For retyping of the original manuscripts, thanks are owed to Karen Phillips and Lesa Mansfield. Rahul Kumar is thanked for his electronic dexterity in preparing scanned materials. For final preparations of the manuscript, the assistance of Shauna Gupta was most valuable. Thanks also to Ken Kehl for the preparation of the index. To the editorial staff at Swets Zeitlinger thanks are offered for their patience and their trust in the interest in, and value of, the work produced by the ISATT community. We acknowledge the generosity of Blackwell Publishing and Teachers College Record for allowing us to include the previously published version of Margret Buchmann's chapter. We also acknowledge past members, the current membership and future generations of ISATT members and the inquiring community at large. Any proceeds generated by this volume will be administered by the ISATT executive to assist student travel to conferences. A special acknowledgement to colleagues no longer with us: the dream never dies, just the dreamer. To our families near and far the final thanks are owed because, without each person's support, patience and belief, the freedom of time and spirit to work and play in scholarly fields would not be possible.

Michael Kompf, PhD
Brock University, Canada

Pam M. Denicolo PhD
University of Reading, UK

October 2002

Introduction

The History of ISATT

ISATT first met in Tilburg, the Netherlands in October 1983. The first meeting was initiated by Rob Halkes, John Olson, Alan Brown, Christopher Clark, Erik De Corte and William Reid. This gathering and the symposium that took place marked the beginning of a paradigm shift that welcomed a range of alternative approaches to gaining deeper and more meaningful understandings of the processes of teaching and learning. A selection of the papers presented at that meeting are included in this volume.

The second ISATT meeting was also held in Tilburg in 1985 with researchers and thinkers from 12 countries in attendance. Contributions from these authors and presenters make up the balance of this volume. Subsequent ISATT meetings have taken place in many wonderful places with gracious hosts, hospitable surroundings and the stimulating environment that scholarly exchange produces. Venues have included Leuven (Belgium), Surrey (England), Nottingham (England), Gothenburg (Sweden), St. Catharines (Canada), Kiel (Germany), Dublin (Ireland), Faro (Portugal) and the upcoming meeting in Leiden (Netherlands).

The acronym ISATT referred to The International Study Association on Teacher Thinking until broadened in scope by changing a couple words to The International Study Association on Teachers and Teaching, as it is currently known. During the debate over the name change, one sage colleague commented that "it doesn't matter what we call this group, it's still ISATT". Those comments ring true because ISATT is not about names, labels or categories, ISATT is about the academic spirit of unfettered inquiry and the generation of possibilities. Many organizations respond to shifting trends and educational change. ISATT members anticipate and write about these changes in some cases decades before the broader learning community comes to full grasp. Evidence supporting this claim is contained in this volume through works that have endured in importance, insight and applicability.

ISATT's history, if examined in terms of the lengthy list of luminaries who have given keynote addresses, or the significant numbers of new scholars now leading their respective fields, or those who sought and received academic sustenance from attending meetings, would easily fill pages. As another colleague commented, "When you attend an ISATT conference you meet your bibliography".

Further insights into ISATT's history are available in the Forewords of most of the books we have produced. Yet the most remarkable accounts of ISATT's history can be had from the stories members tell of memorable incidents that have contributed to and shaped their lives and careers in the academy and the collegium.

Purpose of this book

This book was stimulated into being through discussions between the 'old stalwarts' of ISATT who noted with regret that, while the books from the first two conferences were no longer available for purchase, their own copies were so well thumbed by their doctoral students as to be in a sorry state of decay. Our re- examinations of the contents of these books reinforced those reflections on their value to current and future researchers interested in teachers and teaching.

These papers presented as book chapters have historical value, marking as they do both a change in topic focus and a revolution in research practice. Although the purpose of research in this field remained the improvement of teachers' practice and the educational experience of pupils/students, the importance of teachers' and students' views about what was actually going on in the classroom became increasingly evident as the outside observers' perspectives and interpretations diminished. Authenticity, with 'thick' description, became the goal that superseded concerns about reliability and generalisability. When the books were originally compiled, although experimental designs and the collection of quantitative data were yet pre-eminent in the field of research in education, these were mediated and enriched by studies that included the collection of qualitative data. Such studies strove to illustrate, and thence understand, what teachers and students were thinking about when they were involved in education activities, what their intentions and emotional reactions were, how they predicted and reflected on action and why they responded in particular ways.

These chapters have a current practical value in that they provide a large reference resource for, and a wide range of examples of, both topics and methods of research. For new researchers in particular this is useful for alerting them to previous foundational studies and to explorations with techniques that revealed strengths, limitations and opportunities for innovation.

Value for the future can be found in the texts that note lacunae in research and as-yet-unresolved issues, not all of which have been subsequently addressed. Further, since the chapters derive from research conducted in a variety of national contexts, revealing some evidence of common constraints and opportunities impinging on education at the time, questions are stimulated about what has changed and what has remained the same in the interim.

These questions led us, in the process of requesting permission of authors to re-present these chapters, to ask them also for a short retrospective view on their contribution. We asked them to consider, from the point of view of recent and current work, how well the content stood up to the test of time and what significance the issues raised still have, if any. Many responded with some initial surprise. They wondered themselves whether others would be interested in what was now historical research but, on revisiting it, they found that some issues were still prevalent in current practice while others recognised that the chapter had sown the seeds of a lifetime of continuing research. We have added the commentaries we received to the ends of the respective chapters.

Points to Ponder

Whilst reading these chapters against a background of new dominant pressures on education caused by changes in political and cultural imperatives and developments in technology, we suggest that you too consider how well these ideas, theories, methods of research and results from the past hold up in the current context. Consider: how they relate to your own practice as a teacher, researcher and learner; what implications are revealed for teacher training, continuing professional development and research on teachers' thinking and practice.

Section A

Teacher Thinking: Practical Theories

The first section contains nine chapters that deal with teacher thinking and the ideas of practical theories. In Chapter 1 Teachers' Teaching Criteria, Rob Halkes and Rien Deijkers report on their ongoing study of teachers' teaching criteria. They approach these "personal subjective values that a teacher tries to pursue or keep constant while teaching…", with several research techniques. They note the importance of the criteria of "control" and "work ethos" to teachers. Based on a literature study they describe seven categories of teachers' teaching criteria that do not complement each other within actual teaching situations but rather confront the teacher with dilemmas.

In Chapter 2, Alan Brown discusses Professional Literacy, Resourcefulness and What Makes Teaching Interesting by asking teachers what they found satisfying, rewarding or stimulating. The substantiveness of their declarations, and the thought processes that led to them, produced many interesting findings. In Chapter 3 Rainer Bromme and Gudrun Dobslaw's: Teachers' Instructional Quality and Their Explanation of Students' Understanding describes models of instructional quality and stresses the selection of tasks suitable for students. They analyse the cognitive tools used by teachers when thinking about tasks and students. Chapter 4 How Do Teachers Think About Their Craft? is contributed by Sally Brown and Donald McIntyre. They explore the integrated knowing, thinking and action which are assumed to characterise teachers' professional craft knowledge and offer an account of the research process and preliminary steps for a conceptual framework that reflects the ways in which experienced teachers think about the strengths of their own teaching.

John Olson provides Chapter 5. Information Technology and Teacher Routines: Learning from the Microcomputer examines four cases of using computers as a new subject and as a teaching aid. Issues of computer awareness and literacy describe teachers' experimentation with the "new" medium of instruction. An action research approach is argued to enable teachers to learn more about the values and limits in practice and practices. In Chapters 6 Planning and Thinking in Junior School Writing Lessons: an Exploratory Study by James Calderhead, and Chapter 8 Teachers' Thinking – Intentions and Practice: An Action Research Perspective by Christopher Day, each author discusses how research on teacher thinking might help teachers deal with persisting problems they face in the context of action-research. They reconstruct teachers' thought or intentions and compare these with the actual teaching activities of the teachers. Thus, the teachers involved in action research are confronted with both kinds of information. Evolving discrepancies are discussed as a basis for making decisions about improvement of teaching. Calderhead adds that it is essential to probe students' thinking about perception of teachers' activities as well if these change efforts are to succeed.

In Chapter 7, Categories in Teacher Planning, Harm Tillema tested several systems of description for the characterization of teachers' planning statements using loglinear analysis and found that teachers are very conscious of their intentions while planning at the specific and concrete level. Relevant categories for the description of teachers' planning statements are subject matter structure and sequence and prior knowledge and motivation of students. Chapter 9 - What Makes Teachers Tick? Considering the Routines of Teaching provided by John Olson - suggests that teachers might profit from thinking about their actions and the knowledge-in-action that exists in their routine activities. He proposes that routines may be the highly polished results of previous teacher thinking in the context of feedback and evaluation of activities. Further, he considers the notion that the intention of research should be concerned with making articulate what is inarticulate or tacit to the teachers themselves as might be evident in thought and practice.

Chapter 1
Teachers' Teaching Criteria

Rob Halkes and Rien Deijkers

Summary

On the basis of a teachers' thinking perspective we explore teachers' teaching criteria: personal subjective values a teacher tries to pursue or keep constant while teaching. These may have a divergent character such as: personal action rules, principles, and images of teaching-learning situations, personal intentions, and the like. In two initial studies we tentatively confirmed the possibility to distinguish such criteria and their relationship with teaching activities. We describe, furthermore, the research project aimed at determining these teaching criteria, individual differences between teachers concerning their criteria and the relationship with experience and subject matter specialization.

Introduction

The ongoing study we report here, starts from a teachers' thinking perspective to find out why teaching processes go on as they do. We deem information on this to be necessary for knowing how teachers influence learning processes. The perspective's basic assumption as for us, is the notion that teaching behaviour originates from the person: he/she initiates actions, proceeds with them and eventually stops them. The phenomenon of overt behaviour cannot, by definition, be controlled by itself: the person does this. Now, whatever part of the person one might feel to be responsible for the initiating, controlling, etc., of behaviour, 'mind' must have a central function here. How mind relates to action is a troublesome question; how conscious 'mind' must be of the initiating-, carrying through-, or controlling process itself, is a second problem. Whether cognitions, or feelings, or 'states' of mind have a dominant influence over behaviour is still another one. On the other hand, the possible impact of external influences over the person on his behaviour is not specifically known too. Anyway, being pragmatic, we started from a 'cognitive action' framework to be able, in the first place, to study why teachers behave, act, as they do.

This cognitive action framework has some basic notions like the one mentioned above. Furthermore, it conceptualizes teaching as existing of basic units of action, each consisting of two components: an overt behavioral one and a covert 'mind' component (cf. De Corte, 1982). As we stated, above, the interrelation between the two is not known. Like in a reflex behaviour might condition cognition on the basis of an evaluative process. In rationally prepared actions behaviour might be the consequence of cognition. But what to say of the prepared for action of an athlete: he/she prepares alright, but will not be possible to cognitize in detail the flow of movements needed for a winning act, nor might be able to answer the question as to how he/she did it in general. A lot is still to be known here.

What boundaries in fact exist between such cognitive actions, or, what the unity of one specific cognitive action is, is determined, as far as we see it, by the subjective perception of the person him/herself. The first sentence of a lesson might be one to a student teacher. It might be the whole introduction of the lesson to an experienced teacher. The actual extent of a 'cognitive action' will have to do with the development of routines as new entities of cognitive actions into which former ones are 'absorbed', and with the perception of eventual problems with or within these actions. Maybe these subjective perceived cognitive actions are what teachers would call their teaching tasks.

Concerning teaching, we tune to the circumscription that teaching is intended action in a complex situation that to the teacher is as rational as possible, is meaningful, sensible, according to his/her personal intentions. In order to teach, teachers and student teachers will construct some subjective or 'naïf' theory about what to do in the anticipated (pre-active) and perceived (inter-active) teaching-learning situation. On this point we refer to several theoretical orientations or models of the teaching process, for instance: Argyris and Schön's theories of action (1974), Kelly's Personal Construct Psychology (1955), Elbaz's practical knowledge model for teaching (1983), Miller, Galanter and Pribram's Plan theory (1960); as well as any kind of information processing model of teaching, cf. Clark and Peterson (1984), Shavelson and Stern (1981), Huber and Mandl (1984), etc. etc. It is interesting to see that in most of these studies attention is paid to aspects or objects (e.g. the learner, content, strategies) of the thinking *processes* of teachers (cf. Clark and Peterson, 1984) and not so much as to the *content* of teachers' thinking. In other words: what are teachers' intentions; what are their interpretations of events; what value do they attach to certain kinds of student teacher relations; what personal needs do they have; what rules do they follow in their actions; and so on. We conceptualize what might be known as 'practical principles', 'personal intentions', 'images', 'plans', 'attitudes', 'imperative cognitions', and so on, provisionally, as subjective teaching criteria. They all come to a notion of what to do, what to avoid or what to strive for in teaching. They are personal values a teacher tries to pursue or keep constant while teaching. They are not specifically bound to a single teaching situation, but more general functioning as it where as a background for actual teaching acts and action decisions. A teacher will use them in preactive and interactive teaching situations, consciously – or unconsciously, to select teaching acts, to control their proceedings as well as to evaluate their effects.

Initial Studies

On the basis of two studies we got some insight into the phenomenon of subjective teaching criteria and their impact. The first one was a study of classroom discipline in a new primary school in Tilburg.[1] The staff of this school asked us for support in

[1] See for a detailed description of this study: Rijkers-Mikkers, R. (1983). *Ordeverstoringen in de klas*, een onderzoek in het kader van schoolwerkplanontwikkeling, Tilburg University,

their development of the school and classrooms' atmosphere, the pedagogical climate. In consultation with them, we thought the best way to do this was to study teacher's reactions to possible disturbances of classroom discipline. The instance of a disturbance of discipline, such as the sudden shouting of a child, falling of furniture, a child going to the loo unauthorized, fighting, slamming of the door, etc., was conceptualized by us in a four step model of analysis: (1) *observation:* the teacher notices an occurrence of what might be interpreted as a disturbance of discipline; (2) *evaluation* of this event, in which the teacher diagnoses cause, gravity, impact, etc., and decides whether a reaction is needed and what this will be; (3) *implementation* of this reaction and (4) *evaluation* of the effect of the reaction (cf. Miller et al's TOTE-model, 1960).

The research question most relevant here was: what events occurring during teaching, are mentioned by the teachers as discipline disturbances and what criteria do they use in defining an event as such? All teachers of the school (10) were observed in three different teaching situations: instruction, seat-work and a more 'open' situation, e.g. handicraft, music, or social sciences 'project work'. The observer made notes on events that he/she would depict as a possible disturbance. Following each observation period an interview took place in which the teacher was asked to mention occurrences of discipline disturbances that were discussed as to the reason for this event being such a disturbance. Then the observer mentioned his/her noticed events and, again, these were discussed as possible disturbances. The interviews were tape-recorded. The qualitative analysis of observation sheets and interviews was based on Glaser and Strauss (1967). We cannot go into detailing the results here, but several notions concerning teachers' teaching criteria were tentatively found:

- Concerning the notion of classroom discipline, it was found that these teachers did have different images of classroom discipline for different teaching situations: for instance, 'discipline' during instruction was more 'strict' then during seat-work; also for distinguished periods within these situations; for instance, the climate during the beginning of seat-work lessons when children entered and gradually sat down, was far more permissive than during actual seat-work. It appeared to be that teachers did use their different images of periods within the lessons as a reason for defining the same events as disturbing or not e.g. a child walking about through the classroom. So, in relation to these images teachers must have different sets of criteria applying to the related situations.
- Teachers did have a definite image for student activity or 'business': "as long as they are working, it's o.k." they typically said. Actually, students were distinguished as 'hard workers' or 'non worker';
- 'Noise' was a very delicate kind of criterion, because it had to do with a subjective average level of noise which the teacher tolerated;
- 'Prevention' was another criterion for teacher's reaction to 'not-yet-disturbances', preventing excrescence of an event into a serious disturbance, e.g.: a conflict into fighting, talking into shouting.

---

unpublished doctoral thesis; Halkes, R. & Rijkers-Mikkers, R. (in preparation), *Disturbances of classroom discipline, obstruction of teachers' criteria.*

Based on individual differences between the teachers as to the preponderance of certain criteria over others, we concluded that teachers do have a set of subjective teaching criteria that could explain for individual teaching characteristics.

Related to this was the study of student teachers during their teacher-trainee lessons in secondary education, intended to find categories for the description of student-teachers' thinking characteristics.[2] Again, observations of 20 student teachers for two or three lessons, one of which was tape-recorded, were followed by interviews. Observation and interview focused on posture, gesticulation, voice, use of material and ways of dealing with influence. The study was mainly based on Erving Goffman's (1969) theory on the presentation of self in everyday life. His categories of 'front'- and 'back-stage'-behaviour, 'decorum', etc., were very helpful in the qualitative analysis (Glaser and Strauss, 1967) and tentative explanation of student teachers' teaching. The results showed for instance, that Goffman's idea of 'making work', the way in which someone tends to show by his behaviour that he 'works' according to perceived standards, might explain for the fact that these student-teachers did not seem to be able to deal with silence and for quasi-actions, such as constantly moving around with a piece of paper at hand, or for not being able to 'rest' when students were doing seat-work on the basis of information on posture, gesticulation, use of space and voice, it could be hypothesized that student-teachers tend to show, unconsciously, that they are feeling not to meet 'general' standards of teaching as perceived by themselves. The drops back of voice, moving to the side of the classroom, etc. are examples of this. Whether a teacher meets his own subjective teaching criteria, might show in his overt behaviour, we concluded.

The Study of Teachers' Teaching Criteria

Based on the aforementioned ideas we conduct a research project to explore and determine what subjective teaching criteria teachers have, how they relate to teaching acts and what individual differences concerning these criteria might account for teaching characteristics.

The design of the research project is based on a principle of a large-spectrum approach (De Corte, 1979) that states that several methods and techniques should be used in studying the same object in order to overcome technique specific artifacts. Besides an ongoing literature study we proceed for the time being with three research techniques:

- A questionnaire, indicating teachers' adherence to several statements which are based on a operationalization of tentative criteria;
- An interview based on Kelly's Personal Repertory Grid technique (cf. e.g. Bannister and Fransella, 1977) to (re) construct a teacher's teaching criteria;
- A stimulated-recall technique to study teacher's interactive thoughts and decisions to interpret teaching criteria in action.

---

[2] See for a detailed description: Van Wezel-Krijnen, N. (1983). *Op de drempel van het leraarschap*, een studie naar houding en gedrag van aspirantleraren tijdens oefenlessen. Tilburg University: unpublished doctoral thesis; Halkes, R. & Van Wezel-Krijnen, N. (in preparation). *On the threshold of teaching, how subjective teacher-trainee's teaching criteria show.*

By the choice of these techniques, we hope to cover a dimension from general, say, attitude-like criteria to specific action-criteria-in-use. Here we will report results of the literature and questionnaire study, which will be discussed in the framework of some tentative findings from the interviews.

Besides the general research question stated at the beginning of this section, we also wanted to find out how discipline or subject-matter specialization and years of experience relate to teaching criteria. For the time being we restricted the research to the secondary level.

Literature

Literature covered

To get a hold on possible teachers' teaching criteria we studied information on teachers' thinking in a widest sense: we entered literature from three perspectives:

- Innovation and curriculum implementation, e.g.: Fuller, 1969; Giacquinta, 1975; Hall and Rutherford, 1976; Olson, 1980, 1981, 1982; Ponder and Doyle, 1977;
- Attitudes towards education, e.g.: Kerlinger, 1968; Krampen, 1979;
- Wehling and Charters, 1969; Zeiher, 1973;
- Teachers' thought, judgements, decisions and behaviour, e.g.: Clark and Yinger, 1980; Hofer, 1981; Huber and Mandl, 1982; Jackson, 1968; Leithwood and MacDonald, 1981; Lortie, 1975; Marland, 1977; Munby, 1983; Shavelson and Stern, 1981; Shulman, 1981; etc. etc.

Most of these studies did provide us with a number of relevant criteria sometimes rather accurately formulated in preferences, behaviour, or tendencies of avoidance (things teachers try to steer clear of). On the basis of these descriptions of criteria, sometimes being synonyms or sort-like descriptions of the same issues, we induced (categories of) criteria. For instance: Munby (1983) reports the interest of teachers for differences between students, Bussis et al (1976) talk about the relevancy of taking students' need and feelings into account, Kerlinger (1967) finds that teachers attach great importance to needs and interests of students, while Wehling and Charters (1969) make mention of teachers' emphasizing students' needs and interests. These were subsumed under the category: 'adaptation to individual characteristics of and differences between students' (see below). In this way we formulated 7 (categories of) teaching criteria, based on 112 formulations of possible criteria as formulated in literature.

Teachers' teaching criteria from literature

These are the following:

(1) Student activity and participation, which relate to the extent in which a teacher wants her/his students to be involved in his/her teaching and to participate in classroom activities, on the one hand, and, on the other, a teacher does not want ongoing activities to be interrupted or disturbed: 'flow of activity' (cf. Shavelson and Stern, 1981);

(2) Adaptation to individual characteristics of and differences between students, which is mentioned in one way or another in nearly all studies, aspects of which are: needs, feelings, interests, ability viewpoints, and so on;

(3) Subject matter; this criterion depends on the way in which a teacher emphasizes subject matter 'transmission' as 'storing knowledge', and is related to:
(4) Pedagogical aims or educational objectives the teacher adheres to, like attitude development, social skills training, etc.;
(5) The nature of teacher-student and student-student interaction that a teacher prefers or tolerates, characterized by, for instance, students' obedience or compliance (cf. Lortie, 1975), discipline (Munby, 1982) or teacher's authority and prestige (e.g. 'emotional disengagement', cf. Wehling, 1969);
(6) Teachers' needs, a rather large category of criteria in which the following could be distinguished:
    a. Need for *certainty*, gained by 'practical and concrete support' (such as curriculum guidelines, formal tests and regulations on grading, feedback showing students' competence and results and teachers' own professional and subject matter knowledge) but also by the use of personally preferred teaching activities, and 'influence' (cf. Olson, 1981);
    b. Need for *autonomy* and independence of persons (e.g.: colleagues, parents) and curriculum (Cf. Leithwood and MacDonald, 1981);
    c. Need for *respect* and esteem from colleagues, parents and students;
(7) Scientific orientation that concerns teacher's emphasis on 'basic' knowledge and structure of discipline on one hand, and the pursuit of 'objectivity', on the other.

The Questionnaire Study

Development and try-out of the questionnaire

In this study we wanted to find empirical support for these categories of teachers' teaching criteria and to explore individual differences as well as the way they relate with teachers experience and subject-matter (discipline) specialization. The questionnaire consists of 65 statements to be rated on a 5-point Likert-scale, and 10 teaching-learning activities to be rated on a 5-point scale indicating personal degree of difficulty in using such an activity in class. Some general questions on age, experience, subject-matter specialization, tasks or functions at school level, etc., are added too.

Criteria were operationalized into items with the aid of the criteria descriptions mentioned above, on an intermediate level and a pool of 203 ready-made statements on a concrete level, which related to these descriptions. These ready-made items were also collected from literature. In the development of the questionnaire, we preferably used these, but we also transferred respondents' statements from the two initial studies and formulated items by ourselves.

A preliminary draft of the questionnaire was evaluated by about 10 colleagues on clarity of formulation, concept validity (category representation), parsimony of statements' wording, etc.

The first draft was presented as a try-out to 25 secondary school teachers who volunteered to participate. Some of them completed the questionnaire in the presence of the authors in order to discuss eventual problems; others were asked to state problems in the margin. The data from this try-out were rather roughly statistically analysed on discriminatory power of items.

Information thus collected aided in adjusting the questionnaire.

Data collection, processing and general results

In April 1983, 900 questionnaires were sent to principals of 22 secondary schools on pre-university (VWO) and higher general (HAVO) level, who volunteered to disseminate the lists to their teachers. End of June 1983, 249 completed lists were returned. The rather small amount of response could be explained by ministerial activities (school budget cuts) discharges (due to reduction of students-teacher ratio and decline of student numbers), which might account for a non-cooperative research climate, as well as business during examinations and a principal not keeping his promises.

Data were processed by factor analysis using principal components and orthogonal rotation according to the Varimax procedure (SPSS). Based on Cattell's 'Scree-test' and criteria of interpretability and invariance (Jae-on Kim and Mueller, 1978, 44-5) we chose a 6-factorial solution explaining 30.5% of total variance[5)]. This forces us to interpret results only tentatively. But aided by the two other ongoing studies within the research project we are gaining more insight into the nature of the criteria, thus being able to construct a definite draft of the questionnaire.

Five of the six factors were theoretically interpretable:

(1) *Teacher control* consists of nine items with factor load equal to or larger than .40 that relate to teacher's striving for control over the teaching-learning processes. This factor accounts for 26.6% of explained variance. Some examples:

(2) if students co-decide on classroom activities, little will be learned (factor loading .53);
- only the teacher must decide how teaching will be done (.60);
- parental participation concerning the content of education furthers the quality of teaching (-.44);
- if students would participate in decisions on content of teaching, little will be learned (.51).

(3) *Work ethos*, accounts for 20.6% of explained variance, consists of 7 items ($\leq$.40) and relates to student participation, flow of activity and 'keeping students busy' by detailed prescription of students' activities. Examples:
- Teacher's detailed prescription of student's activities is essential to the process of learning (.55);
- in changing classroom activities the teacher must prevent students from getting distracted (.41);
- students must participate in classroom activities without interruption (.51);
- a student will have better results if he follows teacher's instructions and tasks (.53).

(4) *Affiliation,* accounts for 15.1% of explained variance, consists of 7 items (< .40), indicates that the teacher attaches high value to social relationships, needs respect and esteem from students as well as from colleagues, is sensitive to (social) differences between students and tries to be empathic with them. Again some examples:
- to be successful in learning, a student must feel at ease/at home in class (.50);
- a teacher who gets little positive feedback from his students, is not functioning well (.54);

- as a teacher you overlook a student more if his participation develops all right (.54);
- a teacher cannot do without his colleagues esteem (.52).

(5) *Social development,* indicates a non-subject matter orientation to teaching, in which classroom discipline and authority is seen as an impediment; accounts for 13.8% of explained variance and consists of 6 items (< 40). For instance:
- intellectual development can do without factual knowledge (.50);
- students' social development is more relevant than subject matter acquisition (.58);
- nowadays, there is too much emphasis on classroom discipline (.56);
- education is impossible without teachers' authority over the students (-.47)

(5) Perceived (subjective) *difficulty of 'direct instruction'* [3] *activities,* accounts for 12.3% of explained variance and groups six items (≤ .40) asking for ratings on difficulty of teaching activities with a 'direct instructional' character, mainly because the teacher is always 'in control'; examples of which are:
- lecturing (.75)
- explanation by the teacher (.82)
- discussion directed by the teacher (.45).

As mentioned before, the sixth factor (6 items < 40) explaining 11.6% of explained variance was not interpretable in a clear-cut way. It indicated on one side teachers not being in for student participation in decisions on classroom activities (two items with factor loads, resp .41and .45) and perceiving low control teaching activities as difficult (three items: students experimenting, .42; students formulating problems or hypotheses, .52; student tasks of a free or divergent character, .63) while on the other side stating that use of textbooks diminishes teachers' flexibility (.40).

Teaching experience and criteria

We considered teachers teaching for more than 3 years as experienced, less than 3 years as inexperienced. Here again we need to be cautious in interpreting the data from the first draft of the questionnaire because only 25 inexperienced teachers did respond (versus 224 experienced teachers). Concerning averaged factor scores between experienced and inexperienced teachers, a significant difference was found only for individual factor scores on the first factor, teacher control ($T= -4.03$, $p < 0.000$). The general notion that inexperienced teachers are concerned less about their control in classroom than their experienced colleagues, seems to be confirmed here; 14 items gave significant differences on average scores between the two groups. Some examples are:

---

[3] Figure 1. Overview of eigenvalue of factors, % of explained variance and cumulative percentage of variance.

Factor	Eigen value	% of variance	Cumulative %
1	8.097	10.7	10.7
2	4.040	5.3	16.0
3	3.480	4.6	20.5
4	2.409	3.2	23.7
5	2.331	3.1	26.8
6	2.229	2.9	29.7
7	2.026	2.7	32.4

- Nowadays, there is too much emphasis on classroom discipline: x experience 3.7; x inexperience 2.9; T= 3.27; p<0.001;
- Education is impossible without teachers' authority over students: x experience 2.0; x inexperience 2.8; T= -2.91; p<0.007. (1=agree; 5=disagree)

Subject-Matter Specialization and Criteria

To get a general insight in this relationship, we clustered teachers' discipline specialization into four subject-matter areas: (A) language and foreign language teachers (n=90); (B) science teachers (n=70); (C) social science, geography, history and economy teachers (n=59); and (D) teachers in modern arts, music and sports (n=28). Figure 1 reveals teachers' average factor scores for these distinguished subject-matter areas.

We cannot detail the data here, but we will summarize our findings that were based on T-tests of item scores as well:

- We found significant differences within all factors but the third one: 'Affiliation' criteria do not lead to significant differences;

Teacher control, the first factor, especially distinguishes the A and B teachers who adhere more than average to control criteria, from the C and D teachers who adhere less to these. The differences in factor scores between A and B on the one hand, and D on the other, are significant;

SUBJECT-MATTER AREAS*

FACTOR	A	B	C	D
I	-0.10	-0.17	0.17	0.42
II	-0.09	0.21	-0.15	0.08
III	-0.00	0.13	-0.07	-0.18
IV	0.14	0.02	0.03	-0.55
V	0.00	0.19	-0.19	-0.09
VI	0.10	-0.25	0.02	0.22

Figure 1. Teachers' average factor scores for subject-matter areas.
* Explanation in text

- B, science, teachers prefer less than average criteria concerning 'work ethos', the second factor; language (A) and arts (D) teachers score more or less on average while social science (C) teachers score above average to this factor;
- As for 'Social development' (IV) we found that (D) teachers adhere most to these criteria, while the other teachers score more or less on average.
- Subjective difficulty with 'direct instruction' –activities (V) shows a significant difference between teachers in science and teachers in social science (C), the former stating less difficulty than the latter ones.

Discussion

Despite several restrictions to the research results so far, the main conclusion to be drawn is that teacher's subjective teaching criteria are empirically discernable and that teachers do diverge as to the adherence to and impact of different criteria on their teaching activities. Years of teaching experience and discipline specialization seem to relate with these. We do not know for sure what set or categories of criteria cover(s) the range of all possible subjective teaching criteria. Literature and our own findings indicate teacher 'control' and those concerning 'work ethos', such as student involvement and flow of activity, to be at the heart of teaching. For realizing educational intentions a teacher needs some influence over what is going on and students must respond to it. This seems to be the most fundamental issue a teacher has to work out for him or herself: how to realize at least some intentions without loosing students' cooperativeness which is needed to show to outsiders that work is going on. So, stated otherwise, non-involvement of students really is a threat to teaching. Any teacher will deal with control and work ethos in a personal meaningful way related to individual characteristics. By these we do not mean personal, stable traits, rather the effects of individual ways of going about with teaching as precipitants of teaching experience, or as features of subjective theories of action. Thus, subjective teaching theories and individual teacher characteristics will account for the extent to which a specific teaching criterion is used implicitly or explicitly by a teacher.

Taking a closer look at teaching criteria it appears that in an individual case, they do not complement each other but confront the teacher with dilemmas, situations in which different or opposite roads of acting or reasoning might be followed according to different criteria of the same personal value.[4] Olson (1983) believes these dilemmas to reflect the ambiguities of the teachers' work that are brought to consciousness within context of innovation. This relates with the need for certainty within innovative as well as in teacher-trainee situations, explaining concerns with self (cf. Fuller, 1969; Hall et al., 1976). Wagner (1984) found empirical evidence for these dilemmas, which she calls 'knots' in thinking. Olson mentions five of such dilemmas:

- Social versus intellectual dimensions of teacher influence;
- Tangible versus intangible indicators of progress;
- General versus subject centered education;
- Political versus educational considerations;
- Occupational versus educational conflicts (Olson, 1983, 15).

We like to add:

- The striving for control versus need for esteem and affiliation, and
- Autonomy and independence of prescription versus practical support and advice.

But, rather then trying to enumerate all possible dilemmas, we think it is better to reveal the origins of such dilemmas, which, as for us could be conceptualized as teachers' teaching criteria. The way in which the dilemmas they pose are resolved and the effects of this resolution as feedback to the teacher might determine his/her professional development.

Resolving problems of conflicting criteria might be one way in which teaching criteria relate to teaching activities. Though we do not know the specific relationships between, or lines of influence from criteria to activities, we have got a grasp on the potential of the concept. In contrast with classical distinctions, such as

[4] If they would not be valued equally, they would not pose dilemmas.

'subject versus student centered' or 'progressive versus traditional', teachers' subjective teaching criteria could be a more distinguishing concept in accounting for or for description of teachers' individual characteristics.

One of the constraints of the research project relates to the further conceptualization of subjective teaching criteria. We depict these as being somewhere on the dimension of ideological wishes versus practical constraints in teaching, e.g.: a teacher wishing to interact with his/her students in a friendly democratic way, but 'learning' by experience that this does not work, thus posing him/her in another way with dilemmas we try to formulate criteria more near the practical than the hoped for, so as to set conditions for the representation of teachers' activities; a query, as said before, we try to approach through (video) retrospection techniques. Still then, the relation between thinking and action, between articulated and tacit knowledge, will stay problematic and needs more fundamental research. The development of routines is a relevant object of study within this framework (cf. Lowyck, 1984) as well as contextual and situational influences. In the initial studies we conducted, as well as through the REPGRID interviews, we get an impression of the way in which situational influences are processed by teachers.

On the basis of experience, again, they seem to construct specific situational images which will function as conditional frameworks for the selection of teaching activities and for decisions as how to proceed or what to tolerate. Related to the situational constraints then, teaching criteria are specified so as to render the teacher concrete guidelines and security for his activities.

Last but not least we see teaching criteria also as a powerful concept in studying curriculum implementation. Criteria implicit in curriculum and those adhered to by the implementing teachers could be compared. Eventual contrasts could be related to actual use of the curriculum in class, a procedure more or less preceded and proven fruitful by Olson (1981). Combined with efforts to depict teachers' interpretations of the curriculum (cf. Ben-Peretz, 1984) one could have a better insight into the way new curricula are perceived and dealt with by teachers as well as in factors explaining this.[5] Although one has tried to variablize teachers' influences over the curriculum, their real arguments for their ways of handling, as for us, were not revealed.

---

[5] Cf. Halkes, R. & Mijnster, T. (in preparation). Teachers' construction of a curriculum in arithmetic.

Chapter 2

Professional Literacy, Resourcefulness and What Makes Teaching Interesting

Alan F. Brown

Summary

The purpose of this study was to understand what makes teaching interesting. To do this, we simply asked 22 teachers what they found satisfying, rewarding, or stimulating and what was dull, bland, or boring. The analysis of their answers was done in two ways: (1) the substantives of what they said and (2) the unique patterns of their thought processes. The findings of the first analysis were familiar: Kids are exciting and yard duty is tedious. The findings from the second were surprising: Their self-estimates of quality of work life clearly favoured those teachers whose ways of answering these questions showed them to be (a) more resourceful in describing teaching and (b) more articulate and aware of their own practical preferences, i.e., more professionally literate. These findings are encouraging for persons interested in teacher morale in that some previous studies carried out for different purposes have demonstrated that series of workshops may be deliberately designed to change those two variables.

Introduction

Quality of life, or when referenced to the workplace, quality of work life (QWL), is of current interest particularly in management literature and research. Perhaps the interest arises out of the age-old evidence that it is cheaper to make a person feel good than to raise pay. Whether or not either does affect output matters less in the more professional levels of management and in the public service sector including teaching because in these tasks, quality of work life itself becomes an output. If it is no place for you, you change it, exit, or learn to put up with it one way or another. Thus quality of work life as experienced by the teacher becomes as major a purpose of the principal as did its earlier counterparts: dedication, organizational climate, commitment, morale, spirit, atmosphere, feeling tone, vibes.

Designing a program to improve the quality of work life – the main purpose of good management (Stogdill 1965) – has engaged the attention of consultant and researcher alike. In times of economic restraint, university underfunding and the reticent marketplace, it seems that any plan that looks promising is worth the attempt. Unfortunately this is usually not so. What may look promising appears so because its attention is directed more to conditions of the workplace than to the worker. The workplace is visible, is objective. The worker's intentions, wills, feelings are idiosyncratic, are subjective. Being less visible becomes seen as less accessible and thus researchers are less inclined to study it. Accordingly, the design of programs becomes based more upon the external manifestations of quality of work life than upon the cause itself. Inspired perhaps by Peters and Watermans' concept of excellence (1982), school systems may elect only those programs that are readily affected; changing how a person feels about work may mean changing how that person feels more than changing the work place.

There have been several reports on how life at the school affects those who work there, e.g., Goodlad's *A place called school* (1983) and Batcher's *Emotion in the classroom* (1981). The study reported here examines not only what teachers find interesting and exciting or tedious and draining but also why they may find it so.

A Structure for Thoughts Behind Actions

Structured reflection is a term that has been used to describe some of our previous work trying to set out teachers' definitions of teacher effectiveness and teacher promotability (Brown 1984, Brown 1982, Rix 1982). In these studies, it was discovered that it is not at all unusual for persons to have two quite different sets of definitions of the situation, the posted set and the operational set. The posted ones are those that we post as upon a billboard, to see and to be seen. The operational ones on the other hand are the ones which we actually are using, know it or not, as the tacit or implicit guides to our thoughts and actions.

There are three classes of posted constructions of reality, we have found so far. There are some posteds that are our ideals; some that are our thoughts-in-abstraction and some are outright lies, or perhaps at least our attempts at being diplomatic.

Taken in order, the ideals, or the first class of posted constructs, are the should-be's, or the ought-to-be's of Wagner (1984) so that it is not at all discouraging to find posted definitions differing from the operational, that is those definitions that are inferable directly from our actions, or the 'are's.' When there is this discrepancy within our thinking, it is perhaps recognition that your reach must exceed your grasp (or what's a heaven for?). The chasm between the ideal and the real always seems to widen when you get grand visions of the possible but settle interimly at least on the mundane actual.

Posted definitions of a situation may also however be nothing more than our reflection-in-abstraction. Thus, our second class of posted constructs are when we reflect first upon some abstract idea rather than upon an actual event; we have an armchair theory of reality in which we may actually claim to believe. This happens not so much because we are out of touch with reality as because we fail to attend to our own involvement within reality.

An example of the incongruity that may develop between a person's posted constructs which are reflection-in-abstraction and that same person's operational constructs – the ones actually in use at the time – is seen in the supervisor who stands and exhorts all teachers, every one of them in the very same way, to 'individualize instruction.' The possible contradiction between what is said and what is done is apparent only to the listener. This contradiction happens easily in administration so that one finds a superintendent waxing eloquently over a fascination with the intricately complex matrix of programs he/she has just drawn up for teachers' professional development, i.e., the growth of real live people. The necessary preoccupation with the processes for working with people comes to take precedence over the very people they were supposed to be about. The people get posted, the process prevails.

The third class of posted constructs in quite different. Whether it is called diplomacy, persuasion, salesmanship, hypocrisy, or just spinning lies, they all require you to know what you are up to and why. The teacher who claims to post that kids come first and then breaks a leg trying to beat them out of school at four o'clock (yes, one did but it was an icy Edmonton January) is an example. This is quite different from the second class of posted constructs; it is the difference between the smart hypocrite and the ignorant hypocrite: the former knows that he is lying; the latter does not and is therefore a dangerous phony.

It is now possible to suggest that all of us have different levels of what can be termed 'professional literacy,' that is our ability to articulate in some meaningful set of symbols, including language, the nature of our professional experience and what we do with it. This is the extended definition of the term and it is assisted by "From utterance to text" (Olson 1977) and *Literacy, language and learning* (Olson, Torrence and Hildyard 1985). We suggest that quality of work life is a matter of learning and that specific learning is aided by the concept of professional literacy and its development.

Whether or not teacher thinking and professional literacy are interdependent, as is implied here, remains a moot point. It is sometimes held, as did Cvitkovik (1986), that literacy and thought are quite independent. Thought, he argued, is not specifiable in terms of rules and representation while literacy is. Literacy is a special skill, which may, nonetheless, provide some of the base on which this thought is carried out. Exploring that base is the central interest in the present study.

The notion of professional literacy deals mostly with the second class of posteds and their degree of correspondence to the operational sets. It is conveniently similar to what G. A. Kelly (1955, pp. 475-476) calls 'level of cognitive awareness.' At a low level of cognitive awareness, a person is dealing with elements that lie at the outer extremities of the range of convenience of his available constructs, so that the opportunities for validation or invalidation of those constructs is minimized. Perhaps that is a good thing, not being able to see your own errors. However, this person, says Kelly, 'cannot be very keenly aware of what he is up to.'

A high level of cognitive awareness could be thought of as the descriptor for persons whose constructs of what they are up to are indeed, what it actually is that they are up to. Their thoughts and actions are integrated; they exhibit 'integrity' in at least the conceptual sense. Similarly, their priorities would correspond so that the constructs they *demonstrate* to be 'of most importance' to the quality of their work life would, at perhaps some other time, be the same as the constructs they would upon reflection *claim* to be 'of most importance.' If you keep saying (even to yourself) that 'working with people' is the most important thing that makes teaching interesting, then it should follow that 'working with people' ought to be a construct that would differentiate strongly between those elements of your work you choose as interesting and those that are dull. You would then have a 'high level of cognitive awareness', perhaps are called 'professionally literate', or even demonstrate integrity.

Thus the first question I wish to pose here is whether the level of one's cognitive awareness, or professional literacy, is a cognitive determinant of feelings of job satisfaction, itself related to sense of quality of work life. The question would seem to call for a linear relationship. The lower the level, the less chance there is for validation of your ruminations about work and thus the less likely you are to find the workplace satisfying. The higher the level, the more frequently you have been able to check back successfully and build new thoughts about the place. Thus, you can come to feel better about it.

The next question uses the concept of cognitive complexity (Bieri 1955). According to this, the greater your cognitive complexity the more ably you cope with the tasks at hand. Many operational definitions of cognitive complexity now abound but the simplest estimate allowable would be your resourcefulness, or the number and variety of constructs you possess and are able to draw upon to confront a topic. In earlier studies on developing school principals' ability to make effective staff decisions (e.g., Brown, Rix, Cholvat 1983, for recommendations for promotion or selection) we found a wide range in their repertoire of operative criteria. Despite what they said or posted on the topic, some principals found their repertoires of operative criteria would be quickly exhausted when actually selecting among available candidates. They had to resort to repetitions or synonyms. On the other hand were the more discerning principals who could keep on finding unique characteristics in their candidates; they were thus able to make choices that are more sophisticated for more specific purposes. They had greater cognitive complexity.

The concept of cognitive complexity means being more resourceful, or having more strings for your bow when attacking a problem or reflecting on reality. Parry (1978) made the discovery that those principals who were regarded as more effective by their superintendents were those who used a significantly greater number of criteria than did the less effective; their superintendents had still more. Cognitive complexity also refers to construct interrelatedness but, because these two measures were once found to be highly correlated (Brown, 1964), we will now use the simple number of different constructs used by a subject as a crude estimate of cognitive complexity. The question posed here is whether possession of this should turn out to bode well for the way you feel about your job, i.e., quality of work life. Again, a linear relationship will be looked for.

The Study

This study has three phases, one for each of the variables discussed above: (a) resourcefulness or estimate of cognitive complexity, (b) professional literacy estimated by level of cognitive awareness and (c) quality of work life. There were 22 experienced teachers and we met on several occasions in different groups in early 1985. I did not consider that demographic variables such as age, gender or ethnicity, would have any effect on a study of organizational dynamics. Indeed, they did not. Probably other variables such as level of education and length of experience would not have had any effect either but the subjects studied were relatively homogeneous on those variables. I knew these people well, had established rapport and they responded with enthusiasm. This point I mention because of the crudity of some measures.

For resourcefulness or estimate of level of cognitive complexity, a variation of our Discriminant Perception Repertory Test (Brown, 1964) was devised. This is distantly related to Kelly's Role Construct Repertory Test (Kelly, 1955) but has no roles or poles. Several measures of the reliability and validity of our test have been carried out (for summary report see Brown, 1982, pp. 17-18) when used for organizational decisions but none when used for job satisfaction. Because the format is quite similar and because our subjects were familiar with its personnel decision use, we infer its use to be sufficiently reliable and valid for the present. The instructions given to my teachers this time were:

> Think of three elements of your work or workplace in which you are very interested, find stimulating or exciting, derive satisfaction or reward from, or make you want to stay with it or are otherwise a positive element in what you do and why you do it.
>
> These elements should now be identified below as Element A, Element B, Element C. Note that these elements may be any tasks, persons, challenges, groups, committees, staffs, work routines, events, duties, responsibilities, opportunities, good times, bad times with adults or children or both or neither so long as – to you at least – they are thought of as part of your work or somehow associated with it.
>
> Think of three elements of your work that are bland, dull, boring, tedious, irritative, annoying, uninteresting, a nuisance to bear, make you want to avoid it, cause bother or are otherwise a negative element in what you do and why you do it.
>
> Now enter these as Elements X, Y, and Z below noting the same range of conditions as above.

The purpose of these instructions was to elicit a representative sample of the respondent's positive and negative elements about job satisfaction. Following these instructions was a series of forced comparisons among these elements. There were 18 such comparisons, which exhausts the combinations of six elements taken three at a time, excluding the two homogeneous sets. The intent of this instruction was to exhaust their repertories of constructs attributable to these elements. Because my teachers were already familiar with this format, I allowed them to process their own protocols thus avoiding any problems of scorer verification.

Their 18 constructs, understood at least to them, now became a simple 18-item Likert scale against which in turn each of the six elements was rated. Processing the protocol meant subtracting, for each construct, the summed weights of its X, Y, Z elements from it's A, B, C elements, to arrive at an estimate of the power of that construct upon the critical decision made earlier: what makes teaching exciting and dull? This may be termed the decision power of each construct, showing its strength or weakness as a criterion for job satisfaction, at least for that idiosyncratic respondent. The respondent's complete set of criteria may now be compared to each other to show a scale of the priorities of their constructs, ranked according to the strength or weakness of each construct to discriminate between positive and negative environs at work.

The second phase of the study was approximately three weeks later, with no intermittent reference. The purpose was to obtain an estimate of what can be termed professional literacy that is similar to Kelly's level of cognitive awareness. The estimate is arrived at by making a simple comparison of a person's posted and operational criteria. It is done this way: from the foregoing first phase, each teacher was able to develop a scale of priorities among their operational criteria. They did this by rank-ordering their own criteria according to the decision power of each criterion. To arrive at a comparable scale of posted criteria requires a person to contemplate concepts-in-abstraction.

Contemplating concepts in abstraction is what we do all the time, whether reading journal articles or the daily newspaper. Making preferences among these concepts, we also do regularly. That is particularly so of teachers, school administrators or others in helping professions. The larger problematic of this phase occurs when our concepts in abstraction come out-of-synch with our concepts in use; that happens when the abstractions become disconnected from that from which they were to have been abstracted: the real events, actual people, specific acts. Alternatively, perhaps some never were connected: like the *pater noster*, we acquired it in the abstract and kept it there.

It involves an inferential leap of perilous dimension to thrust the value-laden term 'character' onto the innocent test-taker whose posted and operational definitions of reality happen to coincide. Then, for the son of a Presbyterian minister, Kelly was admirably agile in keeping morals out of cognitive theory. Guilt, for example, is only that response to the discovery that what you did does not congrue with your set of personal constructs. Therefore, in this paper there is the assumption that the congruence of one's priorities in the abstract with their priorities as shown in actual behaviour is an indication that they are integrated. Lack of cognitive integrity, or low level of cognitive awareness, simply means you 'better smarten up.'

The procedure for this second phase was to give the teachers back their own lists of 18 statements. As a point of procedure, the actual processing of the first phase, described above, was not completed until after the second phase so as not to contaminate the task. They were simply asked to:

> Read all your 18 statements and decide which of your criteria are most important to you and your work. Try to decide the relative importance to you of the effect on the quality of life at work of each of these 18 and rank-order them from 1st to 18th or from 1st to however many different criteria you have, not counting synonyms, antonyms.

This was not a memory-check. With the elements having been removed, they were studying attributive statements in the abstract, not identified directly with a concrete referent such as their six elements. The likelihood that these rankings of posted attributions would correspond to their earlier operationally – derived rankings of element-referenced attributions is the likelihood that the terms they talk about lie within their broad range of convenience, or are the operational ideas that they are easily able to articulate meaningfully. The likelihood they would not correspond is greater when the teacher has a lower level of cognitive awareness, has less connection between ideas forced out of experience and terms taken off the shelf. That is, he is 'less keenly aware of what he is up to.' The correspondence between the two sets of rankings is the second variable.

For the third variable, quality of work life, the crude global estimate was used. This was sought on a different occasion. No claim can be made for its reliability although at that moment for that person it was probably a valid answer to the question of "where, on a scale of 1 to 10, would you place your own quality of work life?" I would like to have attempted this, which is the dependent variable here, in a more definitive fashion but was convinced none exists.

The Findings

The main directions of this study were confirmed: the bigger your resourcefulness or level of cognitive complexity, possibly the higher too is your correspondence between what you do and what you talk about (among school administrators: goal-directed behaviour), and the better the both of these, the more likely you will be to find yourself sitting on a comfy cloud when it comes to estimating your quality of work life. That is indeed an overgeneralization but at least there is enough to encourage a more penetrating study of the inner dynamics that help to turn chaos into challenge or to make a nightmare just a nuisance. The encouragement seems to come from the cognitive nature of the determinants. The encouragement lies not in the idiosyncratic nature of these determinants; it is encouraging because we are much more likely to be successful in changing cognitions than in working directly on people's wills and feelings.

Why this possibility is thought to be more likely is because of the reasonably temporary nature of results of inspirational professional activity days on the one hand and the successes we have had with professional development projects resulting in an increase in the number of personal constructs (Brown, 1982), change in the structure of professional criteria which changes held for a period of eighteen months (Rix, 1981). An interesting example of personal change through cognitive structuring is in Hill (1988), "Understanding the disoriented senior as a personal scientist: the case of Dr. Rager," where three months of intensive personal attention brought back an older man's constructs, poetry and will.

Table One sets out the cognitive structure findings and seems to say 'the more strings to your bow the more fun is the hunt' but less clear is the relation with level of cognitive awareness. This ambiguity cleared during interviews with those showing anomalous results. Some seemed to say, 'we're quite happy with how we think we see things so don't confuse us with reality.'

In the last column of Table One, Anomalies, attention is drawn to those teachers whose results are 'anomalous' to the general trend posited earlier. The As are anomalous in that they showed a high degree of cognitive awareness but a medium, low or very low estimate of quality of life. Their comments were the classic ones found in studies of worker satisfaction implying or stating 'of course we know the way it is and even if it's not that bad it could be very much better.' The thinking of the 'A' teachers is that of the hardheaded idealist who, without any cognitive disjunction, lives with realistic visions of improvement. What would a staff be without the divine discontent?

The Bs comments however were to the contrary and equally believable. These teachers all displayed, if not quite the classic 'bovine contentment,' at least equanimity with reality and a willingness to accept cognitive disjunctions as part of the taken-for-granteds of life in complex organizations especially schools with its comings and goings of unpredictable people.

The Cs remind us against temptations to go running gleefully into predicting personal behaviour: individual personal problems and challenges set them askew.

Looking at the substantive content of teacher thinking provided no surprises. The what and the why of interest in the workplace are familiar to educators: children are challenging and paper work is boring. One of them, Don (3), has high cognitive awareness, high cognitive complexity, and high quality of work life. He especially finds rewarding the personal relationships developed through coaching sports team and is frustrated by school board trustees because he has no control over them (note, as compared with his team members).

Several negative attributions of teaching – yard duty, futile trustees – were seen as not enriching or even related to the essence of being a teacher while others – grading, report cards – were seen as detractors from joy but were nevertheless something that just has to be done.

By defining their attributions to the task of being a teacher, our 22 participants re-emphasized the finding (Blackshaw, 1982) that personal constructs underlie satisfaction.

The purpose of this paper was not to identify or propose some new way of making teaching more exciting. Teachers do that themselves. Indeed from what we see, those highly resourceful and professionally literate teachers will derive a rewarding satisfaction from it and, what is more significant, if they do not they will know why not.

Table 1: 22 Teachers Grouped by Self-Estimates of Quality of Work Life (QWL), Measures of Cognitive Complexity (CCX) and Level of Cognitive Awareness (LCA).

	QWL	CCX	LCA**	Anomalies
1. Nel	10 v. high	12 high	11 high	
2. Bob	9 v. high	13 v. high	8 med.	B
3. Don	8 high	14 v. high	11 high	
4. Pete	8 high	12 high	10 med.	
5. John	8 high	11 h. med.	10 med.	
6. Len	8 high	10 h. med.	13 high	
7. Jim	8 high	7 low	4 low	B
8. Mal	8 high	6 v.low	10 med.	C
9. Jayne	7 med.	12 high	10 med.	
10. Lois	7 med.	12 high	10 med.	
11. Geri	7 med.	8 l. med.	12 high	A
12. Candy	7 med.	9 med.	9 med.	
13. Jan	7 med.	8 l. med.	5 low	
14. Donna	6 med.	8 l. med.	10 med	
15. Guy	5 low	8 l. med.	12 high	A
16. Gaye	5 low	10 h. med.	9 med.	C
17. Fred	5 low	7 low	13 high	A
18. Alf	4 low	8 l. med.	6 low	
19. Vern	4 low	7 low	12 high	
20. Lorri	3 low	8 l. med.	6 low	
21. Dona	3 low	6 v. low	5 low	
22. Thom	2 v.low	10 h. med.	10 med	A

Possible range 1-10
1-18 1-14

* Number of different constructs used when comparing work situations.
**Congruence of priorities within sets of posted-operative constructs.
Anomalies: A's know very well what they are up to and claim that as a result their QWL could take improvement.
B's are quite happy thank you with the way they see things now and show little desire to delve deeper.
C's are a mix of personal hopes, problems, opportunities and temporary inconveniences.

Chapter 3

Teachers' Instructional Quality and Their Explanation of Students' Understanding

Rainer Bromme and Gundrun Dobslaw

Summary

'Instructional Quality' models stress the selection of tasks that are suitable for students. This study analyzes the cognitive tools that teachers use when they think about tasks and students. Mathematics teachers were asked how they explain the understanding of mathematical tasks to their students. The teachers' replies reveal that task-related conceptions of the understanding process are present, as well as broader, task independent conceptions of understanding. There are also individual differences between teachers. In addition, the teachers were observed during lessons. Connections between the range of ideas of understanding, and the instructional quality of the teachers were examined. Teachers who were rated to be 'enthusiastic' in their teaching tend to prefer task related or mixed explanations for their students' understanding.

Introduction

It is of interest to explain understanding, since the teacher's most important task is to promote this in her students. To achieve this, the teacher must realistically assess her student's previous understanding. To be able to plan her lessons, and to make allowance for the difficulties some students might have, it is important that she be able to explain to herself the extent to which understanding has taken place.

We investigated explanations in order to learn something about the teacher's professional knowledge. The investigation is based on viewing 'teachers as experts' (Fogarty et al., 1983; Leinhardt, 1983; Bromme, 1986). Like other experts, e.g., doctors, chess-players, or engineers, teachers use a network of professional concepts and concept relations. These concepts structure their conceptions of the tasks they have to cope with, and of the conditions they have to take into account in doing so, as well as the 'image' or 'model' they have of the situation in which their professional activity takes place (Schön, 1983).

It is helpful to record professional knowledge while it is being used, that is in connection with concrete situations taken from the experts' field of experience. Then there will be less bias within verbal reports and the answers will be richer than if one tries to record a lexical or systematical representation of their knowledge. Asking for *explanations* for the understanding of mathematical tasks will create a situation for the teachers in which they may use their own knowledge about 'understanding', and thus be able to explain it.

The professional knowledge with regard to pedagogical-psychological teaching problems, such as classroom-management, student aggression, etc. has been extensively examined (e.g., Krause et al. in this volume). On the other hand, the teachers' concepts of more subject-specific issues i.e. problems linked to the subject matter taught, are less well understood (cf. Clark in this volume).

The models and findings in recent research on efficient teaching show that almost all teacher variables, which were found to be important for the students' learning, refer directly to dealing with the subject matter. They require at least subject-matter-relevant judgments from the teacher, such as allotting tasks according to the students' level of learning, evaluation as to which are unequivocal forms of presenting the subject matter, etc. Judgments about the students or his/her level of learning and his/her learning potential must be linked to the respective details of the task in order to be able to make a decision as to which action should be taken. This is only a hypothesis, taken from an analysis of the demands the teacher has to cope with. It must be investigated empirically whether this linking really takes place in teachers' minds.

Our question is whether evidence for such a linkage of subject matter components with observations and assumptions about the students' characteristics or students' behaviour in the lesson, are to be found in the teachers' explanation, or whether these explanations refer to person characteristics and behaviour not related to the subject matter.

The first type of explanatory concepts are called in the following, task-related concepts, because the subject matter in our study is represented as a mathematical task. All concepts about students' thinking, knowledge, behaviour, etc. directly dealing with the subject matter are called task-related concepts.

The second type of explanatory concepts are called task-independent. Examples are giftedness, general motivation, student's memory, absenteeism, attention, support at home, etc. Sometimes we will call the second type of causes 'purely' psychological explanations. Obviously, this is a quite narrow notion for 'psychological', but a little exaggeration may help to clarify the distinction.

A similar differentiation is made in the empirical research into conditions of students' learning at school. In this field, a division is made between instruction independent predictors such as intelligence, anxiety, and sex, and more instruction dependent predictors such as instructional quality and previous marks, when trying to predict student performance. It has been revealed that the causal effect on learning achievement of instruction independent factors becomes increasingly weaker the longer the students are at school, while the actual lesson quality, and the student's accumulated knowledge become increasingly important for explaining and predicting school performance (Simons et al., 1975). For example, Schwarzer

(1979) showed that the students' subject-specific previous experience formed the best single predictor of students' performance, when compared with intelligence and anxiety. Therefore, the question arises, whether the causal importance of such subject-related factors is also mirrored in the teacher's professional explanation.

A first study on students' understanding from a teachers' point of view has shown that the subject content occupies a prominent space in teachers' explanations of understanding (Bromme and Juhl, 1984). The objective of this study is to find out whether this result can be generalized to other teachers and another type of school. Additionally we will investigate whether teachers differ in the extent to which they include the subject matter in their explanations. Finally, we will inquire whether there are connections between the explanations and the quality of the teachers' classroom activity.

For this purpose, the quality of teaching is assessed according to certain research variables into teacher effectiveness. As these models stress the presentation of and the working with the subject content, it will be of interest to see whether teachers who give particularly 'good' lessons with regard to these variables, will also extensively refer to the subject content in their explanations.

Questions To Be Studied

1. Which concepts are named by mathematics teachers in order to explain their students' understanding of mathematical tasks?
2. Do teachers differ in their preferences for task related vs. task independent (pure psychological) concepts?
3. Is there a connection between the type of explanation and the quality of classroom activity in the teachers' subsequent teaching unit?

Method

Teachers' explanations were obtained by using a table in which the teacher had to list a mathematical task that she had presented in her preceding teaching unit (cf. fig. 1).

task	name	reasons for understanding
understood	name	reasons for understanding
task not	name	reasons for not-understanding
understood	name	reasons for not-understanding

Figure 1. Tasks

For this task, each teacher had to select two students who had understood the task and two who had not. In the columns of the table, the teacher was asked to write down the causes of understanding resp. not understanding this task. In constructing this table, we adopted a basic idea of Kelly's (1955) repertory-grid test, that linking two given persons and/or facts requires the application of the subject's existing system of concepts (Ben-Peretz, 1984).

This method of attaining data does not impose predetermined items on the teacher, but permits us to obtain free answers. We considered this necessary as until now there are so few research results, that the presentation of items would be rather arbitrary (for this see also Huber and Mandl, 1979). Using concrete mathematical tasks taken from the classroom assures that teachers will be able to base their answers on actual experience instead of "theorizing." Finally, the tables, by their standardized form offer the possibility of comparative and summarizing evaluation of the answers.

In each case, the teachers had to select the students from a list of 15 students chosen at random. Additionally, they had to note four tasks that had been actually worked on in preceding teaching unit. Thus, we obtained eight explanations per teacher pertaining to understanding and pertaining to non-understanding. The teachers were permitted to give several reasons per student. The explanations were obtained immediately after the end of a teaching sequence – which had dealt with geometry for almost all of the teachers.

Lesson quality was then assessed by observers in the subsequent teaching series. Each teacher was observed in at least four lessons, her teaching being recorded in connection with another study. An observer who was familiar with the mathematical subject matter of this teaching series, and with the variables for the quality of teaching, assessed the teacher's behaviour according to six variables:

1. The teacher's control of the course of the lesson.
2. Clarity of organizational orders and subject matter instruction.
3. Clarity when using blackboards and overhead projectors.
4. 'Warmth' of teaching style.
5. The teachers' interest in the subject matter and emotional participation (enthusiasm).
6. Anticipatory approach towards disruptions and discipline problems ('withitness').

Active control was characterized by an active as opposed to a reactive content-related and organizational lesson control.

Assessment of the clarity of teachers' remarks related both to organizational and content-related instructions. This variable was one of the five variables in the review by Rosenshine (1971) which relate strongest to learning success.

In assessing the *clarity of blackboard and overhead projector use,* behaviour closely related to communicating subject matter was taken into account.

The item *'warmth'* of teaching style concerned the extent of which the teacher created a climate of understanding and personal acceptance in her lesson.

The variable of *teachers' interest and emotional participation* described the degree to which the teacher was able to convey the impression that she personally found the subject interesting and the course of the lesson exciting. This variable is also labeled 'enthusiasm.'

The variable *anticipation of disciplinary problems* was introduced to assess classroom management. This variable described whether the teacher only intervened to deal with problems or events which had already been developing for some time, or whether she acted at the first signs of a disturbance ('withitness', as described by Kounin, 1979, cf. Rheinberg and Hoss, 1979).

Data Collection and Sampling

The study was carried out as part of a larger project on stochastic education. Twenty-six teachers participated in the overall study, which included pre-tests and post-tests and the recording of additional data. The average professional experience of these teachers was 92 months.

For our purposes, the interest of these tests is merely that they show that our questions and our observation had no influence on performance of the classes concerned in comparison to a control group of ten other teachers who were not observed.

Thirteen teachers working with fifth and sixth year classes and belonging to three different schools were included in the study of teachers' explanations. *

Data Analysis

The free explanations were coded with a previously developed and tested category system. They were subdivided into units of meaning by one of the authors and subsequently processed by two trained coders. The inter-coder reliability was 92 percent. Before further analysis, agreement was reached on the remainder of codings after discussion with the coders. The categories (see table 1) can be assigned to two different fields according to our main questions:

- Categories describing various stages of the process of solving mathematical tasks, or referring to subject-matter-related student behaviours or performance. These were named task-related categories.
- Categories referring to pedagogical-psychological content such as student motivation, attention in the classroom, students' self-confidence, etc. These were name task-independent categories ('purely' psychological categories).

* Teachers' explanations were collected by G. Dobslaw (1983) and have been reanalyzed for this paper.

There were also three remainder categories, the most frequent being 'mere description of the course of work'.

The coding system was constructed after a survey both of the literature on understanding as a cognitive process, and of studies of teacher explanations that discuss personal, task, and teaching variables (Bromme and Juhl, 1984).

There is no unified model of conceptualization and explanation in cognitive psychology for the process of understanding mathematical tasks (cf. Greeno, 1978; Resnick and Ford, 1981; Michener, 1978). Many approaches, however, have been based on a problem-solving model previously developed by Polya (1966) for working on mathematical problems.

Solving a task requires the following: previous knowledge, in order to be able to understand the task; recognition of that which is sought; knowledge concerning the proper solution algorithm and related problems. The particular importance of previous knowledge for understanding tasks is emphasized both by Ausubel and by the constructivist theory of memory.

The significance of the connection of the task with both other tasks and the conceptual system of mathematics is due to the nature of mathematical knowledge, as well as to the educational function of tasks as special instances of concepts and rules which are to be acquired.

Categories 6.1 to 6.7 are based on these elements of understanding. They refer to the process of understanding as a cognitive process and are used where task-related explanations appear. Explanations for the understanding of mathematical tasks can be given on various dimensions and with different explanatory depth. This is why causes like motivation, working behaviour, etc. were included twice – both task-related and task-independent. Furthermore, we included explanations that are task-independent, such as influences external to school or the students' general speed of work. (The further evaluation steps are presented together with the results).

In the task-related categories, the teacher's predominately considered their previous knowledge. This can be seen in the frequency with which categories were named which involve the relationship to other tasks, and the work behaviour when working on the task. In the task-independent categories, the teachers' predominately named style of thinking (e.g., 'has a good imaginative ability'), attention, concentration and general work behaviour in the lesson (e.g., 'always cooperates well').

Results

The first question: The frequency distribution across all categories (cf. Table 1) produces nearly the same number of task-related and task-independent explanations (44% and 38%). Explanations of understanding and non-understanding were complied for the table.

In the task-related categories, the teacher's predominately considered their previous knowledge. This can be seen in the frequency with which categories were named which involve the relationship to other tasks, and the work behaviour when working on the task. In the task-independent categories, the teachers' predominately named style of thinking (e.g., 'has a good imaginative ability'), attention, concentration and general work behaviour in the lesson (e.g., 'always cooperates well').

Table 1. Relative frequencies of the content analysis evaluation of teacher explanations referring of understanding and not understanding of the mathematical task. As far as the relative frequencies of the categories named can be used to draw a picture of the processes considered important for the students' understanding, we get the following result:

Understanding mathematical tasks requires from the students both specialized previous knowledge (knowing the subject matter treated in previous lessons) and general, cognitive skills such as imagination, the ability to draw conclusions, reasoning, etc. (which were summarized under category ten as 'style of thinking'). Likewise, students must be able to remember the special mathematical solution strategies necessary for a task, and must be able to apply them.

Categories referring to tasks	
2. Motivation ref. to task	3
6.1 Previous knowledge of the task related subject matter	12
6.2 Recognizing the solution sought	1
6.3 Recalling or recognizing the idea of solution	4
6.4 Elaboration of an idea of solution by the student	1
6.5 Implementing the idea of solution	3
6.6 Recognizing the connection with other tasks	6
6.8 Other items of the process of problem-solving	2
8. Observable working behaviour with regard to task	5
12. Giftedness ref. specifically to math. Topics	1
13. Task characteristics of teaching whilst treating task	4
	--- ca. 44%
Categories extending across tasks	
1. General motivation	3
3. Attention	9
4. Self-Confidence	3
5. Memory	1
9. Speed	3
10. General styles of thinking	11
11. General giftedness	1
16. Homework	1
17. Gen. Evaluation of achievement	4
18. Non-school influence	2
	--- ca. 38%
Rest	
6.7 Mere description of the course of work	11
7. Observable working behaviour in general	6
14. Gen. Characteristics of teaching	1
19. Impossible to code	3

The second most frequent task-related category is that of recognizing the connection of the task with other tasks. Among the task-independent categories, the most frequent is that of styles of thinking, followed by attention, cooperation, and concentration.

The second question of the study was: Do certain teachers prefer task-related or pedagogical-psychological causes? The teachers were allowed to give several causes for each student, and did so. Combinations of causes were also given; the teacher stating, for instance, that the student did not pay attention, was therefore unable to retain a certain mathematical theorem, and consequently could not understand the task. These combined causes can be purely task-related, purely task-independent, or mixed, as in the example (attention and lack of mathematical knowledge). Decisions were reached by coding with the categories described above (see table 1).

When a combination of cause contained at least one category belonging to the task-independent and at least one from the task-related category, it was classed as a mixed combination of causes. Table 2 contains the frequencies of the preferred explanations under the headings 'understood' and 'not understood', broken down for each teacher.

The frequency with which different types of combination were used was significantly different, as shown by Friedman's rank variance analyses ($x^2 = 147$, p = .000 understood, $x^2 = 258$, p = .001 not understood). Mixed explanations were most frequently given, with purely subject-related explanations coming second. This holds for the explanation of both understood and not understood tasks.

No teachers provided exclusively task-related or exclusively task-independent explanations over all of their eight students. Rather, we found that they prefer either mixed and task-related, or mixed and pedagogical-psychological explanations or mixed explanations only.

This means that three groups can be clearly distinguished: Group A gives for the majority of its student's task-specific, and for some students mixed explanations. Group B gives only mixed explanations for all students, and Group C consists of teachers who prefer task independent causes for the majority of their students, giving mixed explanations for the rest.

Table 2. Teachers' Explanations

Teacher	task-rel. A	ped.-psy. B	mixed C	task-rel. A	ped.psy. B	mixed C
08	2	5	1	1	3	4
09	5	1	3	2	--	6
10	3	--	5	4	1	3
11	3	--	5	3	--	5
12	--	6	2	--	7	1
26	3	3	1	1	5	2
30	2	--	6	2	4	2
31	--	5	1	--	3	5
32	1	1	6	3	--	5
33	1	2	5	5	2	1
34	5	1	2	4	2	1
36	2	--	5	5	--	3
37	3	1	4	2	--	4

Sometimes a teacher has mentioned three students per table only. Therefore the rows' frequencies do not sum up to 8.

Hence, there are not teachers (on exception) giving, for some of their students, only task-related explanations, and only task-independent ones for others. It ensues that explanations are given according to individual emphasis, which is influenced, but not determined, by the facts (task/students) to be explained. The preferences for explanations for understanding and non-understanding differ only slightly. To illustrate the explanation preferences, we shall provide an example for each group of teachers; A, B and C.

In Group A (predominantly task-related explanations), the special previous knowledge with regard to the task concerned is frequently names in combination with recognition of the task's connection with other tasks.

In Group B (predominantly mixed explanations), teachers combine the deficits in previous knowledge with lack of attention and lack of self-confidence for most of their students who do not understand tasks. In the case of understanding, combining task-related categories with the student's imagination and thinking styles is frequent. In Group C (predominantly task-independent explanations), a teacher will name, for instance, general working behaviour in the classroom, use of time, and influences external to school.

Consideration of the individual cases, however, does not only show that teachers place their emphasis differently, but it also additionally reveals that there are great differences in detail between the combinations of causes named by the teachers. The explanations thus do not adhere to a pattern, which is stable across all behaviours to be explained, and all teachers, but contain a variety of perceived combinations of causes.

The third question under investigation: As the quality of classroom activity is described in models of effective teaching, by variables that mainly concern the handling of the subject matter, we investigated the question as to whether a connection can be found between the teaching and the preferred explanation. In view of the small number (N=13) of teachers who were further interviewed regarding their explanations, such a connection can only be determined by considering the individual cases. The basis for this is a factor analysis, which was performed for the entire sample with the above listed observation variables. The classroom observations were averaged across the four lessons and then submitted to factor analysis.

Using an obliquely rotated main component analysis with correlated factors, a factor structure was obtained that allowed clear interpretation. The assumption of a two-factor structure was previously tested using an exploratory analysis without a preassigned number of factors.

This also produced two factors with eigenvalues greater than 1.0 (Kaiser criterion). For theoretical reasons, an independence of possible factors was not assumed. However, a low correlation was obtained between the two factors (r=.23).

The two factors explained 77% of the total variance, and explained the six items concerning lesson quality to a satisfactory degree (communalities between .65 and .89). The contents of the two dimensions can be differentiated as follows: Factor 1 described aspects of subject-matter-related and student-related engagement, while Factor 2 concerns aspects of the organizational and content structuring of the course of the lesson.

In order to test whether it was acceptable to take an average across all lessons, we inspected factor solutions at specific time-points. These produced the same factor structure with only slight variations.

The factor values for each observed teacher on each factor were known at the time of factor computation. We then tested the extent to which each individual teacher's values on both factors related to her career experience. No significant relationships could be found.

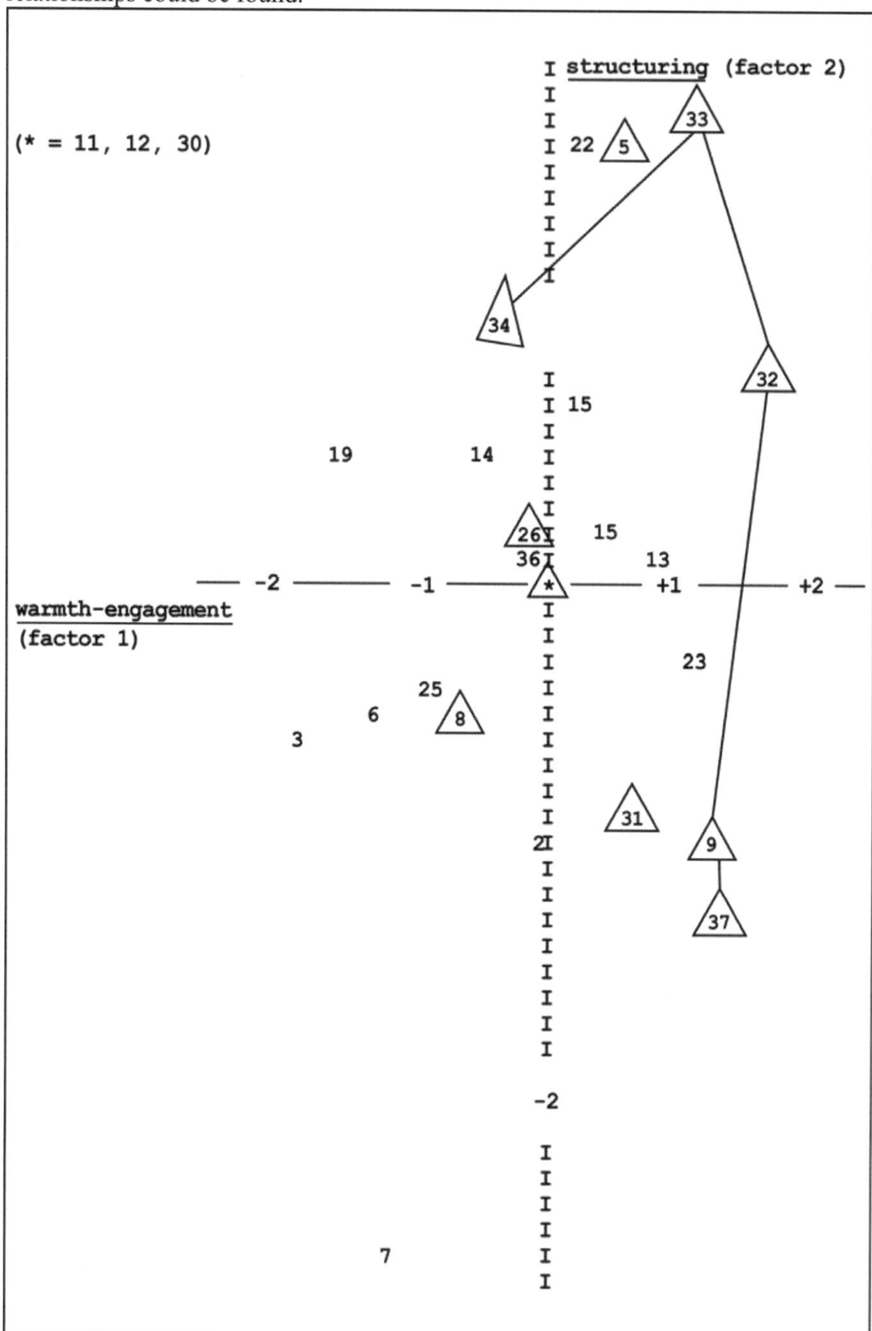

Figure 1. Factor scores of all 26 teachers.

Figure 1 shows the factor values of all 26 teachers participating in the study. The teachers whose explanations of understanding were collected, and are described in this paper, are marked by triangles. The letters serve to designate their preferences of explanation. To obtain these, the frequencies of understanding explanations were compiled from Table 2. Grouping as to A, B, C was performed according to the most frequent type of cause named.

A study of individual cases should focus on those teachers who have relatively extreme factor values compared to their colleagues. They are situated at 'the margin' of the diagram and have been linked for purposes of illustration. It is evident that these 'good' teachers belong either to A (predominantly task-related explanation) or to B (predominantly mixed explanations). The four teachers 33, 32, 9 and 37 are among the teachers in the overall sample who show the most 'enthusiasm and warmth' (factor 1).

For the other dimension, no such connection can be determined. Here, teachers 33, 34, 32 show rather high values, whereas 31, 9, and 37 show rather small ones.

This would seem to justify the cautious conclusion that the 'enthusiastic' teachers prefer task-related and mixed explanations. The opposite conclusion, of course, does not hold; i.e., there are also teachers having these preferences of explanation who show only small or negative factor values on the dimensions of quality.

Discussion

The teachers' explanations of causes make allowance for the cumulative character of learning in school. Mastery of the previous subject matter is considered a prerequisite for understanding a task. Tasks cannot be considered in isolation in the classroom. In order to understand them, it is necessary to recognize their relationships to other tasks and areas of the subject matter taught. This item is also frequently named by the teachers. The prerequisites of understanding, however, do not only lie within the student's knowledge and immediate cognitive processes. Understanding also requires activity and attention. Both the student's observable working behaviour and concentration are frequently named. The second most frequently named item is the style of thinking (for example: is able to think logically or spatially), which is a field-specific ability of the student that is relevant for the task in question.

This image of 'understanding' from the teachers' point of view is quite consistent with basic assumptions of modern memory research and teaching/learning research, for instance in its emphasis on the importance of previous knowledge for learning, or in the frequent naming of task-related activities. The analogies to the importance of the 'time on task' variable of empirical teaching research are obvious.

The frequency distributions of causes confirm the findings of Bromme and Juhl (1984) obtained from another sample taken from a different type of school (Gymnasium), but matched for age group and school subject. There is, however, a difference. While 69 percent of all items were task-related, field-specific explanations in the Gymnasium, there were only 44 percent in our present sample. In the case of the Gymnasium teachers, there were only 21 percent task-independent, general categories compared with 38 percent in the present study. It is possible that the difference can be explained by the fact that in the Gymnasium particular emphasis is traditionally placed on teaching subject-matter specific, curricular knowledge.

The relatively high proportion of task-related concepts of explanation, however, remains notable, even if it is smaller than that in the Bromme and Juhl study (1984). The high proportion is possibly due to the method of obtaining data, as student and task were equally salient. This situation, however, is realistic, in that it conforms to teachers' everyday teaching experience.

The frequency distribution of the causes named does not yet clearly define the individual patterns of explanation. For the second question of the study, the explanation was thus evaluated for each teacher in order to see whether it was purely task-related, purely task-independent, or mixed. One half of the teachers used mixed explanations for most of the eight students they had to evaluate; the other teachers either preferred purely task-related or purely task-independent explanations. All teachers, however, gave mixed explanations, so that we have to assume teacher preferences rather than strong differences between teachers.

The variability of the individual patterns of explanation is remarkable. It indicates the presence of comprehensive and organized professional knowledge about 'understanding' (cf. Rheinberg and Elke, 1978). Among these explanations, the mixed ones best conform to that which is understood as an explanation according to what is called 'understanding' within the philosophy of science, namely the linking of a specific observation to a more general assumption.

For the third question of the study concerning the connection between the explanations and the quality of teaching, we have behaviour observations from a teaching unit at our disposal, which was held after the explanations had been given. This enables us to obtain those connections between explanation preferences and quality of teaching that are very stable. In view of this difficulty, it is remarkable that the four teachers who show – compared to the overall sample – a relatively high amount of subject-specific interest and warmth (factor 1), prefer task-related or mixed explanations rather than task –independent ones.

This result can only be taken to be a first indication toward a possible connection of this kind. At least it is a hint that a connection between the quality of instruction and the teachers' explanatory concepts might actually exist.

Author Reflection Twenty Years On

The main message of this chapter is the importance of subject matter related issues in teachers' thinking. Analyzing the integration between pedagogical/didactical information and subject matter related information in teachers' views of classroom events is still a challenge for research on teacher thinking. This chapter is an example of study that has this integration in its focus.

Chapter 4

How Do Teachers Think About Their Craft?

Sally Brown and Donald McIntyre

Summary

Research is described which attempts to explore the integrated knowing, thinking and action which are assumed to characterize teachers' professional craft knowledge. Its approach gives priority to: (i) investigating how teachers think about what they do well in the classroom, (ii) letting those in the classroom (i.e. teachers and pupils) decide what is done well, (iii) using pupils' judgements about teachers to decide with whom the research is done and (iv) persuading teachers to articulate what they do well and how they do it. As a research strategy, it fits into the heterogeneous grouping known as 'case-studies', but it also has a primary concern with the development of generalizations across teachers. An account is offered of the research approach, collection of data, and some preliminary steps in the generation of a conceptual framework to reflect the ways in which experienced teachers think about the strengths of their own teaching.

An Investigation of Teachers' Professional Craft Knowledge

Our interest is in the practical part of experienced teachers' professional knowledge, which we refer to as *professional craft knowledge* (PCK). By that, we mean the knowledge, which is:

(i) Embedded in, and tacitly guiding, teachers' everyday actions in the classroom;
(ii) Derived from practical experience rather than formal training;
(iii) Seldom made explicit;
(iv) Related to the intuitive, spontaneous and routine aspects of teaching rather than to the more reflective and thoughtful activities in which teachers may engage at other times;
(v) Reflected in the 'core professionalism' of teachers and their 'theories in use' rather than their 'extended professionalism' and 'espoused theories'.

The plans for our research have drawn on the ideas of Deforges and McNamara (Deforges and McNamara, 1977 and 1979; McNamara and Deforges, 1978), and our aim is to generate theory, which is grounded in the practice of classroom teaching. Our starting point is to assume that most experienced teachers already use a wide range of sophisticated skills about which it would be most valuable for us to learn. It is their strengths and not their deficiencies in which we are interested. This approach has implications for the use, which might later be made of our findings for in-service staff development or the design of more appropriate educational innovation. At present, staff development and innovation tend to operate in our country to a deficit model: problems or omissions are identified and the gaps are plugged. Our research implies a different model in which those developments and innovations would be built on existing strengths. Furthermore, an understanding of what experienced teachers do well would have clear benefits for pre-service teacher education.

Research Approach

Decisions about which of the many aspects of teachers' PCK the research will be concerned with, are made by those inside the classroom: pupils' statements about their teachers' strengths determine which teachers we approach with an invitation to become involved (we look for high levels of agreement among pupils about specific teacher characteristics and for a variety of characteristics among teachers); and teachers' perceptions of their own strengths determine the particular aspects of their teaching to which we pay attention. Our concern is with what *teachers* see as necessary and desirable to do in their classroom circumstances, with understanding how *teachers* construe their own teaching, and not at all with evaluating that teaching. Every effort has to be made to minimize the influence on our data collection and analysis from the well-known models of teaching and teachers generated by 'outsiders' (e.g. process-product or teacher as decision-maker, classroom manager, information processor, facilitator, dilemma-resolver).

The context in which we have chosen to work is primarily that of teaching in the early years of secondary school (with pupils aged 12 to 14) and, to a lesser extent, in the later years of primary (with pupils aged 10 to 12). All the classes we work with are mixed ability within comprehensive schools. Sixteen teachers have been involved: in secondary, the twelve teachers come from art, computing, English (2 teachers), French, geography, history, mathematics, physical education, outdoor education and science (2 teachers); and the four primary teachers have responsibilities across the curriculum. No assumptions can be made about the validity of the findings for other contexts (e.g. for the teaching of other age groups).

The research has to find ways of stimulating teachers to articulate and elaborate aspects of their tacit professional craft knowledge. How that will be done will depend on the nature (presently unknown) of that knowledge. The investigation has been planned, therefore, in two stages. First, we are attempting a straightforward exploration of the ways in which teachers construe those features of their existing patterns of classroom teaching, which give them satisfaction. These patterns are highly adapted to the circumstances in which they find themselves and the purposes to which they find they have to give priority. The second stage will use the findings of the first to help to decide a strategy for persuading teachers to elucidate as fully as possible *how* they do the things, which they do well. It is the first of these stages, which is reported here.

The demands that our research makes on teachers are substantially less than, for example, those which action research would call for. We do not ask them to become researchers but rather to talk about their classroom practice; the responsibility for the theorizing remains with us. Nevertheless, our anxiety at the start of this empirical research was high. We were to be concerned with an exploration of: the complexities and uncertainties of teachers' professional craft knowledge and practical expertise; their intuitive performances (which some believe are not open to explicit description and must remain implicit in their actions); the plethora of decisions, judgements, interactions and behaviours of which experienced teachers may be unaware and which they carry out more or less spontaneously; and the ways in which they make use of their previous experiences in dealing with new and, strictly speaking, unique situations which they encounter in the classroom. Although we started from the assumption that there is such a thing as teachers' professional craft knowledge, we knew that for the most part this knowledge is not articulated. Was it sensible, therefore, to plan to undertake an investigation of what pupils and teachers construed as good teaching?

We knew that pupils could talk in general terms about what their teachers do well; the data collected in our preliminary work had satisfied us of this. Whether pupils would be willing and able to identify specific instances of the qualities displayed by teachers in particular lessons remained to be seen. We knew from previous experience that teachers would be able to talk about their teaching, but we could not predict whether they would do so only in terms of their formal teaching knowledge (e.g. as formulated in most teacher education courses), how talking about their teaching might affect their classroom performance (e.g. increasing self-consciousness or engendering paralysis) or if our probing would stimulate *post hoc* relationalisations of a kind in which we were not interested.

The Collection of Data

Before the teachers finally agreed to work with us, we showed them summaries of the statements made about them by pupils. However, we requested that prior to seeing this summary each teacher should give us a brief account of what they saw themselves as doing well in their teaching. This they all agreed to do and so provided us with a preliminary set of data to help us formulate comparisons between teachers' and pupils' constructs of the teaching.

Some aspects of our approach have been common across all the teachers with whom we have worked. We have endeavoured to retain a pattern of observing and recording two to four hours of teaching for each teacher; each teacher has made the choice of the unit of work to be observed; we asked each teacher to tell us in writing what aspects of their teaching they were particularly pleased with (i) at the end of the first observed part of the unit and (ii) as they reflected on the whole unit of work at the end; we interviewed each teacher after the second and any subsequent observed lessons; we asked pupils at the end of each observed lesson and at the end of the unit of work to tell us (in writing) about what their teachers did well; we gave the teachers copies of the audio tapes of the lessons; and in the light of them having had an opportunity to listen to the tapes, we carried out two further and longer interviews some two weeks after the end of the unit of work.

There were differences, however, among the different cases. In the primary schools, we usually observed two full mornings with a variety of different aspects of the curriculum intertwined. In the secondary school, timetabling and time allocation differed among subjects: 'two to four hours of teaching' in English and French was a week's work including several double periods; in art it was two weeks of single periods; in physical education there were two double periods of basketball over two weeks; in outdoor education we had a whole day expedition; and in computing there were five weeks of single periods. This has consequences not only for practical arrangements but also for the unity in a 'unit of work', for how pupils and teachers construe the lessons as having continuity and for organization, tactics and strategies which teachers use to provide introductions to, and linkages between, the classroom works of different lessons.

We are not yet at the point where we can offer an account of our 'results'. All we can do is to give some tentative impressions of what appears to be emerging from the analysis of our data.

Analysis of the Data

Data from Pupils

Some general ideas about the teachers we worked with can be gleaned from the pupils' statements. These were of two kinds: first, some pupils (about 350) were asked to comment on general qualities which they admired in their teachers; and secondly, others (about 400) were asked to comment on what they thought was good about the teaching in a particular lesson. In relation to the first of these, we can say that with very few exceptions the statement we collected were sensible, positive and useful. Qualities of teachers which were cited included the ability to create a relaxed and enjoyable atmosphere, classroom control, presentation of work so that it is interesting and understandable, clarity of explanations and instructions, helping with difficulties, treating pupils as mature individuals and encouraging pupils to raise expectations of themselves. An exploratory study with fifty-seven teachers reassured us that their categorization of pupils' statements agreed with our own. It was these statements that were used to select the teachers, and so our sample is biased towards those teachers who have good relationships with their pupils.

Our collection of comments from pupils about what had gone well in each lesson or selection of work was quite productive. They almost always commented on how these teachers explained clearly so everyone understood and knew what to do, the ways in which they were ready to help pupils when they were stuck and the classroom control that ensure the work would be carried out in a relaxed and industrious atmosphere. Beyond this, the statements reflected the diversity of the qualities displayed by the different teachers such as an individual's use of humour to sustain an enjoyable atmosphere, a readiness to take account of pupils' opinions, a supportive way of dealing with pupil's public errors, efforts to involved everyone in the activities and to develop self-confidence, kindness and warmth, effective demonstrations, specific subject related qualities, taking the trouble to provide background information and choice of helpful and enjoyable lesson content or activities.

We became aware, however, of some of the limitations of this approach. We are convinced that we should not use it with pupils younger than our youngest group (age 10). The texture detail and elaboration provided in the written responses becomes very thin at this stage, and the approach which at all stages is biased towards the views of the competent writers becomes even more suspect with younger pupils. It also seems that pupils have a low tolerance for repeating the same task at the end of each lesson.

Data from Teachers

The data we have collected from the classrooms and subsequent interviews with teachers are astonishingly varied and enormously rich. The accounts they give of what they themselves do well are striking in their individuality but demonstrate high levels of internal consistency. In one sense, the diversity of the coherent 'stories' of good teaching is unnerving when we face the prospect of trying to reach widely generalisable constructions of teachers' craft knowledge. Nevertheless, we are currently stimulated by what the different teachers say about such things as:

- The ways in which they manage information about, and introductions to, the work of the lesson;
- Their approaches to taking account of differences among their pupils;
- The ways in which they deal with the errors which pupils make;
- Their attempts to build up confidence and trust with pupils;
- Their distinctive concerns for the characteristics of individual pupils;
- Their strategies for diffusing potential discipline problems and for making sure that recalcitrant pupils do not become alienated from the work or the experiences;
- The nature and purpose of their 'discussions' with pupils;
- Their efforts to ensure that everyone is involved in the work and all achievements are recognized;
- Their ways of managing group activities;
- How they change tack when pupils' interest or attention may be flagging;
- How they endeavour to ensure that pupils' creative efforts will not be hindered by technical expertise that is lacking;
- The ways in which they create a relaxed and enjoyable but, nevertheless, disciplined atmosphere.

Towards Generalizations: A Framework of Concepts

Since our approach uses sixteen case studies, one of our concerns must be to understand how each teacher, as an individual, evaluates and talks about his or her own teaching. However, if our aim is to understand, communicate and ultimately to use those aspects of PCK that we are investigating, then we have to formulate our findings in ways that are potentially generalisable across teachers. This implies an attempt to establish generalizations and economical concepts across the individual cases. Such generalizations will not be of the probabilistic kind, which arise from the application of statistics to large data sets in, for example, survey research or experimental studies. Our generalizations are better described as naturalistic, and as forming hypotheses to be carried from one case to the next rather than as general laws applying across a population. Together, the generalizations make up a provisional theoretical framework which is intended to expose and make explicit some aspects of the tacit knowledge of expert experienced teachers (who may well manage and teach their classes in such taken-for-granted and routinised ways that they are unconscious of the craft knowledge embedded in their actions).

We have used a careful systematic, case-by-case, interactive, inductive process to identify a set of concepts that appear to be generalisable among the criteria which teachers use in evaluating their own teaching as satisfactory. Outsiders' concepts (such as 'objectives' or 'decision-points') have been eschewed unless they arise directly from the teachers' accounts. Ideally, the generalizations formulated should: be directly supported by evidence; be falsifiable; not discount any of the teachers' accounts; go beyond evidence from one individual or one occasion; demonstrate fully the logic and rationality of the teachers' actions; and be recognizable and acceptable to the teachers.

At the time of writing, we have carried out a detailed analysis of at least one interview with each of the sixteen teachers. Using the data from the secondary teachers, we have inductively generated a tentative framework of concepts, which appears to encompass and reflect the ways in which these teachers talk about the strengths of their classroom teaching. Only a tiny proportion of their statements cannot be so accommodated. The data from the primary teachers are also fully accounted for within the framework, but these were not used in the inductive framework-building process. (The sample of primary teachers is too small to predict whether the same framework would emerge if we had started with a consideration of their data.)

The framework is still in its infancy, but we are finding it useful to organize our ideas about the ways in which teachers perceive their own teaching. The rest of this paper outlines some of the control concepts, which seem to be helpful in characterizing teachers' thinking and are sufficiently simple to reflect the immediacy and spontaneity of classroom actions.

In the first place, it seems that most teachers evaluate their lessons in terms of what we have called *normal desirable states of pupil activity* (NDS). A lesson is deemed satisfactory so long as pupils continue to act in these ways, which are seen as routinely desirable. Different teachers may have different criteria for what counts as an NDS. For example, they may expect pupils to be (i) interested, involved and asking questions, or (ii) working independently, using worksheets and doing everything pretty much on their own, or (iii) thinking about what they are doing. These NDS's may vary according to the phase of the lesson or the particular pupils being taught.

In addition to (and occasionally instead of) criteria of NDS of pupil activity, some teachers evaluate their lessons in terms of concepts of *progress*. The concepts of progress, which are used, fall into three broad categories: development of pupils' attributes (e.g. knowledge, confidence); progress through the work (e.g. the syllabus, the teacher's plan for the day); and the production of something (e.g. an artifact, a performance).

The concepts of *NDS* and *progress* are distinguished by the former involving a pattern of activity being maintained without change over a period (albeit sometimes a short period) of time, and the latter introducing a development aspect in contrast with the steady state of the former. For example, where a teacher talks about 'pupils understanding' we would categorise a reference to 'pupils understanding what is going on in the classroom or what the teacher is saying' as *NDS*; if the emphasis is on 'pupils developing an understanding what is going on in the classroom or what the teacher is saying' as *NDS;* if the emphasis is on 'pupils developing an understanding of something', then this would be seen as an example of *progress*. Unambiguous categorization, however, is not always possible: where teachers talk about 'pupils picking things up' (in the sense of understanding), we have some difficulty in deciding how this should be classified. This is no great problem since the framework of concepts is intended to communicate ideas about teaching and not to enable every statement to be reliably coded.

It will be necessary to examine more closely than we have been able to do so far the relationship between the concepts of NDS and progress for individual teachers. A cursory look suggests that sometimes an NDS and an example of progress may simply be different ways of looking at the same thing; or the progress may be a necessary development before the establishment of an NDS; or the NDS may provide the necessary conditions for making the progress.

The third set of generalisable concepts relates to the standards which teachers apply to their criteria for the first two kinds of concepts (NDS and progress). These standards depend on the *conditions*, which impinge on the teaching. It seems that there are several sub-sets of concepts here, which refer to conditions of *time* (usually a constraint and includes timetabling factors), *material conditions* (space, equipment, class size, weather), *pupils* (enduring characteristics or behaviour on the day), and *teachers* (personal characteristics or feelings/behaviour on the day). We have yet to unpack this complex of concepts, but we are aware that conditions of one kind can affect conditions of other kinds. For example, a constraint of time may result in pupils, or teachers feeling under pressure; in that event, time may be the basic cause of, say, a disruption in the NDS of pupil activity, but the immediate causal factor may be seen as some aspect of the pupils' or teacher's behaviour.

A fourth set of concepts is concerned with the teachers' evaluations of their own activities. These evaluations appear to have two facets: first, the *extent to which their actions maintain their NDS*; and secondly, the *extent to which their actions promote progress*. In addition, it appears that the *conditions* (time, material conditions, pupils, teachers), which influenced the standards applied to NDS or progress, are also seen to impinge on and lead to variation in the teachers' actions.

At the moment, therefore, our emerging conceptual framework for teachers' thinking about their classroom teaching has to take account of our evidence that:

> The criteria and goals teachers use for evaluating their teaching are concerned with: first, maintaining some *normal desirable state of pupil activity*, and secondly, promoting some kind of *progress*. The teachers evaluate *all* their own actions in relation to these purposes. The standards they apply to achieving these goals in practice, and the single-mindedness of their own actions, are influenced by various *conditions*, which impinge on the teaching.

It is also of interest to note some of the things, which do *not* characterize our teachers' evaluations of their classroom teaching. For example, the words 'learning', 'objectives', 'aims' and 'decisions' are rarely used and, as yet, we have not identified examples of teachers evaluating any aspect of their teaching as inherently desirable, as characteristic of good teaching, or indeed in any other terms than as instrumental towards some kind of *NDS* of pupil activity or some kind of *progress*. However, we would emphasize that these are not conclusions, only ideas that we are exploring.

Our next step is to see whether our tentative generalizations hold up when we analyze the rest of our data. Then the crucial questions for the teachers: Are the abstractions we have made from their accounts intelligible to them? Do they distort what the teachers are telling us? Do they capture the major emphases of the ways they think about their classroom teaching?

Acknowledgement

We gratefully acknowledge a research grant from the Scottish Education Department (SED) for this research. The views expressed in this paper, however, are those of the authors, and are not necessarily shared by the SED.

Chapter 5

Information Technology and Teacher Routines: Learning From the Microcomputer

J.K. Olson

Introduction: Computers as a Subject

We have undertaken case studies of forms of microcomputer use in elementary and high schools using a variety of data collecting methods (interview, journal, stimulated recall, observation, repertory grid and questionnaire).

The schools we studied were selected from a list of about 30 schools in a large suburban school board, which were conducting action – research projects. We selected these schools in order to include different levels of schooling and different aspects of the curriculum. In each school, we interviewed the principal teacher and one or more teachers involved in the action research. For each school we collected information about the history of computer use in the school and the background of the teachers. The information we collected was used to produce a policy-oriented report to the Ontario Ministry of Education (Olson, J. and Eaton, S., 1986).

From the data we collected, we constructed eight accounts of computer use, which we have divided into two groups. The first group comprises four cases in which the computer is used as a way of teaching computers as a new subject in the elementary school curriculum. Within the subject domain, we find activities like: programming, low-resolution graphics and analysis of software. Our second group comprises four cases in which computers are being used as a teaching aid. Here we find drill and practice, tutorial, simulation, and word processing activities carried on within the framework of existing subjects. The way the computer is used here is heavily influenced by existing plans for the scope and sequence of the subject and ways of teaching the subject; but there are interesting ways in which computer use affects these curriculum elements.

As we do not have space here to treat both of these areas, we will discuss how teachers use the computer as part of a new subject, which they call computer 'awareness', or computer 'literacy' (see also Ragsdale, 1982; Amarel, 1984). Although some distinguish between these two terms, the common thread is that these teachers are experimenting with a new subject and with ways of teaching this subject within existing constraints of curriculum and resources. They have volunteered to do the additional work of teaching about computers and have been granted the use of at least one Apple 2e, one green screen and a disk drive; most of these teachers have printers and colour monitors, and some have peripherals like joy sticks and Koala pads.

There is no formal curriculum for the new subject. Teachers are making it up as they go along according to their own interests, and they express the uncertainties that come with such 'unlicensed' practice. There are many questions that must be left aside here although they were discussed in the report of this research: Why these teachers? Why have they been allowed to experiment so freely? Why the computer promotion by their school board? Why choose the particular activities they have as definitive computer awareness?

All four teachers we are considering adopted a similar strategy for using computers in the classroom that we call the 'teach yourself' routine. We will outline the main features of this routine and describe the difficulties teachers had in maintaining it. We hope to show why teachers use the 'teach yourself' routine and to see what their response to the difficulties of the strategy tells us about their approach to '*newness*' itself as an element in working life in the classroom.

The Teach Yourself Routine

The routine has these composite features:

a. The computer subject goes on all the time. Students go to the computer based on a rotary; students are there while the teacher is teaching the rest of the class.

b. Students engage in learning the subject mostly by programming the computer, but also through running assorted software.

c. The teacher teaches the rest of the class and aims at minimal contact with the 'class' working at the computer.

d. To ensure minimal disruption and delay, the teacher may offer whole class instruction in key computer moves or refer students to manuals, and the teacher relies on computer 'Whiz-kids' to help those who are 'stuck' and to tutor their peers. The teacher also asks students to rely on the manual to help them 'debug' their own problems.

e. The teacher relies on certain students to preview software.

f. The teacher chooses software and activities that require minimal teacher support. Thus Basic is favoured over Logo as a programming activity because they think there are more and more simpler steps in Basic.

g. Access to the computer is part of the classroom reward structure.

These are the mean features of the strategy as teachers described it. Why such a strategy?

First, we should say, and the teachers would say, that this strategy is not new. It is an extension of self-directed seatwork with which teachers are quite familiar. There are many similarities with sending children to the library to do library research using the librarian as a support rather than 'peer tutors', except that instead of using software students use reference materials. These teachers have taken a familiar strategy 'off the shelf' and used it to support student self-directed work on the computer.

Given that normally only one machine is available, it is not surprising that teachers have chosen to 'teach' the subject in parallel with the teaching they have to do anyway, and given their idea that doing the subject is doing programming, access to the machine is critical. There are, of course, other ways of conceiving of doing computers as a subject, but these teachers all chose doing programming, or becoming familiar with *types* of software and peripherals as their way of defining it.

We can think of reasons why it is not surprising that teacher proceed this way. They find that some students are teaching themselves anyway, and these few students stand out in the eyes of these teachers. They are interested in computing and eager to explore software. They seem to know how to 'debug' it. Teachers see them as members of the computer generation; it is a subject that is part of their culture. Doing computers, one might say is a recess and lunchtime activity that now has a legitimate place in the classroom. The new subject is one that has come from the student culture itself, so these teachers seem to think. Some teachers feel they are on the outside of it in some ways, and this feeling depends on whether or not the teacher is a computer 'buff' or expert, but even so the sophisticated teachers are prepared to let students teach themselves computer skills.

Teachers are also short of time. How do you add another subject and a non-licensed one at that on already heavy work? (See also Shinegold, 1983). The teachers let the students work mostly on their own thus freeing them to teach the rest of the class what had to be taught. In any event, teachers found themselves in the middle of two lessons rather than teaching the main class, with the computer 'class' teaching itself. They were called upon to do two things at once; much more so than they had hoped; and more so than normal when the teach yourself method is used in other contexts, say in library work or using work stations.

We thus became interested in why the 'teach yourself' routine had not worked as well as in other contexts. We found a number of reasons.

1. The programming required more support than teachers had imagined it would. Students could not handle messages like 'syntax error'; no teacher had ever said 'syntax error.' Teachers give positive feedback, which has a 'warming' affect, but the machines gave negative feedback that could not be interpreted, and was read by students as unhelpful and cool.
2. Computer literate peers did not tutor. They would 'debug' and move on, and the students they 'helped' were not any the wiser about why they had run into problems.
3. Some students found that handbooks were difficult to use.
4. Some students had difficulty reading text on the screen.
5. Students became bored with software they could not control, and some types of software were not that interesting as such. Logo was cited as a poor performer in the 'teach yourself' strategy by teachers.
6. Students were posing management problems that could not be quickly dealt with like other seatwork situations.

Teachers were not able to do two things at once without flaw, and we saw examples of 'slippage' that came about because the strategy did not work as planned. These incidents give a sense of what 'slippage' was like:

1. One student entered programming lines incorrectly, and was left for up to an hour before receiving corrective feedback by the teacher.

2. Another student entered a poem with a particular shape but did not know how to paragraph and ended up with a printout quite unlike what she had hoped to get. What she saw on the screen was not what was printed because she lacked control of the word processing.
3. A student tried to produce a graphic without her coordinates that had been left at home; she guessed at the coordinates.

Teachers were aware of the 'slippage.' They could see it in the videotapes we showed them; yet they said they were satisfied with their efforts at running the 'teach yourself' approach to computers as a new subject although they were frustrated that the strategy had not worked as well as they had hoped it would; that is that students would teach themselves. Because the students at the computer needed more support than anticipated teachers found themselves having to reconstrue the teach yourself routine. It became important to understand how the teachers construed the modified teach yourself routine. We shall call this modified approach the 'two things at once' routine. It is, of course, common for teachers to do two things at once in the classroom, but we were interested in a special variant that involves the use of computers.

How Teachers Construed the 'Two-Things-At-Once' Routine

We wanted to explore in greater depth how teachers construed their having to do two things at once and to do this we asked them to construe elements of computer use which involved episodes of delay and interruption and which required to intervene in student activity. An example of elements is: "One student does not want to work with another student at the computer." We asked them to sort these episodes (elements) into categories, and to subsequently construe all of the episodes using their own categories (Olson and Reid, 1982). We found the following constructs to be common:

1. Teacher-resolved vs. student-resolved episodes.
2. Student vs. technology problem episodes.
3. Quickly resolved vs. slowly resolved episodes.
4. Existing rules applied vs. new rules required.
5. Routine responses adequate vs. judgements called for.

From our analysis of the constructs, we began to appreciate the problems teachers hoped to avoid in adopting the 'teach yourself' routine in the first place, and the problems caused for them by its partial breakdown into their having to do two things at once. We could see that the 'teach yourself' routine was intended to minimize interruptions from the computer 'class,' and that their views about software and machinery were based on their constructs of speedy resolution of classroom problems using existing routines. Indeed, for these teachers complex software was not the issue in the two things at once routine, complex student reaction to it was. These teachers did not want to have to invent new procedures for teaching; they hoped that the existing procedures would work. Ironically, the very way they defined doing computers as a subject ensured that their existing routines would not work well. The level of teacher support demanded by programming simply went well. The level of teacher support demanded by programming simply went well beyond what they were used to giving in other 'teach yourself' situations, and although they thought that the peer tutoring would sustain the routine, peers did not tutor each other; computer literate students helped 'debug' situations where they could, when they could, and the teacher had to be called in often as well.

We can see now how these teachers have 'installed' the computer subject in their classrooms. The teacher reaction to newness might be put this way. The subject was taught using a well-tried approach: the 'teach yourself' strategy. The effect of the unforeseen elements of the new subject (for example, negative feedback of an imprecise kind) was that the well-tried methods did not work well, and although the teachers professed satisfaction with them, we were not so sure they really were neither satisfied, nor sure whether they ought to be satisfied.

Expressive and Instrumental Elements of Teaching

Why did they express satisfaction? We have here an apparent paradox that we felt ought to be resolved. In spite of the fact that the teach yourself routine did not work as well as teachers hoped, they were prepared to continue with this conception and way of teaching the subject. They found that students enjoyed the new subject in spite of the difficulties of the 'teach yourself' strategy. The teachers felt that the difficulties were worth the trouble because they enjoyed the children's pleasure at having a computer in their room. The teachers construed the using a computer as their way of being 'modern,' of expressing something about the kind of teacher they were. This purpose seemed to overrule the evidence that their chosen way of teaching about a new subject had not worked out that well. Their mild disquiet about the 'slippage' was countered by the thought that their students had made significant gains in computer literacy. Evidence for these claims, however, seemed sparse, and negative evidence was ignored while claims about progress were made with little positive evidence available.

Why did the teachers act this way? Following the work of Harre (1979), we believe that teaching activities can be considered from two points of view: the expressive and the instrumental. Expressive teaching acts convey messages about how the teacher wishes to be seen by his/her students. As Harre suggests it is through the expressive dimension of our actions that we convey messages about the sort of person for which we wish to be taken. Instrumental teaching acts are those we commonly associate with the classroom – the 'processes' of the 'process-product' paradigm; acts which are directed as fostering learning, what we normally attend to in studying the influence of the teachers on the outcomes of classroom activity.

We argue that exclusive focus on instrumental acts of teaching is an insufficient basis from which to make sense of what happens in classrooms. One must understand both the expressive and instrumental dimensions of teaching activities. In our case, we could not make sense of the paradox we had uncovered without appreciating the expressive dimension of teaching. We believe that the teachers value the computer as a symbol to be used *expressively* by them to enhance their standing in the eyes of the students, parents, and principal. The minor difficulties in achieving computer literacy could be ignored for the time being since they were not as important as the expressive process. Let us consider how this is so.

These teachers spoke about the positive impact having a computer had on the class; how students were eager to use it; how much they enjoyed having it in the room. By having a computer in their room, these teachers were able to say something about their interest in modern teaching methods and in the needs of their students, and about the relevance of their classroom to life outside it. The eight case studies reveal a common concern about the expressive dimension of using computers in the classroom. In the end, of course, expressive and instrumental issues will receive more attention. How quickly the process occurs and how it might be facilitated are important further questions for research and policy (see Berg, 1983; O'Shea, 1984).

This study raises questions about how well existing routines can be modified to take advantage of new technologies; it also raises questions about what sustains these routines, what teachers have invested in them and how teachers can become more critical of the costs and benefits of the routines that they use.

Assessment of Routines and Paradoxes of Teaching

Working with these teachers in the way we have involved them and us in a process of reflection-in-action (Schön, 1983). In confronting teachers with their practice, we have asked them to consider the expressive and instrumental functions of their teaching; to bring to the foreground matters which often remain tacit. It has been said that action research is not an appropriate activity for teachers. Construed as 'research' in the usual way, this claim is probably well founded. It is unrealistic to expect teachers to do research studies of the experimental or quasi-experimental type in their own classrooms but it is essential that they subject their practice to critical, systematic reflection. We think this reflective process is in fact how action research ought to be construed – the careful consideration of specific elements of teaching, and that we ought to broaden how we construe research itself in education.

The systematic consideration of practice itself can yield the theories-in-action that are immanent in practice. As Schön (1983) has suggested tacit knowledge is not prepositional knowledge, but is knowledge no less for that, and the recovery from practice of the theories implicit in the practice are what we take action research to be; thus verifying theory and practice in one process. In this study, such collaborative action research has led to the identification and dissolution of a paradox. We think that apparently inconsistent or unusual practices might be usefully construed as paradoxical; that is 'contrary to received opinion or expectation; sometimes with a unfavourable connotation as being discordant with what is held to be established truth' (OED). Thus, sometimes teachers appear to be doing things in ways contrary to orthodox opinion, and even their own espoused view.

Often teachers are viewed by outsiders as practicing inconsistently when compared to certain ideologies. The teachers are said to be in the grip of a dilemma. In fact, what has happened is that the outsider often has perceived an apparent paradox and has failed to resolve it, or has not discovered the real paradoxes. It is not at all clear that teachers actually experience the dilemmas orthodoxy tends to make us think they experience.

Close attention to how teachers think about their practice is a way of surfacing and understanding paradoxes of practice. Some paradoxes are real; for example, the teacher may not be acting in accord with his/her own intentions. Other paradoxes may be only apparent; the teacher may be pursuing goals that are not those that are espoused, but valued nonetheless, at least by the teacher in relation to whom the practice is not paradoxical. Only orthodoxy renders the practice so. Thus, the unusual practice seen in instrumental terms may appear paradoxical, but seen expressively is not paradoxical, at least not in relation to what the teacher is trying to do, and in relation to a holistic account of what teaching entails.

It is here, we believe, that the notion of expressive and instrumental dimensions of teaching is especially important. Teachers are conditioned by their education and by professional journals to pay attention to the instrumental dimensions of teaching as a way of assessing their practice; yet their practice cannot be other than paradoxical unless the expressive dimension is considered. As we saw in these cases, upon analysis of their practice, it became clear that expressive elements of their actions allowed us to account for the apparent paradox of persistence with ineffective teaching routines – that is with the doing two things at once routine in spite of the difficulties that gave rise to. Such a routine left the teacher frustrated by interruption, yet in the absence of means to overcome these interruptions and wanting to continue to have students attend the computer 'class' for expressive reasons, the teachers were prepared to continue with the routine. The paradox has disappeared.

Action research in the classroom enables teachers to explore paradoxes of practice and to learn from them more about what they value in practice and about the limits, and limitations, of existing routines. In the cases we have been examining here the limits of the teach yourself routine can be clearly seen; such a routine does not effectively extend to teaching computers as a subject when construed as learning how to program. In discovering the limits of the routine, its limitations come to be known. As practice is recovered and analysed in this way there exists the possibility of a body of practical wisdom which could become the rational basis for practice itself – a true educational theory based on practical wisdom derived from the interpretation of practice. We attach this significance to the research.

Author Reflection Twenty Years On

This paper urged researchers and teachers themselves to look at the expressive and instrumental actions of teaching when trying to understand what teachers do in practice. It suggested that teacher routines are more complex than often noted. It should also be said a better understanding of teacher culture is needed if school reforms are to succeed.

Missing in many innovation strategies is an adequate understanding of what norms of practice prevail in teacher and student culture, and how teachers view the risks of disengaging from those norms sufficiently to risk engagement with innovative practice. Existing change models tend to be focused on management issues and to underplay the importance of established classroom practices. Research is needed to find out more about teachers' work culture and the technologies that sustain it, and the implications of new approaches for those technologies. *This process involves a high degree of negative feedback.* For example, how well does stating curriculum policy in the form of *outcomes* work as a way of curriculum improvement? How well are teachers able to implement *constructivist* ideas in practice?

These questions are not only about research. They have implications for in-service teacher education and professional development. The teacher voice is currently not central in school change and thus negative feedback is lost. Indeed school systems are lobbied by powerful stakeholders – such as business interests –

which promote frames for discussion taken from their practice – notions such as, for example, *best practice* or *value added* whose nature – for them and those who would adopt these ideas – beg many questions about values. Lacking negative feedback and with a limited language to express educational purposes, the danger arises that curriculum policies may be mandated without adequate tests of their validity and efficacy.

The potential for curriculum planning and for research to enhance many worthwhile innovations through critical feedback is great, especially when based on a morally adequate discourse amongst the key players, not the least of which are teachers themselves.

Chapter 6

Planning and Thinking in Junior School Writing Lessons: An Exploratory Study

James Calderhead

Summary

Eight junior schoolteachers in three schools took part in an exploratory study to analyze their thinking in the preparation and implementation of writing lessons. It was intended to assess whether such an analysis would enable teachers to understand better their classroom practice and to identify more easily the difficulties involved in teaching writing. Data was gathered using the methods of stimulated recall and interview, and field notes were made by the researcher. It was found that in order to conceptualize some of the difficulties of teaching writing and consider strategies for change, teachers also had to draw upon evidence concerning pupils' performance and the constraints upon their own practice. It is suggested that in order for research on teacher thinking to inform issues of teaching practice it may have to combine the investigation of the mental life of teachers and pupils and also the school context in which they work.

Recent research on teachers' thinking provides numerous insights into the nature and development of classroom practice. In the preactive phase of teaching, before teachers come into contact with their pupils, it would seem that teachers engage in a problem-solving process, designing activities which suit the interests and abilities of their pupils, the demands of curriculum guidelines, and the material and procedural context of their classroom. Teachers make decisions about who is going to do what, when, how and with which materials, and in doing this may give thought to a variety of factors (see Morine 1976, Yinger 1980, McCutcheon 1980). In the interactive phase, when teachers are in face-to-face interaction with pupils, there is little time or opportunity for such reflection and problem solving. Instead, teachers think mostly about implementing the mental image or *plan* that they previously devised (see McKay and Marland 1978, Peterson and Clark 1978, McNair and Joyce 1979). Comparisons of experienced and beginning teachers have demonstrated that experienced teachers possess memorised repertoires of plans and a series of classroom routines for their implementation. This reduces the demands upon their preactive and interactive thinking (see Calderhead, 1984).

Although the concepts used in research on teacher – thinking 'routine', 'activity' and 'plan', for example – deserve further refinement and clarification, they contribute to a general cognitive model of teaching which offers a conceptualization of classroom practice in terms of designing, implementing and maintaining activities. The research described here set out with the pragmatic objective of exploring the contribution of such a model to helping teachers analyze and reconstruct their practice in their attempts to achieve solutions to persistent problems in everyday classroom life.

The particular problem area examined was the teaching of writing (or composition) in the upper junior school. Difficulties in teaching writing at this age level were found to be commonly expressed. Teachers frequently reported being dissatisfied with their own classroom practices in this area of the curriculum, yet were unsure how to improve them. Generally, they felt they were implementing the procedures that they had learned at college and which they had read about in prescriptive texts on the teaching of writing, yet at the same time these procedures seemed unsatisfactory. In particular, teachers reported that their children lacked ideas and imagination in their creative writing, that little evidence of improvement in ability to use writing skills or conform to writing conventions appeared over the course of a year, and that there was little transfer of skill from writing exercises (in spelling, grammar, sentence structure and punctuation) to children's own free writing. Teachers felt that their approach to the teaching of writing was inadequate though they were unclear about the precise nature of the problem and how they might alleviate it.

The aim of this study was to investigate how a cognitive analysis of teachers' practice might enable the problems of teaching writing to be better conceptualised, allowing teachers to perceive more clearly the nature of their practice and how it might be further developed. The focus of the study was on the analysis of how teachers prepare and organise their pupils and the classroom environment for writing work. The research did not attempt to evaluate the effectiveness of alternative teaching strategies against particular objective criteria, or to analyse the children's composition process in any detail, although these are themselves obviously related and important areas of research.

An Exploratory Study

The research involved 8 teachers in 3 schools who volunteered to explore the nature of their classroom practice in writing instruction. The study initially set out to describe the nature of the plans and classroom routines that teachers adopted. Teachers' preactive thinking was studied through interviews, and interactive thinking through stimulated recall. Lessons were also observed and general notes were made of the teachers' and pupils' actions. The data collected from each classroom were reported back to teachers individually and on other occasions to groups of teachers within each school. The data became the subject of discussions about classroom practice.

As a result of these discussions, it quickly became apparent that other data were necessary in order to evaluate writing instruction. In particular, pupils' responses to the teacher were thought to be valuable in indicating whether teachers' expectations were actually being fulfilled. Several pupils in each class were consequently interviewed at convenient points during writing lessons. They were asked to explain what they were doing, what they had been thinking about during particular parts of the lesson and what examples of good and poor work in this subject area were like. The teachers were actively involved in the research, contributing to decisions about which data to collect, and interpreting and evaluating the findings. In this sense, the research undertaken was both exploratory and collaborative.

Summary of Findings

In initial discussions about writing in the junior school, all of the teachers in the study acknowledged that writing entered into most areas of the curriculum. However, four areas in particular were regarded as being centrally important:

1) Creative writing lessons (the term *creative writing* – was often used simply to refer to writing on a given topic – creative or otherwise)
2) Science/project work
3) Writing skills work (grammar, punctuation, exercises etc.)
4) Incidental writing (for wall displays, school radio, in school magazine etc.)

The latter was often regarded as a 'fill-in' or supplementary activity. The teachers varied in their emphasis on writing skills activities, most believing that there was little if any evident transfer of learning from such activities to free-writing. Nevertheless, they usually followed a textbook that was prescribed within the school. It was in science/project and creative writing work those teachers regarded the main, and most valuable, writing activities to occur, and it was on these that the research focused. Teachers viewed science work as providing the opportunity for pupils to become acquainted with experiments, to develop experience in recording and tabulating data and in writing up their observations. Project work was typically viewed as encouraging pupils to refer to a wide variety of books and to search out relevant information; project work frequently involved an equal amount of artwork (pictures, maps) as written work in producing a folder or project book. Both project and science involved teachers in a considerable amount of planning, and preparation, particularly in making work cards and ensuring that appropriate materials were available. In designing work cards teachers' concerns were with providing tasks that would generally be understood by the children and would suit the materials available. Reference to teachers' guides was common in the case of science though not in the case of project work. The questions on the work cards usually required straightforward descriptive replies. E.g.: "What can you find out about the Eiffel Tower?" "Describe how cheese is made?"

Individual differences amongst the pupils were not catered for. Teachers generally claimed that the more able pupils would simply make more of the task, giving fuller answers to the questions asked. Teachers also reported that because of the amount of time involved in making work cards they would be used over several years with different classes. During the interactive phase of science and project lessons, teachers devoted most of their time to managerial concerns, coping with pupils' difficulties, misunderstandings, lost materials, and dealing with inappropriate levels of noise and movement. Science and project work did in fact tend to consist of fairly noisy, bustling activities.

Teachers viewed creative writing lessons as occasions on which the greatest demands were made upon children's writing skills. Teachers' planning for creative writing usually involved thinking about a topic that would interest and involve the pupils. Topics that had been successfully used in past years were frequently adopted again. How the topic might be developed and how it might be made interesting to the pupils were also commonly considered in advance. In the interactive phase of writing lessons, teachers' concerns were mostly with developing pupils' enthusiasm for the topic and helping them to organize their ideas about it.

The more data that were collected, the more it became apparent that there were several mismatches between teachers' intentions and the more immediate thoughts and actions of teachers and pupils in the classroom. In both science/project and creative writing work, teachers perceived themselves as providing open tasks to pupils, and the central aim of these lessons was often viewed as being the encouragement of children to develop and organize their ideas or observations and to report them in written form. Superficially, it seemed as though this was the case, but in fact pupils frequently interpreted the tasks as closed ones and the teachers did not discourage, and in some cases unwittingly encouraged, this interpretation - a finding in common with several other classroom studies (see Doyle, 1979).

In project work, the children were generally required to search out information on given topics, to explain particular problems or processes, or to draw maps, pictures, etc. This might seem to be a relatively open task, yet in operation it was quite limited. Most of the writing consisted of copying out of books. After observing the classes, it was estimated that only about 10% of children in fact read the books and provided descriptions or explanations in their own words. In talking to the children about their work and how they approached it, the majority took it for granted that this was simply what 'doing project' involved. A few did report that they thought they should really read several books and then write about what they have read, but admitted that they usually copied because it was easier. An estimated 10% did not appear to understand what they were writing, but learned from others that were the relevant sections to copy. When the children were asked about what distinguished good project work, they most commonly reported *neatness* and *attractiveness* as the important features.

In science work, there was a similar dependence on copying. When writing up, the children were typically asked to report what they did, what they saw and what they found out. In fact, they based their writing on the instructions given on the workcard, adding a few brief statements at the end to describe their findings. A few able children expanded the latter section. Some of the less able children copied the directions from the workcard, without even changing the tense and form of the verbs, adding little else. When asked about their work, the pupils again mentioned the importance of *neatness* and also expressed concern over *obtaining the right answers*. The pupils often knew what the results of their experiments should be even before carrying them out and achieving this result was often regarded as the most important part of the exercise. For example, on one occasion, a group of children was observed to carry out an experiment using three candles with three different sized jam jars placed over the top. They knew from talking to other children that the 'correct' result of the experiment was to find that the smaller the jar, the less time the candle burned. However, their own finding was that the candle in the smallest jam jar burned longest. This led them to repeat the experiment and eventually to realize that a chip in the rim of the jar might be allowing air in. They repeated the experiment again, replacing the faulty jar, and found the expected result. In writing up their experiment, however, none of them made any mention of their own particular problems and discoveries, reporting their procedures as if the experiment had 'worked' first time. When asked about this, they suggested that it would not have been "right" or "proper" to report their actual findings and also that it would have "taken too long".

In creative writing lessons, the introductory phase of the lesson was viewed by teachers as a time to direct the children's attention to the chosen topic, engender interest and enthusiasm, and encourage the children to develop and organize their ideas ready for writing. During this discussion phase, many of the ideas and much of the vocabulary was in fact provided by the teacher, interestingly, pupils perceived this phase of the lesson quite differently. They viewed it as a process of tuning in to what the teacher wanted or expected, noting the ideas and the vocabulary that would be relevant and acceptable. When asked about their approach to the task, they appeared to attach little importance to coming up with their own ideas. When asked to suggest what made a good story, the most common criteria mentioned were *neatness* and *length.* Only on one occasion were pupils found to be greatly concerned with communicating their own ideas in writing. On this occasion, the class had been hill-walking with the headteacher and some volunteer parents the previous day. Their own class teacher had stayed in school. The children were enthusiastic in talking about their own experiences and when asked to write about the walk, they generally perceived the main purpose of the task to be one of recounting these experiences to their teacher.

It was also noted that the feedback provided to pupils about their writing did not appear to match teachers' objectives of helping pupils to develop, organize and express their ideas. Feedback on pupils' performances was provided both during the lesson and afterwards, and in both oral and written form. Over 90% of all feedback statements referred to spelling, punctuation or grammar. Most of the remaining comments were general in nature, such as "good" or "well done". Hardly ever did teachers make any comments about the pupils' ideas, the content of the pupils' writing, or their use of language.

Teachers' Interpretations and Reactions to the Mismatches

These observations together with teachers' and pupils' reports of their activity were presented to teachers for their comments and reactions. In the case of science and project work, pupils' different interpretations of the activity and their reliance on copying in carrying it out did not surprise the teachers. They reported that their ideal aims for project and science were probably ambitious for most of the children, and whilst their aims were not being fully met they regarded science and project work as providing pupils with some basic valuable experiences, such as referring to library books, tabulating information and working in groups when carrying out experiments. The teachers were very much aware of the heavy managerial demands upon them during science and project lessons. In fact, they could not imagine taking a more actively instructional role in these activities, and were generally prepared to live with the situation as it was.

In the case of creative writing, discussion of the mismatches between teachers' curricular intentions, teachers' and pupils' actual behaviour led teachers to consider why such practices have come about. They suggested a number of contextual factors, which they thought exerted considerable influence upon them. For example, although several of the teachers claimed that they did not themselves place the highest value on neatness and length in writing, they were aware that parents and head teachers did expect to see full, neat notebooks at the end of the year, with regularly completed essays, and felt that this was an expectation to which they had, to some extent, to comply. Inevitably, this was communicated both implicitly and explicitly to the children. The timetable was also viewed by some teachers as a constraint. Since writing was regarded in the schools as one of the basic elements of the curriculum, it was generally timetabled. It was expected that creative writing would occur during the same period each week. Yet, teachers were aware that this made it difficult to capitalise on events occurring at other times that might provide useful topics for writing. They also felt that it introduced artificiality into the task itself, writing becoming a weekly ritual in which pupils were expected to write (often imaginatively) on a topic selected by the teacher and about which the pupils might have little if any knowledge or enthusiasm.

Teachers were also well aware that the majority of tasks given to pupils in the class, and perhaps throughout their school life, were closed. In consequence, children might become used to school tasks being ones in which they have to find right answers or in which they must give those answers which they think the teacher wants. Teachers reckoned that to encourage pupils to abandon this approach to school work, and to develop communication in which pupils played a more active role, would be very difficult. A further constraint became evident to teachers in one school incidentally as a result of the research. Both of the participating teachers in the school noticed that when the researcher visited their classrooms, the children's writing improved – becoming longer, containing more ideas of their own, and, according to the teachers, revealing more effort to express themselves. The teachers attributed this improvement to the children attempting to impress the researcher with their work. As a result of the research, the teachers came to realize the importance of the pupils possessing a sense of audience when they wrote, and they developed strategies for attempting to achieve this in their normal classroom work.

In discussing the evidence concerning their classroom practice, it was apparent that teachers had a very limited conception of the processes in which the children were involved in writing. There appeared to be considerable mystery about the composition process itself and teachers were unsure of how to facilitate it. For example, one teacher commented:

> I can give them ideas and suggest vocabulary, but I can't write the essay for them. It's up to them. If they are having difficulty I sometimes get them to tell me their story and after we've talked about it they go and write it down. But I don't see what else I can do. They've got to write it on their own.

Another similarly claimed:

> The children have to think about what they're going to write. Then for some of them everything seems to click into place and off they go, but for others they wait and wait and it never seems to click.

After being confronted with the evidence collected from their own and others' classrooms, most of the teachers in the study did report making some changes in their practice. These were generally fairly minor, such as changing the nature of the feedback they gave, selecting writing topics closer to children's own experience, thus attempting to make the pupils less dependent on ideas from the teacher, and developing opportunities for pupils to write for other audiences and encouraging pupils to develop a sense of audience whilst writing. In addition, other changes were perceived by teachers as being desirable but were regarded as beyond their own control.

Interestingly, the changes that teachers made in their practice were all in the direction of conforming to recommendations contained in the Local Education Authority (LEA) Guidelines for writing in the junior school and which are commonly contained in prescriptive texts on the teaching of writing. Consequently, the study did not result in any new solutions to the problems of teaching writing, but it did appear to enable teachers to better conceptualize the nature, causes and effects of their own practice, to perceive more accurately the need for change and the direction in which this could be made. In discussion with teachers over the LEA guidelines it was apparent that they had difficulty understanding the relevance of many of the LEA recommendations and difficulty in relating the LEA guidelines to the problems they experienced in the classroom and to their own understanding of their practice and its effects. As one teacher pointed out when discussing the importance of audience:

> "The children either have the skills to write or they don't. Changing the audience isn't going to give them any more skill."

This teacher's own way of viewing the problems he faced prevented him from perceiving the relevance of the recommendations. One of the outcomes of this study is to offer some support to the notion that by enabling teachers to conceptualize their own practice more fully, they can relate it more adequately to curriculum guidelines and to discussions, recommendations or models concerning classroom practice, as well as to their own curricular intentions (see Olson, 1982).

Conclusions

It is impossible to draw firm conclusions from this small-scale, exploratory study, though a number of observations seem worthy of mention. Firstly, it appears that teachers sometimes have vague or inaccurate conceptions of their own practice. Numerous mismatches occur, between teachers' intentions for practice and their actual practice, between teachers-expectations/interpretations of pupils' performance and pupils' actual performance, and between teachers' expected learning effects and pupils' actual learning. Secondly, the study suggests that information concerning teachers' thinking can contribute to teachers' analysis and reconstruction of classroom practice. However, the value of this information alone would seem to be rather limited, certainly as far as teachers' identification of difficulties in teaching writing are concerned. Although a change in teachers' practice requires changes in planning and in interactive routines, an analysis of these plans and routines was found not to be sufficient to identify the difficulties of practice. Teachers' planning and thinking is geared towards producing certain effects on the pupils and it occurs within a network of constraints that shape its possible form. In order to evaluate their practice, identify difficulties and consider possible strategies for change, teachers had to think of their teaching in relation to its context and effects. If the findings of this research were more widely generalized, it would suggest that for research on teacher thinking to contribute to a better understanding of teaching and point towards solutions to everyday teaching problems, it may have to broaden its scope of enquiry to include pupils' thoughts and actions and the context in which teachers and pupils work.

One further outcome of the study is to suggest the value of small-scale exploratory, collaborative work with teachers in the development of our understanding of classroom practice. Exploring the relationships between existing models of teaching and problems of actual practice can both illuminate the nature of classroom difficulties and indicate the limitations of our theoretical frameworks. If research on teacher thinking is to contribute to a theory of teaching that is of value to teachers, researchers must not lose sight of the everyday classroom problems that teachers experience.

Chapter 7

Categories in Teacher Planning

Harm Tillema

Summary

A variety of planning decisions have been found in a growing number of studies on teacher planning. However, they are often framed under different labels or categories that seem to have considerable overlap. In this paper, it is advocated that we need better descriptive labels or categories that must be derived from the actual planning behavior of teachers.

In this study 356 planning statements, made by teachers, are analyzed for the degree of specificity with which teachers make these planning decisions. Based on a loglinear analysis it is decided which system of categories fits the data adequately.

It is concluded that teachers make planning decisions in a specific and even concrete way and are concentrated on the planning categories: structure and sequence of subject matter, the presentation of subject matter and the diagnosis of prior knowledge of pupils.

Introduction

A growing, international number of studies have investigated teacher-planning behavior (Clark and Yinger, 1979; Yinger, 1977; Peterson, Marx and Clark, 1978; Bromme, 1980; Ben Peretz, 1981). This interest in planning has probably its origin in a conclusion made by Jackson (1968) that teachers do not use planning prescriptions from the curriculum. The first descriptive planning studies (Zahorik, 1970; Taylor, 1970) investigated whether curriculum-planning categories, such as objectives, learning activities, organization, and evaluation (Tyler, 1950) were used by teachers. These studies showed that subject matter is the most important planning category for teachers, followed by teaching learning activities and materials. Specification of objectives and evaluation were least frequently observed categories. Planning as done by teachers is now commonly seen (Clark, 1980) as an activity in which a number of categorizable decisions are made, therefore it becomes important to know what actually constitutes the planning decisions of teachers. In more recent studies (Peterson, Marx and Clark, 1978; Marx and Peterson, 1981; Yinger, 1977) several planning categories have been found: motivation, subject matter sequencing, diagnosis, organization of materials, teaching-learning activities, also these studies confirmed the observations from previous descriptive studies.

Most descriptive studies about planning on a general level agree what are the planning categories actually used, but this results from generalizing about subject matter domains, periods of time and differences between grades. It could, however, be argued that it is more interesting to know how teachers plan in concrete cases and for well defined subject matter domains (Morine, 1976) in order to know how specific and detailed is the teacher plan for individual lessons.

A more differentiated picture results when planning of well-defined subject matter domains is taken as a starting point for investigation. Morine (1976) studied the planning of a reading and arithmetic lesson and used several methods (observation, simulation, thinking-aloud procedures, questionnaires) in two different grades. She found that teachers plan on a concrete level, which could be related to higher learning gains. Teacher planning focused on subject matter presentation and the cognitive aspects of learning. Teachers did not considered objectives, alternative teaching procedures or subject matter per se. Also, there were found some differences in planning between grades. Unnoticed in previous studies is the observation that teachers place great interest in the use and making of materials. This study is in accordance with research that studied planning in general, however from the study of Morine it can be concluded that planning categories are clearly interdependent, which can influence the importance of a planning category in particular.

Other studies have added several conflicting findings. Mintz (1980) studied the planning of a reading lesson with a simulation method and found that diagnosis of reading difficulties was one of the most important categories. Lydecker (1981) studied a social science lesson and found the lowest frequency for diagnosis. Also these studies differ in the importance attributed to the textbook and materials used. Subject matter is the most important category in simulation studies (Peterson, Marx and Clark, 1978; Mintz, 1980), in which a teacher plans for non-regular topics and mostly for unknown students (also diagnosis becomes an important category in these studies). Studies that used other methods have frequently yielded other results. With thinking-aloud procedures, subject matter presentation and pupil's activities become important (Ben Peretz, 1981; Yinger, 1977). Written planning protocols (Mintz, 1980; Ben Peretz, 1981) stress the importance of sequences (of activities, subject matter) but this category often does not appear in other studies.

In general these studies indicate that teacher planning consists of the following categories: subject matter, teaching activities and pupil related categories, but between the studies there are substantial differences in the importance of categories and even some contradictions have been found. This does not necessarily mean that the way teachers plan is responsible for these differences. Sometimes these differences can be attributed to the methods conducted in research. Several problems can be identified.

A first problem in most studies of teacher planning is that the researcher a priori selects which categories will be used to describe the planning statements made by teachers. Frequently, the categories are derived from a prescriptive planning model. However, results from earlier studies show that prescriptive categories are not an adequate basis for describing the actual planning decisions made by teachers. Because statements are molded in the a priori selected categories, a discrepancy could exist between the researcher's model of planning and the actual decisions made in the planning by teachers. Therefore it should be investigated which system of planning categories best fits the planning statements of teachers.

Morine (1976) and Berliner & Tikunoff (1976) have used a procedure that could bring the researcher's description of planning categories more in accordance with the actual categories used by teachers. In this procedure, common themes or topics are selected through a stepwise and iterative analysis of protocols. These topics are then clustered under appropriate labels that describe the statements in terms of the teacher's comments.

A second problem is the amount of overlap in the categories researchers use. Most categories are often not clearly defined. In the study of Yinger (1977) the category activities could be regarded as part of subject matter presentation in the study of Peterson, Marx & Clark (1978). Taylor (1970) identifies a category 'teaching context', which is facetted in several categories in other studies. There is also less agreement about the number of categories used for describing teacher planning. Taylor (1970) selects four categories; Morine (1976) identifies 45 of them.

A third problem is the specificity of teacher planning. Most studies do not indicate how specific teachers plan; this information however is of potential value because it can be argued that the more concrete teachers planning decisions are, the more it indicates the perceived importance of a category in the teacher's planning. For the same reason the more general or global remarks need not be related to the actual planning decisions. Morine (1976) analyzed the degree of specificity in lesson plans. Ben Peretz (1981) also judged the specificity of lesson plans, but concluded contrary to Morine that protocols largely consisted of general statements.

A fourth problem is the unit of analysis. The early studies concentrated on planning over large periods of time and not specifically on subject matter domains. Later studies concentrated on the planning of individual lessons. There seem to be substantial differences in teacher planning, depending upon the planning period (semester, weekly or lesson planning - Yinger, 1977) and also depending upon subject matter domains. Therefore, the analysis of teacher planning must be specific to these factors.

This study seeks to answer the following questions with respect to planning categories teachers use in their daily lesson planning of social science instruction: What is the degree of specificity in the planning protocols of teachers; which system of planning categories adequately describes these planning statements, when a distinction is made between a descriptive and a prescriptive system.

Method

Participants

Fifteen teachers participated in this study on a voluntary basis. They were teaching in the 5th and 6th grade of primary schools and all of them had at least 2 years of experience in their classes. The teachers in this study used the same social science curriculum. Teachers were asked to plan a lesson on the topic: 'rich and poor countries'. The curriculum was used as material and teachers were asked to plan in their usual way. It was assured that this topic had not already been treated in their classroom. The researcher was present during the lesson planning.

Procedure
In this study, a combination of thinking-aloud and stimulated recall procedures was used. In the instruction of the thinking aloud procedure teachers were asked to plan their lesson as they normally do and say anything that comes into mind. The researcher registered the verbalizations made by the teachers. In assuring that teachers make statements, not only about the cognitive processing of lesson information, but also on their deliberations and decisions, the thinking-aloud statements were used in a stimulated recall procedure afterwards.

After completing the planning task during the thinking-aloud session teachers were confronted with the statements made and were asked to elaborate on them. In this stimulated recall session, the researcher used a disruption technique (Wahl, 1981) to discuss the statements with the teacher to assure the precise meaning of a planning category and the degree of specificity of the statement. For instance: when a teacher mentioned the motivation of pupils in the thinking-aloud protocol, the researcher after confronting the teacher with this statement, asks typical questions as: how did you realize this, how is it related to, how do you know it will work.

Scoring
1. Category systems.
The protocols of teachers were analyzed with a stepwise procedure to find statements with a common content. The protocols were seen as a text (Bromme, 1981) and the propositions in the text were compared between protocols to find teacher formulated clusters of same planning comments. This rough material (unique but comparable descriptions of planning statements made by teachers) consisted of 15 different descriptive topics:

1. Background of pupils: social status, outer school influences, as well as general ability and intelligence of pupils.
2. Prior knowledge: the diagnosis of pupil's knowledge during instruction
3. otivation: an estimation about the involvement of pupils and a willingness to use ideas of pupils and the activation of pupils during the lesson.
4. Interest: the affective and emotional prerequisites of instruction.
5. Teaching strategy: the choice of a teaching model with respect to the intended learning processes.
6. Teaching procedure: the kinds of teaching activities used.
7. Materials: the choice of materials and media.
8. Organization: the sequence of activities and use of materials.
9. Learning difficulties: the problems or difficulties in the subject matter for pupils.
10. Sequence: the manner of presentation of subject matter.
11. Structure: the content of instruction and outline of the subject matter.
12. Prior instruction: already given instruction and related topics.
13. Intentions: teacher's intentions and own perspective on instruction.
14. Objectives: a description of intended content, activities.
15. Aims: pedagogical deliberations about instruction.

The description in clusters of planning statements is not mutually exclusive; it merely indicates distinguishable formulations in the protocols of teachers.

In a next step, these statements were coded in two different category systems. The first category system was derived primarily from prescriptive planning- or curriculum models. Four categories were identified:

1. Objectives (clusters: 13,14,15)
2. Subject matter organization (clusters: 5, 10, 11).
3. Activities (clusters: 6, 7 and 8).
4. Evaluation (cluster I and 9).

It was hypothesized that this prescriptive planning category system is only one alternative to describe the planning statements adequately. A rival system of planning categories was constructed by selecting a more specific subset of categories, which were frequently mentioned in the planning protocols. Teachers mentioned at least the following categories (see table 1 from which the categories were derived):

a. Prior knowledge (cluster 2) and motivation (cluster 3).
b. Teaching procedures (cluster 6), materials (cluster 7) and organization (cluster 8).
c. Subject matter structure and sequence (cluster 11 and 10).

Different to the prescriptive category system, this descriptive category system is more in line with theories of instruction (al design), (Glaser, 1976).

2. Degree of specificity

Statements made by teachers may differ in the degree of specificity. General remarks in the protocols need not be related to actual planning decisions. Four levels of specificity were identified:

1. Deliberation - general statement about the importance of a planning decision but not necessarily related to the specific lesson topic. 'I think it is important that, you have to watch that'.
2. Focus – statement about the importance of a planning decision specific to the lesson planning. 'In this lesson I have to take into account, I want that to happen'.
3. Specification – statement about the elaboration of a planning decision. 'I will do the following (e.g. motivation, learning activities)'.
4. Concrete statement - statement about the elaboration of a planning decision on an operational level. 'I start with a story, I will ask that question'.

Every statement in the protocols was scored into one of the levels of specificity.

Analysis

To determine the category system that best fits the planning statements (goodness of fit) a loglinear analysis of the data was performed. In loglinear analysis, frequency differences between categories are fit and compared with a test model that describes all relevant categories. It is tested in how far a simpler model of categories fits the observed frequencies without deviating significantly from the test model. Loglinear analysis is especially suited to find patterned frequencies (or categories) in the obtained data and to determine a pattern that fits these data. The parameters to be tested are: prescriptive category system (P), descriptive category system (D) and degree of specificity (S).

Results

Three hundred fifty-six distinct statements could be classified as planning decisions in the protocols. They are used here for further analysis. The mean number of statements in the protocols is 23.73. Table I gives the percentages for each cluster of statements.

Table 1: Percentages of the Planning Statements.

1. Pupils' background	02%
2. Prior knowledge	15
3. Motivation	10
4. Pupils' interest	05
5. Teaching strategy	05
6. Teaching procedures	09
7. Materials	07
8. Organization of activities	09
9. Learning difficulties	04
10. Subject matter sequence	09
11. Subject matter structure	09
12. Prior instruction	04
13. Teacher intentions	05
14. Objectives	05
15. Aims	02

The percentages for the three parameters: prescriptive categories (P), descriptive categories (D) and level of specificity (S) are given in Table 2.

Table 2: Parameter Percentages.

P - prescriptive category system:	
- objectives	12%
- learning activities	25
- subject matter organization	23
- evaluation	06
D - descriptive category system:	
- prior knowledge/motivation	35
- organization/procedures/materials	25
- subject matter structure/sequence	18
S - level of specificity:	
- deliberation	06
- focus	19
- specification	25
- concrete statement	49

In subsequent analysis, the parameters are analyzed for their goodness of fit in all their combinations (table 3 gives the results).

Table 3: Comparison of Fitted Models.

Model	Df	Likelihood	P	Pearson	Chi	P
D	60	183.25	.00	172.17	.00	.00
S	60	94.26	.00	105.93	.00	.00
P	60	206.50	.00	199.55	.00	.00
D,S(additive)	57	65.88	.19	69.46	.12	.12
D,P	57	178.12	.00	165.09	.00	.00
P'S	57	89.13	.00	100.97	.00	.00
D,P,S	54	60.75	.24	65.14	.14	.14
DS (interactive)	48	57.40	.16	60.06	.11	.11
DP	48	162.80	.00	151.22	.00	.00
PS	48	77.13	.00	81.80	.00	.00
DS,P	45	52.27	.21	55.51	.13	.13
DP,S	45	45.43	.45	46.59	.40	.40
PS,D	45	48.74	.32	49.04	.31	.31
DS,DP	36	36.95	.42	37.05	.42	.42
DS,PS	36	40.26	.28	40.72	.27	.27
DP,PS	36	33.43	.59	33.15	.60	.60
DP,PS,DS	27	26.64	.59	24.14	.62	.62

This overview of fitted models shows that no parameter by itself can adequately describe the data – they all differ significantly from the observed data. If the parameters are combined to describe the data, it shows that a combination of the descriptive category system (D) and (S) level of specificity can adequately describe the data. A combination of all parameters would also be acceptable, but the decline in likelihood ratio with respect to the degrees of freedom is not substantial enough. This means that adding the prescriptive category system gives no more extra information than already available in the two other parameters.

An interactive model consisting of the descriptive category system and level of specificity is also acceptable. It is concluded on the basis of the likelihood ratios that a combination of the descriptive category system and level of specificity fits the data adequately.

Next, an item analysis on the protocol statements was performed to determine the correlations between categories. It proved that some categories had no variance, because they were mentioned by all the teachers and were excluded from further analysis: prior knowledge, motivation and sequence of subject matter (however, this result gives further support for the descriptive category system). Table 4 gives the correlations between the remaining categories.

The table shows several interesting aspects. Pupils' background and interests are strongly related to other categories, so is subject matter structure, however teaching procedures and activities are strongly negative related to other categories. The category: prior instruction contributes much to all other planning decisions and formulation of objectives is negatively related to the planning process. This table shows that several categories hardly can be separated categories; this indicates that planning decisions are highly interrelated.

Table 4: Correlations Between Clusters of Statements.

	1	4	5	6	7	8	9	11	12	13	14	Item/ rest
1. Background	-											.34
4. Interest	.03											.36
5. Strategy	.47	-.28										.13
6. Procedures	-.25	.10	.19									-.39
7. Materials	.34	.20	.21	-.16								.17
8. Organization	.02	.15	.14	-.10	.14							.02
9. Difficulties	-.09	.28	.40	-.18	-.10	.14						.05
11. Structure	.28	.68	-.18	-.07	.44	-.10	-.19					.46
12. Prior instruction	.34	.20	.10	-.16	-.02	.21	.21	.44				.26
13. Intentions	.20	.29	.01	-.13	-.30	-.20	-.10	-.13	.07			.14
14. Objectives	-.28	.10	.19	.07	.16	.10	.19	-.07	.16	-.54		-.13
15. Aims	-.19	.37	.09	.25	.04	.02	.09	.25	.04	.13	.28	.17

The critical value for the correlations to be significant is: r = .456 for P (.05 and r = .645 for p < .01.

Discussion

The results of this study indicate that teacher planning can be categorized as decision making about prior knowledge and motivation of pupils and the organization of teaching procedures and activities, taking into account the structure and sequence of subject matter. Teachers are aware of their intentions and plan on a specific and even concrete level. Also this study indicates that level of specificity is an important factor in describing teacher planning, even more than differences between systems of planning categories.

Some categories have special importance for teachers: sequence and structure of subject matter and prior knowledge and motivation of pupils but also the results show that these categories are strongly interrelated with other categories. This could mean that the categories imply each other to some extent and are interdependent. For instance, a strong interrelation was found between the categories: structure of subject matter, pupils' interest and prior instruction; also between the categories: teaching strategy, background of pupils and learning difficulties. This could mean that teachers think in terms of larger clusters of decisions than is accounted for in some planning models.

Teachers in this study did not isolate specific categories and then elaborated upon them, but deliberated what kind of planning decisions affect each other; this was done on a concrete level of specification. This could imply that it may not be fruitful for a theory on teacher planning to identify distinct categories. It is probably more realistic to view teacher planning as composed of a number of central questions for the teacher or decisions, involving planning categories that are not separable but highly interdependent. According to recent research on teacher planning these central questions in planning could be formulated in the following way: What is the central information in the subject matter to be learned and how can I (the teacher) best present it to my pupils? In the process of finding an acceptable solution to this design problem, teachers think in terms of clusters of categories that imply each other.

Another typical finding that could be derived from the data in this study is that teaching procedures/activities are negatively related to other categories - a finding also present in the study of Marx and Peterson (1981). Probably this could be another focus in planning. After having decided about the planning of the subject matter presentation in relation to pupils' interests and prior knowledge, another and distinct focus in planning is deciding about the realization and organization of the plan. This last kind of planning need not be performed during the preactive phase of teaching, but could well be undertaken in the interactive phase, during instruction. Marx and Peterson (1981) found that teachers who were less productive in decision-making during the preactive phase were planners that are more productive in the interactive phase. Zahorik (1970) stated that decisions in the interactive phase are more reactive to students' interests and problems and are positively related to attitudes of students about instruction.

Taken together this could mean that teachers, after having decided about the content of instruction and the way to present subject matter are interactive planners about the activities in which they are to be involved with their pupils. This hypothesis asks for further exploration.

Chapter 8

Teachers' Thinking – Intentions and Practice: An Action Research Perspective

Christopher Day

Summary

This paper attempts to address three questions fundamental to research on teachers' thinking – in what ways do teachers learn, why do they change (or fail to change), and what is the role of the researcher as intervener in this process of teachers' thinking and behaviour? I will argue that if research is to make a significant contribution to teacher learning and change researchers must move away from the notion of themselves as prime designers and interpreters of the motivations, thoughts and actions of others towards a more interdependent role in which collaboration, consultation, and negotiation are first principles. If they are to achieve success, they must be prepared not only to talk with teachers about practice, but also to observe and work with teachers in their behavioural settings. This is based on the assumption that, 'in order to understand teaching, teachers' goals, judgements and decisions must be understood, especially in relation to teachers' behaviour and the classroom context …' (Shavelson and Stern 1981).

Introduction

The paper is divided into two parts. The first part deals with the social, psychological and institutional constraints upon teacher learning. The second part describes briefly the reflections of teachers whose thinking and practice were affected by research in which they played central collaborative roles with the researcher; and describes principles for researcher intervention in relation to a client-centered theory of professional learning and change.

Constraints on Teacher Change

The work to be described in this paper focused on identifying discrepancies between intentions and practice of teachers and attempted to move these closer together on the assumption that this would result in increased teacher satisfaction and teaching effectiveness. While there is a good deal of recent and well documented research on various aspects of teachers' thinking – their pedagogical thoughts, judgements and decisions - there is as yet very little empirical evidence concerning the relationship between this thinking and classroom practice (behaviour). Whereas recent studies agree that the behaviour and thinking of teachers are guided by a set of beliefs that are often unconscious (Clark and Yinger, 1977) and that teachers' decisions are often 'intuitive' (Stenhouse, 1975), we also learn that not all teachers consistently employ practices which directly reflect their beliefs (Duffy, 1977). Furthermore, there is as yet little evidence concerning how teachers assess their plans and accomplishments and so revise them for the future.

We do know, however, that teachers take very little notice of research findings from outside the school - because of the specialist language in which reports are couched, and because they are perceived to be irrelevant to the 'practicality ethic' held by many teachers (Doyle and Ponder, 1976); we know that in schools the general attitude towards introducing and using new knowledge is discouraging (Miles, 1981; Elliott, 1982); and we know that although teachers' learning has as its most significant context, the classroom, for this is the focus of the teachers' professional life where 'practice' confronts 'theory', still most of the opportunities for learning occur outside this context. It is not unreasonable to hypothesize that a significant proportion of the learning associated with any change in practice takes place in this context of use, and that this minimizes the problems of transfer of learning (Eraut, 1982). If this is so, we may move some way to developing and validating a theory of professional learning for teachers, and professional practice for researchers in which a dialectical relationship between researcher and subject, theory and practice is central to the research enterprise. Fenstermacher (1980) has argued that, in order for teachers to adopt research findings, a chain of events should occur. First, teachers must become aware of their subjective beliefs about teaching. Second, these beliefs should be held open to empirical investigation. Thirdly, a subjectively held belief becomes an objectively held belief if it is verified empirically. Disconfirmation of the subjective belief constitutes grounds for a change in belief, consistent with the empirical evidence. And fourth, objectively held beliefs constitute reasonable grounds for action. Though written in a different context, this notion of rational man is not dissimilar to that postulated by Elliott in describing teachers' self-monitoring processes (Elliott, 1976). However, these views of rational man take no account of, for example, how teachers' thinking about teaching is affected by perception, socialization, and time and energy constraints. Take, for example, the way in which new teachers cope with and react to their initial experiences of schools. This process has been depicted as a kind of two-way struggle: student teachers try to create their own social reality in their work by attempting to make it match their personal vision of how it should be, whilst at the same time, the school's organization and social forces attempt to compel the student to conform to its own version of reality (Lacey, 1977). While this analysis, like that of an earlier study (Fuch, 1968) is plausible, it nevertheless lacks any indication of how, over time, the socialised teacher may shift from one position to another. Can he do it on his own? Is help needed? Who should provide this, and what kinds of help are most appropriate?

Towards a Theory of Professional Learning and Change

If teachers are to extend their knowledge about practice (and thus gain the possibility of increasing their professional effectiveness) they will need to investigate both their thinking and their practice. Only by evaluating the compatibilities or incompatibilities that exist within and between these two elements of a teacher's framework for action may we be enabled to increase our knowledge of his thinking. It is not enough to talk to the teacher about teaching; we must also observe him in the behavioural world of the classroom. He may be unwilling or unable to think or behave differently until both thinking and practice have been made explicit.

The problems of increasing knowledge about practice have been highlighted by Keddie (1971) in her distinction between teacher as educationist and teacher as practitioner:

> '... While, therefore, some educational aims may be formulated by teachers as *educationists,* it will not be surprising if 'doctrine' is contradicted by commitments which arise in the situation in which they must act as teachers...' (Keddie, 1971).

In other words, there may well be a difference between what people say and what they do. In the staff room setting, for example, talk about teaching is governed by tacit assumptions about the nature of talk about teaching; whereas in the classroom setting, teaching actions are governed by tacit assumptions about the nature of teaching actions. Our explicit actions, both as educationists and practitioners, are therefore often based on our implicit, unstated knowledge of the nature of practice in any given setting. (Polyani, 1967). Practices are in sense rules of action that allow us both to maintain a stable view of, for example, the classroom or the school, and give priority to certain kinds of information while ignoring other kinds. They are theories of control. A new teacher very quickly develops assumptions about practices that allow him to cope with the complexities of teaching and being a member of staff. He develops 'routines' (Yinger, 1979), and since it is rare for these to be made explicit or tested, the possibilities for evaluating values, assumptions and expectations that underpin his teaching are minimal. It would seem that under normal conditions teachers' thinking is limited by both perceptual and contextual constraints.

Argyris and Schön (1976) characterise the normal world of learning as 'single loop', in which, ... we learn to maintain the field of constancy by learning to design actions that satisfy existing governing variables' (goals). Evidence of 'single loop' learning is often to be found in in-service courses where, because of the audience, teachers feel unable to justify their practice. They claim that it is the only one possible under the circumstances.

While single-loop learning is necessary as a means of maintaining continuity in the highly predictable activities that make up the bulk of our lives, it also limits the possibilities of change. It is argued that if we allow our theory of action to remain unexamined indefinitely, our minds will be closed to much valid information and the possibilities for change will thus be minimal. In effect, if we only maintain our field of constancy we – become 'prisoners of our programs' and only see what we want to see. A second complementary kind of learning is suggested and this is characterised as 'double loop learning'. This involves allowing things that had previously been taken for granted to be seen as problematic, and opening oneself to new perspectives and new sources of evidence. One has to be prepared to see oneself as others see one in order to better understand one's behavioural world, and one's effect upon it. In effect, there are two problems to be faced. The first is concerned with self-confrontation and the extent to which an individual can engage in this, and the second related problem concerns the extent to which the consequences of self-confrontation can be accommodated in thought and action by the teacher without assistance.

The problem with moving towards 'double loop' learning is that attention is once more drawn to myriads of additional variables of information which are normally 'filtered out' by teachers through the development, for example, of routines and decision habits in order to keep mental effort at a reasonable level. (Eraut, 1978). They may no longer respond only intuitively to situations, but are forced, through confrontation with self, into a critical and rational response. Stenhouse (1975) called this, 'disciplined intuition' rather than 'rationalist abstraction' and perceived the need for external support for change which must grow from within the individual. However, McDonald and Rudduck (1971) have pointed out that if teachers take the risk of departing from their niche in the social and organizational structures of the school into which they have been socialized in order to innovate, they risk taking on 'the burden of incompetence' where the approved certainties they have striven to construct since their own days as beginning teachers are laid to one side and they once again become vulnerable to socialisation pressure to return to the norm, much as when they first began. Clearly, changes that affect institutional norms and routines will only take hold if accompanied by a degree of resocialization (Eraut, 1982), and this is likely to require support.

It seems clear that for the individual, once in the system all the incentives lie in one direction - to stay a functional part of the system: socialization forces hold the structure in a steady state. Change is difficult.

Argyris and Schön (1976) highlight the necessity for support in their description of the process of change. They suggest that a teacher faced with a difficult situation in which his theories-in-use (behaviour) do not allow him to achieve what he wants may either change his theories-in-use in order to achieve his espoused theories (which are used to justify or describe behaviour), or continue to use them even though they no are longer perceived to be compatible with his espoused theories.

So far, I have attempted to describe teachers' thinking as it is, constrained by factors of perception, socialisation and time and energy. I now move on to reporting briefly on the changes in thinking and practice which were perceived by teachers as a result of engaging in an extended process of self-evaluation collaboratively designed and supported by a 'research-consultant'; and to suggesting a client-centered classroom based model for professional learning and change.

A Client Centered Model of Researcher Intervention

The research was not traditional, for the researcher did not maintain a role distance from the actor and his environment. Instead he provided the necessary moral, intellectual and resource support for teachers engaged in a process of self-examination. It is suggested that by this means the 'researcher-consultant' achieved access to more valid information concerning how teachers learn and why they change (or fail to change), – and thus teachers' thinking, than had he adopted a more neutral or naturalistic stance.

Briefly, the research set out to test a model of classroom based in-service education. Members of a particular English department volunteered to participate and agreed to the filming of two sequences of six lessons each with Fourth year CSE classes. By reviewing these video films and discussing them with the teachers at length, by interviewing pupils and examining the interactions within the lessons, the researcher was able to offer each of these teachers information by which it was possible to re-examine and reflect upon espoused theory and to generate new personal theory, – what Elbaz (1981) in a study of one teacher's thinking, referred to as 'a practical principle which is the outcome of reflection and deliberation'.

Each teacher's aim was to increase his professional effectiveness in the classroom, and five sequential stages were found to be necessary in order to achieve this: 1. Identification of inconsistencies within his prevailing theory of action through self-confrontation and reflection; 2. evaluation of this confrontation as a means of informing future decision taking; 3. the planning of new theories-in-use; 4. the implementation of those new theories; 5. internalization of new theories of action and further confrontation or, return to confrontation of initial theory of action. (Day, 1981)

The research was thus client-centred, where the researcher intervened in the teacher's life in order to seek questions that are perceived by the client as relevant to his needs, to investigate answers to these questions collaboratively and to place the onus of action on the client himself. The notion here is that where work is related to personal experience and perceived needs and occurs in the context in which this experience and needs occur (i.e. the school or department), the client's personal investment in the learning enterprise will be maximized.

All four teachers achieved change at classroom level in different ways, according to their particular intentions. However, it was the changes in attitudes towards themselves as teachers and towards their teaching that were more significant. Within the constraints of this paper, it is possible to report only briefly on their reactions to their engagement in a process of learning with the support of a researcher, and the changes that they identified in their thinking. Further detail is recorded elsewhere (Day, 1981).

One teacher stated that the work had provided her with the time 'to think about, question and even change my methods, and the frame of reference in which to make decisions and formulate ideas'. She now had a cohesive theory of action. Additionally, she had been able to increase her range of teaching strategies and she now used more individualized and small group project work than previously. She also stated that one year after the work with the researcher had been completed she still had a more generally questioning attitude. This critical awareness had been a direct result of the work; and she had changed as a result of what the study had shown her about herself.

The statements by another teacher showed that his view of himself as a teacher and the roles he believed he should play had been clarified by the research. They also revealed a lack of self-confidence in trying out alternative teaching strategies. The work had, 'highlighted the incompatibility between encouraging self-searching and transmitting a body of knowledge...'

All the teachers commented that without the presence of the researcher, they were unable or unwilling to find the time and energy to continue with the detailed and systematic process of self-evaluation. However, all had already transferred what they had learnt into their work with other classes. One said that it already had some effect on the work he was pursuing with a third year class, 'insofar as they're doing some very successful self-directed work now'; and in his work with a sixth year class where, 'I've been able to extend the amount of material and our approach to it'. Another had negotiated the task rather than imposed it with pupils in first and second year classes, and had changed what had previously consisted of predominantly whole class work to small group work; and another had 'changed the content' (i.e. the negotiation of content with pupils) in order to allow new methods with other classes. A fourth also had employed small group work with younger pupils, but was as yet unwilling to negotiate content with pupils in examination classes because of the responsibility he felt for producing work for external assessment.

The researcher held informal conversations with all the teachers during the two years after the research with them was completed. During the course of these conversations, which were often on matters quite unrelated to the research, they often informed the researcher of how the changes they had made in attitude as well as practice were being sustained. They felt that they trusted much more their own ability to not only find, but also evaluate and modify, their personal solutions to the teaching problems that they encountered. In effect, they felt that they had achieved a new critical standard with regard to themselves as teachers.

In the case of one teacher, Steve, evidence of the long term changes in thinking was able to be collected both during the supported research process, and at intervals up to five years following its completion.

His detailed written evaluation contained what was for me a fascinating account of how his attitudes towards the research and the researcher had changed over time.

> '... over two years it has progressed from feelings of aloofness, caution and occasional cynicism to ... real professional regard and genuine interest and concern with every aspect of your study - both in your terms and mine...'

He commented on the positive value of self-confrontation and the researcher's role as 'consciousness raiser' rather than direct 'change agent':

> 'Your change agency can't change our personalities ... but what I think it can do is reveal to a teacher the nature of his personality in so far as it does or does not elicit responses and promote a learning environment for his group of kids …'

I asked Steve if he had felt any positive gain because of the research:

Steve: 'Yes, two. The first one is ... talk, and I allow far more latitude in a constructive way as far as group and individual's talk is concerned. And I try and participate in it more, and steer it more …
The second thing is, the main gain I think, has been the setting of a higher self-standard, I can't think of any other way in which you'd be so compelled to examine yourself and force yourself as high as you possibly can in the classroom. You try it, and if it works, then you've reached a level to which you must always afterwards aspire, and compare whatever else you do with that. I know when I've fallen short a bit, and I usually know why …'

He stated that the self-confrontation process, assisted by the researcher, had taught him to ask questions, structure conversations, value inter-personal relationships in the classroom, re-assess the value and use of teacher produced resources, and re-assess his teaching role. He had undergone a fundamental change of attitude, and a re-adjustment of his self-concept. He believed that his attitude to the process of teaching and learning had changed permanently from 'mechanistic' to 'humanistic':

He summarized the value of the research both in terms of its process and in terms of product.

'The biggest turn around for me'
'I feel much surer of what the fundamental attributes of a good teacher are in my own mind – without all this navel gazing I would have stayed in my neatly ordered, mechanistic universe for a good while longer. I think I may well have saved myself many errors in approach and the blundering up of a good many blind alleys in starting a new department by looking primarily at its teachers' skills and attitudes rather than its administrative and curricular super-structure ... being a part of this research cost me a lot of time and energy. It was worth it ... there is no communication in learning without feeling – can there be a product which is economically feasible, within the capacity of change agents who don't share your personal attributes, which is a true synthesis of research techniques and human correspondence?'

Long Term Value: The Teacher's View

Some five years later, the same teacher wrote about the changes that had occurred in his thinking as a result *of* engaging in the research process:

> 'I can attest to the validity of this model of reflection and theory building. There has been shift in pedagogy that had come about through the critical evaluation of current practice in the light of both personal and public theory. The gains were hard won and difficult to hold in the sense that once a personal theory had been revalued and aligned with public theory, any slipping back induced some feelings of guilt. Close reflection upon practice became an eradicable habit. The implications of this model of teacher change are powerful ones. It is not important that my experience of the process focused only upon one aspect of my work within 'English' for I discerned a ripple effect which caused me to employ the evaluative methods learnt in one area upon others. Additionally, I had found a way of making public theory my own and this aroused the appetite to search out more that may be of relevance and use. What is significant however is the degree and intensity of external support that was required to engender these effects: there exists no army of researchers who can institute the process in a wide scale. For me this experience accelerated a process that I hope, but cannot be sure, would have taken place anyway. I was enabled to move out rapidly from my 'comfortable routines' and from the 'coping' strategies that often mark the plateau of many teachers at an early stage in their careers. The process gave me renewed access to public theory in the sense that I could use it: prior to that I was aware of such theory, but could not employ or affirm it because my personal theory was too tightly in the grip of my current classroom practice. By taking a risk and letting go of accustomed practice, by becoming theory-less for a time, I was able to address and assimilate the public theory to which I aspired.'

The point being made is that by recognizing the learning and research potential of teachers and engaging them actively in the investigation of their own theories of action we may promote more learning about both teachers and research.

The Role of the Researcher in Research Into Professional Learning

Professional Learning Theory and Change

The six main principles that resulted from the work with Steve and other teachers were: 1. effective learning occurs in response to the confrontation of problems by the learner (This reveals discrepancies within and between thought and action); 2. decisions about teaching should stem from reflection on the effects of previous actions (This results in a reassessment of thinking by the teacher about his thinking and behaviour); 3. effective confrontation of problems requires the maximizing of valid information (This results in a desire by the teacher to change); 4. video-recordings of lessons provide direct evidence of theories-in-use, and are therefore especially effective promoting confrontation (which results in attempts to change); 5. effective professional learning requires internal commitment to the process of learning and freedom of choice for the learner; 6. teachers need support in achieving changes - partly because old routines dominate and new routines need support to develop.

Important assumptions underlying these principles are: a) much knowledge about practice is implicit rather than explicit, teachers (and indeed researchers) have a limited understanding of theories-in-use; b) in so far as assumptions about the influence of external factors and the nature of classroom practice remain unquestioned and unproblematic, these are likely to limit a teacher's capacity to evaluate his work; c) terms such as 'reflection' and 'informed choice' are taken to imply the need for explicit examination of both espoused theories and theories-in-use; d) problems in learning new theories of action stem from existing theories which already determine practice, and unless that practice can be related to exist theory explicitly, it is unlikely that any new theories will be developed.
(Detailed evidence for each of these principles can be found in Day (1979))

In the work described above, since the teacher is seen as an active causal agent in his own learning, the research design cannot either masterminded or unilaterally controlled by the researcher-consultant. The work must be collaborative, with a maximum flow of information between the two and the teacher and consultant's hypotheses being openly stated as they develop. The conceptualization of the problem (comes) through the empirical work, not as its prelude (Shipman, 1976). If this kind of work is to be developed, channels of communication must be established by the research community that enable teachers and researchers to engage in a continuing dialogue about the nature of teaching and learning within the classroom. For both perceptual and practical considerations of time and energy the active support of an outside agent is essential in work concerned with the thinking-practice dimension of teachers. His presence is necessary: a) to establish and sustain a responsive, mutually acceptable dialogue about classroom events and their social and psychological context; b) to audit the process rather than the product of possibly biased self-reporting; c) to create a situation in which the teacher is obliged to reflect systematically on practice. This is unlikely to happen in the crowded school day; d) to act as a resource which the teacher may use at times appropriate to the needs which he perceives; e.g. to relieve the teacher of the task of data collection; e) to represent the academic community at the focus of the teacher's professional life.

The researcher thus becomes a part of rather than apart from the teacher or client. Thus a mutually acceptable language of discovery will develop. Problems of transfer (of knowledge), validity and credibility (of research findings) and 'barriers to change' will be minimized. In research such as this the two main principles for intervention theory and change are that: 1) the perceived needs of the client(s) are of paramount importance; 2) the consultant's role is collaborative and co-equal, but not necessarily neutral.

The assumptions underlying these principles are that:
a) teachers have the capacity to be self-critical; b) teachers are motivated to learn by the identification of a problem that concerns them; c) some teachers may need affective as well as intellectual support; d) the interventionist must help the teacher in processes of internalization, rather than compliance or identification (Kelman, 1961).

Others have reached similar conclusions, though not always supported by empirical work.

Throughout the research, I assumed a model of a teacher who, given particular circumstances, is able to distance himself from the world in which he is an everyday participant and open himself to rational influence by others. I believe that this distancing is an essential first step towards self-evaluation.

Ideally, external agents would not be a necessary part of this process. However, most situations are not ideal, and teachers are not afforded the opportunity for reflection on their teaching. This is not to say that teachers do not reflect in the normal course of events, but that the conditions in most schools for most teachers prohibit any detailed consideration of the complex factors which contribute - often in conflicting ways – to teachers' decision-making in the classroom.

Teachers, then, are likely to operate on a model of restricted professionality. Once they have developed a personal solution to any problems of teaching that they perceive – and this is usually achieved without any systematic assistance by others – it is unlikely that this solution will again be significantly questioned.

This pattern of teacher development has been characterized as single loop learning, where theory making and theory testing is private. The environment is competitive rather than collaborative; and the pursuit of an objective rationality, in which feelings are suppressed, is highly valued (Argyris and Schön, 1976).

The traditional researcher also engages in single loop learning by defining his clients' needs according to what his techniques will allow him to provide. These professional techniques are then reinforced by the institutions that have evolved to make them work (e.g. schools and training establishments) and behaviour is constrained to suit them. Technique thus becomes an instrument of control rather than a means of serving the clients' needs.

Researchers who assist the individual who is attempting to explicate and build his own theory of practice are, as Steve recognized, engaged in 'a synthesis of research technique with human correspondence'. The essential elements in this process of learning are diagnosis (of the world in which he acts), testing (of his theories and assumptions) and the accepting of personal causality (taking responsibility for what he does); and these elements refer both to technical theories (techniques which the practitioner uses in his work), interpersonal theories which state how the practitioner interacts his clients during his work.

Researchers too often operate in their world of restricted professionality and, like other professionals, protect themselves and their colleagues by speaking in abstractions without reference to direct observed events. This has the effect of both controlling others and preventing others from influencing oneself by withholding access to valid information about oneself. The time is ripe for more open access by the research 'subject' to the researcher's thinking and practice to the ultimate benefit of both.

Author Reflection Twenty Years On

It was fascinating reading my own work published more than a decade ago – for two reasons because it caused me to reflect on i) how far my own understanding have moved since I first asked the questions, 'In what ways do teachers learn? Why do they change, (or fail to change)?, and what is the role of the researcher as intervener in this process of teachers' thinking and behaviour?' and ii) the contributions made by other researchers since then.

Within the limited space available, the answer to the first is in the affirmative – and my ongoing research on teacher learning and development school leadership and change reflect this. It probes variations in the work, lives and effectiveness (VITAE) of 300 teachers in 100 schools across England. It too seeks to answer the original questions on learning and change, but take account of the seminal research of Michael Huberman and Milbrey McLaughlin in particular. For since the original paper, there have been advances in thinking – for example, work on situational and organizational and leadership effects on learning (McLaughlin, Leithwood, Lave and Wenger), on teacher efficacy (Ashton), on the importance of teacher biography and history (Goodson), on the role of emotion in cognition (Damaio) and, on change (Fullan). There has also been much more small-scale action research work with teachers, the development of learning networks and the much-publicized work of the school effectiveness and school improvement researchers whose agendas seek to iterate with policy.

Yet, this new knowledge has served to inform further the important effects of context upon teacher learning rather than to answer directly the questions posed in the original paper. It is perhaps an unavoidable characteristic of educational research that is still, by and large, underfunded, that it is still, therefore, predominantly small scale and that, though reaching for an holistic view of the teacher, it illustrates the inescapable problems of capturing complexity and being unable to produce generalisable solutions to perennial problems. With regard to the ongoing issues of establishing structurally supported link between the Academy and schools, well, there are more of them but in the grand scheme of things, relatively few. So cultures are still separate rather than connected.

Chapter 9

What Makes Teachers Tick? Considering the Routines of Teaching

J. K. Olson

Summary

Teachers are often thought to do things in an unconsidered way; the very idea of teacher routine activities, for example, suggests for some that consideration is absent. I would like to explore the possibility in the paper that although teachers may not always be able to tell us “what was in their mind” when doing this or that we would be wrong to assume that something unconsidered was going on. On the contrary, it might be thought that routines are the highest expression of what teachers know how to do and teachers have good reasons for acting as they do in the way they engage in what we call routine activities. We ought to think of routines more as the centre of what teachers do rather than activities of a simple or unconsidered kind.

Introduction

In order to develop the idea that teacher routines ought to be a central focus of study of teaching I will consider first the well known distinction between knowing how and knowing that as it relates to routines. Secondly I want to consider how we can investigate what teachers know how to do, and try to draw a parallel between understanding the actions of teachers and historical research. In doing this I will suggest methods of investigations that focus on the reasons why people act as they do. In this way we can come to appreciate the meaning of teacher actions in classrooms as we see how what teachers know is brought to bear in their practice. Such a way of thinking about teacher thinking has promise for school reform in the long run.

I begin this discussion by referring to a major study of science education that was conducted by the Science Council of Canada (Olson and Russell, 1983). The study had its origins in a general concern about how little young people know about the achievements of Canadian science and about the people who made them.

Teachers were seen to be the problem because they were unwilling or unable to deal with science in Canadian context. The Science Council launched a study aimed, in part, at describing just what science teachers did in their classrooms, what they said they did and how these doings might be understood. It was hoped that such data might cast light on the assertion that teachers were unwilling or unable to deal with science in a particular way. Case studies at eight sites across the country were launched after the researchers had agreed that we would observe teachers teaching and ask them to tell us what they were doing.

When all the cases were in two of us tried to make sense of the actions of the teachers whose cases we had in front of us. We sought to understand the meaning of teacher actions within a framework controlled by their purposes. This might be called a reflexive view of the role of the teacher (Olson, 1983). In this view change occurs as teachers reflect upon their purposes and their ability (their know how) to pursue those purposes and to alter them. To understand what happens in schools we tried to understand the purposes of teachers. We found a number of things that teachers did which we think we understand but are nonetheless troubled by them. Our response to these findings gives a sense of our approach to data of this kind.

First, we found that teachers tended to ignore optional units in curriculum plans. Such optional material often contains socially relevant material involving teaching styles in which pupil opinion is stressed. Teachers said they did not have enough time - we suspect they do not know *how* to handle these options. We were unhappy about the sidelining of this material.

Second, we found that teachers emphasized socialization in their teaching goals - the inculcation of good habits through careful recording of scientific fact. We found that teachers stressed diligence. We found them very suspicious of inquiry and using inquiry approaches infrequently. We suspect they are nervous about their know how here.

Third, we found teachers teaching science without reference to its processes or to its economic/political context. We found teachers using science to develop good habits - good behavior -through instruction in and practice of doing problems, and making notes. We find that teachers know how to do these things and that the apparent limited 'know how' of these science teachers was a cause for concern.

Thinking back, I now ask how did we come to these conclusions? What were we doing when we approached the research this way? One answer might be that we looked at what teachers did as historians might look at the behavior of their subjects. We sought to understand motive; we were concerned about teacher judgements; and how we ought to appraise what we found had happened. We had, of course, the texts of their own commentary, as do historians, and we had some understanding of the context of action. Our analysis of the data was a form of reflection on the actions of the teachers; what did these actions mean, we asked? What was their significance? In taking this approach we have placed a strong emphasis on how teachers think about their work, that is, how they *explain* what they are doing.

How are we to react to these expressions of their approach? Are we to say how reactionary teachers are? For here again we find teachers not achieving those ideal potentials for science education imagined by outside planners and theorists. Were we to lament the inadequacy of their grasp of the pedagogics which could be used? In short were we to once again accuse teachers of unconsidered and limited action? Clearly not, but how should we view their actions and how they talk about them? This was a fundamental problem for the way we interpreted the case study data.

Knowing What and Knowing How

Teachers often have been accused of rather simplistic conceptions of what they do. Their "technology" has been seen to be weak and in need of bolstering especially by extracting solid prescriptions from the social sciences. There is much doubt now about such a diagnosis and prescription. I hope this paper will contribute to an increase in that skepticism (See also Calderhead, 1983).

Dan Lortie (1975), for example, says that the ethos of the profession is tilted against pedagogical inquiry. Teacher theories are simple and uncritical, but is their *practice* itself so bereft of intelligence? Is their know *how* so deficient? This is another matter, and I think the answer is no. The practice is much more skillful and intelligent than teachers talking about it might indicate. What teachers know is embedded in their know *how*. It is only because of what Gilbert Ryle (1949) calls the "intellectualist legend" that we tend to assess the intelligence of performance on the basis of the quality of the supposed antecedent internal operations of planning. If we find that the operations are poorly articulated we assume that the practice itself is also poor; not so, says Ryle. These are two different things. Ryle argues that contrary to the intellectualist legend, efficient practice precedes the theory of it. The intelligence is in the practice, not in *thinking* about it. Abilities are played out in the practice itself – in the know how.

If this is generally correct, we ought to consider carefully studying *practice* to decide what teachers can actually do. That teachers may not be able to give a well articulated propositional account of their practice is another problem. They may or may not. Teachers are very likely making many adjustments as they carry out familiar routines in the classroom. However, I find it difficult to understand how these could be based on conscious decisions based on the weighing up of propositions moment by moment and thus I have some difficulty seeing how an information-processing model would be useful in understanding routines – where teacher craft is most finely tuned and capable. Not "thinking" about teaching does not stop teachers from efficient practice and while it may be a good thing for them to be able to articulate well what they are doing, that doesn't stop the doing it well, or at least better than we may think judging from their own accounts of what they do.

The capacity to talk about it and do it are different, although related, and if we want to study teacher practice – what teachers know how to do, we have to observe what they do. Teacher thinking in this sense is teacher practice; what teachers know is in the practice not only in the cognitions that accompany practice.

We may be interested in how they explain or justify what they do, how their knowledge is articulated retrospectively. Let us not mistake one for the other or collapse the both into the one. Their know "that" performance is one thing; their know "how" performance is another. Both are teacher actions we would like to know more about so that we might understand teachers and teaching better and to understand how teachers come to change their practice.

This line of thinking leads us to be concerned about what teachers do as well as what they say about what they do and to be careful about how we judge their work as a whole. It could be dangerous, for example, to stress some theory about how teachers do what they do and then go ahead and look for evidence of such a theory in action. We might end up by proposing that teachers ought to do it the way theory says, and we would only be assessing what they do in reference to the hidden norm put up by theory. Take the case of teacher planning. As Ryle (1949) suggests "People do not plan their arguments before constructing them. Indeed if they had to plan what to think before thinking it they would never think at all; for this planning would itself be unplanned" (p. 30). Yet some of the research on teacher planning seems to assume that this is exactly what teachers do or ought to do. Perhaps the research is really a norm in disguise. If teachers are not planning before action then they should be is the message of the research, and moreover they should do this according to particular theoretical models for planning. Yet why should they plan before acting when they already know how to act?

Polanyi (1958), in his book *The Study of Man*, picks up Ryle's concern about knowing how and knowing that in his distinction between *tacit* knowledge and *articulate* knowledge. Tacit knowledge is akin to know how. As Polanyi says our powers of knowing "operate widely without causing us to utter any explicit statements and even when they do issue in an utterance this is used merely as an instrument for enlarging the range of tacit powers that originated it" (p. 27). Prearticulate knowledge, or tacit knowledge, we gain through experience; we learn about a new region by walking through it without maps; our knowledge may never be articulate yet we can find our way around. We may learn to perform many activities without any articulation of them. Our knowing preceded our thinking about our knowing. Tacit knowledge precedes articulate knowledge. This is also Ryle's point. Efficient practice precedes the theory of it.

Polanyi goes on to make another distinction about human knowing, i.e., the idea of *focal* and *subsidiary* awareness. This distinction is crucial to his idea of how we come to understand other people, and to how we ought to view the details of their behavior in that process. He argues that the particulars of behaviour make sense only if we understand their purpose. He notes that words, graphs, maps and symbols are never objects of our attention in themselves, but pointers towards the things that they mean. Symbols can serve as instruments of meaning only by being known indirectly or *subsidiarily* while we fix the focus of attention on their meaning and their purpose. The same is true of tools and machines. Their meaning lies in their purpose.

So while we attend to their purpose we are only subsidiarily aware of the things themselves; and so it is for understanding human beings. Comprehension of what another person is doing cannot be had through mere examination of the particulars of their behavior; we have to appreciate what the person does in terms of what he/she is trying to do. We have to understand their behavior as pointers towards the purposes that they serve, and in terms of those purposes. The meaning of what people do lies in the purposes served by those actions. Hence, these actions are not objects of attention in themselves but indicators, known not in them, but subsidiarily to the purposes they serve. In themselves they mean nothing. On this view, atomistic accounts of what teachers do seem bound to fail for lack of a context to make sense of them. Such a focus guarantees we will not really understand the meaning of isolated acts. Only a molar view will do. Such a view is largely absent from research on teacher thinking (See Hills, 1984).

What Ryle and Polanyi are saying is that what people *know* how to do is bound up in *what* they do, and Polanyi goes further to say that the meaning of what people do is to be understood in terms of the purposes they are attempting to achieve. Without understanding those purposes we cannot understand what the actions mean. Later I will consider how such understanding might be achieved and where the pitfalls lie, but turn now to recent and important work by Donald Schön

(1982) which deals with understanding practitioners, and has considerable relevance to our discussion of the nature of teacher practice.

Studying Teachers' Tacit Knowledge

Schön's work owes much to the work of Ryle and Polanyi. In his book The *Reflective Practitioner* he criticizes the idea that theory precedes practice, and that the intelligence lies in the theory not in practice. Essentially, he mounts the same critique of this idea as Ryle and Polanyi. He takes issue with what he calls the "model of technical rationality". This model, he says, assumes that professional activity consists in "instrumental problem solving made rigorous by the application of scientific theory and technique". Science, he says, does not tell us about problems or ends, and problems of ends cannot be reduced to problems of means – to problems of technique. There are no techniques for dealing with problems of ends. Procedure here is tacit, as Polanyi would say.

Schön adopts the idea that knowing is inherent in action and that the challenge for understanding and improving professional practice to make articulate what is inarticulately embodied in action itself; the starting point for improvement is practice, not norms derived from social sciences, although the findings of social sciences might have some bearing on the critique of practice once articulated. Practice itself, he says, is knowledgeable. More is done in our practice than we can say. For Schön to educate the professional is to engage him/her in making the tacit articulate and to subject that to criticism. But how should we go about this? Schön gives some examples. They seem to me to suggest an *historical* approach to understanding practice; a critical reflection on the meaning of one's actions in their context.

Rather than fundamental differences between the study of history and natural science, Polanyi sees simply differences in the degree to which the student has to come to identify with the subject matter being studied. He sees a progression of identification from the inanimate to the animate to man. The need for identification is greatest when the object of study is man: greatest in the study of history; in the study of human action which is what we are concerned about here. I will not rehearse his arguments here, as I want to pursue the significance of the idea of *identification* with teachers as it bears on research into teacher thinking and action.

Polanyi sees three pitfalls that lie in the way of those who would study human action in an historical context, that is, try to understand human actions in the past with reference to reasons. Briefly we might be inclined to judge these actions by our own standards without allowing for differences in setting. This he calls the *rationalist* fallacy. Judging the actions by standards of their own setting renders any *criticism* of the standards of other settings meaningless. This is the fallacy of *relativism*. Reducing man's moral scope to the control of impulses or any other cause and thereby voiding action of moral meaning is the *determinist* fallacy. How do we avoid these fallacies?

Polanyi suggests, first, that we recognize the biological and cultural rootedness of free action; second, that we assume that all humans have free access to standards of truthfulness and are obliged to limit actions in relation to them in all times, and that, third, we assume that the human mind is the seat of responsible choice. While people may fail, they have the capacity to succeed and they have access to truth that will help them succeed. This is where Polanyi wants to lodge his study of man. I think it may be a good place to lodge the study of teacher thinking.

Practically speaking how do we come to understand other people's actions? Polanyi suggests that we have to participate in the life of the person we want to know. We have to identify with the person: "This we do anyway when we study other things, but it reaches a point with man that when we arrive at the contemplation of a human being as a responsible person and we apply to him the same standards as we accept for ourselves, our knowledge of him has definitely lost the character of an observation and has become an encounter instead" (p. 95). Instead of the atomistic approach to understanding people's action that I spoke of earlier, we see here reasons why an holistic view of action is needed.

I believe this is what we have to do in order to understand what teachers are doing in their classrooms if we are concerned to improve what happens there. If one has educational worries at heart, our judgements that we apply have to be based on some defensible *conception* of education. In our analysis we subject the actions of those we encounter, and ourselves as we pursue our study, to these same standards. We assume that we, like they, are free to know those standards and that we seek to understand human action by exploring the mind not just the environment.

Polanyi uses the term "in-dwelling" to suggest the kind of attention to the other person that we should strive to achieve. This view has much in common with an approach to understanding other people developed by George Kelly (1955).

The Search For Reasons

It is interesting to note that George Kelly, through critical reflection on his *own* clinical practice and his know how, produced an articulation of that practice which became his personal construct theory. Kelly's work might be seen as a way of creating the kind of personal encounter that Polanyi considers so important. By rendering the tacit knowledge articulate we can gain the advantage of being able to discuss practice and to subject *know how* that is in practice to critical scrutiny. Kelly himself, as a clinician, was very interested in finding ways to help people articulate what they did and why they did it. He recognized how difficult it is to move from doing things to knowing what one is doing and why. His grids are aids to the accomplishment of such a process; they are the means to achieve the in-dwelling of the clinician of which Polanyi speaks. As Kelly says "If you do not know what is wrong with a person ask him, he may tell you. The clinician who asks such a question will have to be prepared to do a lot of listening" (p. 322-3).

The philosopher Theodore Mischel (1964) points out that Kelly's theory may be regarded as a guide to careful listening. The listening is aimed at uncovering the *reasons* why people behave as they do. Mischel, in fact, reconstructs Kelly's theory in terms of *constructs as reasons*, not as anticipations or plans. Constructs he sees channeling behaviors because they are the rules or reasons which people follow in their behavior. The use of grids and the conversations that go with them might be seen as means to articulate the reasons that are embedded in peoples' actions. In this view rules and plans are not rehearsed in some way prior to action, they are played out in action. Most often, we do not plan then act - we act.

Constructs, I think, do not tell us about our anticipations, they tell us about what is built into our actions as we act; they are our reasons, or the rules that guide and channel our behavior. We simply plunge ahead tacitly acting out what we know until, of course, something gives us pause and we begin to reflect on our actions in effort to make articulate what is tacit - to construe our behaviour. This is the way I would read what Kelly is saying. In this view, Kelly offers us a powerful method for trying to help people articulate for us what the meaning of their actions is, what rules guide and channel behaviour; how they see the world. The actions in themselves are meaningless. To an outsider they could mean anything and thus nothing; unless we get to know how the other person construes himself or herself and us. This is what I take Kelly to be saying.

The significance of this line of thinking for research on teacher thinking is that we need to watch teachers teach and like historians we need also to consult them as we speculate on what their actions mean within the frameworks which give their actions meaning. We need to talk to teachers in order to understand them and to test our ideas about them against their own.

To come back to the case studies I spoke about at the beginning this paper, in our analysis of these studies we did end up judging the actions of the teachers in the same way that history judges the actions of people. We did seek to understand the meaning of their actions by trying to understand what prompted them to act in the way they did in order to see what their reasons were. We did see people doing things we felt they shouldn't be doing, and we said so in our report; but we also assumed that these teachers were responsible for what they were doing, that they were in control of their lives. Where we thought they failed, we did seek to understand why; knowing there are hazards in teaching and knowing that while successes are possible so is failure and that the hazards are great. We did this, we hope, within a shared framework of obligations. Having undertaken this reconstruction I have given here of the work we did for the Science Council, it strikes me now how much like the writing of history was our task in appraising the material presented to use in the eight case studies, and how limited our opportunity was really to get to know the teachers in a way that would leave us confident that we had encountered them, instead of making subjects of them. Perhaps next time we can achieve that.

Author Reflection Twenty Years On

Since this paper was written the notion of novice and expert teacher has become a common way for researchers to describe, but more often implicitly to judge, teacher performance. The term suggests a certain level of cognitive functioning if not awareness. Why use this term "expertise"? For those who advocate such cognitive functioning the term has scientific and efficiency overtones. It derives these overtones from the idea that if only teachers could process information like the expert systems of computers students would learn more effectively.

Over the life of ISATT the notion of novice and expert has been very much in evidence. Research under this rubric involves both getting teachers to tell the researcher something about their expertise: we are puzzled by them and want to understand them better; or there is something about what they do or tell us that makes us want to affirm them by saying they are experts. In these senses the character of expertise is defined by teachers. On the other hand, the researcher might be saying, perhaps not explicitly: we know what expertise is and we are going to examine teacher practice, see who has it and give them a grade. Such a concept of expertise would, of course, serve well as a theoretical basis and hence rationale for teacher testing. Finally researchers may be both describing and evaluating in the same research in a dialectical way – notions of good functioning arising as teacher practice is examined. Somehow standards are evolved by watching what teachers do. The problem is that the normative basis of these evaluations is elusive. What is *good* about good practice, as it were?

The term novice has completely different origins. A novice is a *novitiate* - a term originating from the practice of religious communities. A novice is someone who has yet to have developed the capacity to be a full member of that religious community, who does not know the mysteries of the community, and has yet to be tempered and developed. And so here we are juxtaposing these two terms – expert and novice – that come from such different realms and metaphors of human activity One from the idea of the mind as a computer processing information; the other from the whole person being inducted into the mysteries of practice. It is hard to imagine how the transition from one to the other could occur under such a distinction – such a rubric. This fact alone should give us pause. Are we here being bedeviled by language? Is such a transition possible or are we talking about two different ways of thinking about teaching – different *teloses*?

We need to look at this distinction critically. Are we confounding ourselves by prescribing ways of teaching based on supposed beneficial mental functions and calling them descriptions? Are we reducing teaching to something less than it is by calling teachers *experts*? What should the *telos* of teaching be: expert or novitiate?

Section B

Personal Beliefs: Self as Teacher

The first three chapters in this section the authors, Pope and Scott, Lowyck, and Broeckmans, consider previous research in their respective fields, finding the then dominant reductionist approach inadequate for understanding the complexity of teaching as an activity. They note in particular that teachers' intentionality and the alternative frameworks that they bring to their professional worlds have previously been neglected. Using examples from their own research, these authors illustrate their assertions that more interpretative techniques can enhance understanding of teaching and learning to teach. Maureen Pope and Eileen Scott, in Teachers' epistemology and practice, focus on what teachers count as knowledge – and how resistant such views are to change; Joost Lowyck in his chapter Teacher thinking and teacher routines: a bifurcation?, also addresses the idiosyncratic nature of teachers' interactions within their professional environment, particularly exploring the nature of routines undergone in the preparation for and implementation of teaching. In An attempt to study the process of learning to teach from an integrative viewpoint, Jan Broeckmans addresses the pre-active and interactive phases of teaching but also draws in the post- interactive phase using a combination of observational methods with intro- and retrospective techniques for gathering data. All of these authors support a view that an integration of research approaches and methods is required if we are to gain a rich and realistic understanding of the nuances of the science, art and craft of teaching.

In Chapter 13, entitled Describing teachers' conceptions of their professional world, Staffan Larsson considers paradoxes in the classroom – what teachers think and do in response to what students think and do – in an exploration of teachers' implicit assumptions using the then innovative approach of phenomenography. Again the emphasis is on appreciating the unique aspects of individual interpretations of context and process rather than eschewing such detail to generate higher order generalisations.

There follow two chapters contributed by Jean Clandinin and Michael Connelly. In the first one, What is " personal" in studies of the personal?, they analyse the then recent research literature that purports to address the personal in teacher thinking, drawing three main conclusions. In brief, the first judgment is that, though authors name concepts and approaches differently, there is yet more commonality than might initially be supposed. The second highlights some of the differences that are extant in how we as researchers imagine teachers' thought to be composed, while the third notes an emphasis on cognitive aspects or conceptions to the detriment of consideration of affective processes. Their second chapter, Personal practical knowledge at Bay Street School: ritual, personal philosophy and image, provides an illustration of an enquiry that evades that last problem, being a dialectical examination of Personal Practical Knowledge that, by definition, includes the affective in conjunction with the cognitive.

In the next chapter, written by Miriam Ben Peretz and Lya Kremer-Hayon and entitled Self-evaluation, a critical component in the developing teaching profession, the theme of the cognitive and affective processes combining to influence action is continued. This time the focus is on teachers' self evaluation – learning from feedback generated by self rather than the more typical form of feedback from outside sources. In common with findings from previous chapters, teachers in this study showed great diversity, this time in their ability and willingness to engage in self reflection, each individual reflecting on a personal style and mode of teaching.

Chapter 17 documents Robert Tabachnick and Kenneth Zeichner's study of the interaction between beliefs and behaviour, and their exploration of strategies used by teachers to harmonise differences between them. In this instance it is belief and behaviour that are viewed as inseparable while context plays a considerable influential part. Drawing on a multi-method longitudinal study, the authors of this chapter, Teacher beliefs and classroom behaviours: some teacher responses to inconsistency, use case studies to illustrate how two teachers take different routes to minimising inconsistencies between beliefs and behaviour and how the management structure of their schools influences this.

In contrast to the emphasis on the personal in the preceding chapters, Margret Buchman in Role over person: legitimacy and authenticity in teaching looks at teacher decision making from the two perspectives of 'personal reasons' and 'professional obligations' to develop an argument that the latter should predominate in taking professional action. Although she recognises, like Tabachnick and Zeichner, that personal beliefs guide action, she proposes that these should change if not consistent with professional discipline since she advocates a shift in perspective from self to other when engaged in professional role. Again context is considered as influential but this time it is the prevailing culture of a country rather than of a specific school that is cited. Like Pope and Scott, she considers it important that teachers first identify their personal beliefs as a prelude to comparing them with external standards, national or professional.

While the next chapter, Chapter 19 Forming judgements in the classroom: how do teachers develop expectations of their students' performances? presented by Manfred Hofer, is a study of teachers' reasoning, it begins by assuming that we already have adequate knowledge of the variables that teachers use to evaluate pupils, though we do not know how these variables are inter-related. In contrast to the other chapters in this section,, an experimental approach is used to unravel the interaction of influential factors, though the interpretation of the results was informed by spontaneous comments made by the teachers during the process of the experiment and from their reflections in a subsequent interview. The results again accentuate individual difference, this time in the way that teachers form judgments.

Angelika Wagner's study, Conflicts in consciousness: imperative conditions can lead to knots in thinking, takes us back into an approach that requires teachers to share with the researcher their thinking while engaged in professional activities. The purpose of this research was to illuminate the 'knots' in thinking that occur when imperative thoughts are in conflict thus producing a dilemma. She required teachers and students to recall what was going through their minds at certain points during the class. The results demonstrate interesting differences between teachers'and students' knots and suggest that there may be differences between teachers who espouse different approaches. Significant implications for teacher training are raised.

Chapter 10

Teachers' Epistemology and Practice

Maureen L. Pope and Eileen M. Scott

Summary

This paper draws on the theoretical perspective of George Kelly in addressing the importance of teachers' perspectives on the nature of knowledge. We will suggest that teachers' views on knowledge and theories of learning will affect practice in the classroom. A short summation of current work, describing research findings based on a study of student teachers and practising teachers' views of knowledge, learning and implications for the practice of teaching, will be given.

We contend that theories of knowledge will be integrated into a repertoire of constructs which teachers hold and that the staff developer/teacher educator will need to come to an understanding of the role these perspectives play in relation to other constructs in the teachers repertoire. We will argue that teachers' views may be highly resistant to change and will advocate an approach similar to views expressed by Kelly on Fixed Role therapy as a means of encouraging teachers to reflect on their positions and the implications of their viewpoints for their practice in the classroom.

Teachers' Epistemology and Practice

Theories can be seen as alternative constructions that are reliant on the particular vantage point of the theory builder. 'Facts' are potentially open to reconstruction. These statements represent part of our epistemology and influence how we teach our students. This personal knowledge leads us to conjecture that teachers' views of knowledge, ie. their epistemologies are likely to affect the way they approach their teaching.

Our current research is concerned with the teaching of psychology to student teachers; their epistemologies; their views on teaching and learning and what interactions, if any, there are amongst these aspects which have implications for the process of teacher education. This paper will address the issue of student teachers' epistemologies. Our methodological stance and personal epistemology is such that we find the ideas of G.A. Kelly (1955) particularly useful in providing a framework for our research. Kelly described his philosophical position as that of *'constructive alternativism'* that suggests that people understand themselves, their surroundings and anticipate future eventualities by constructing tentative models and evaluating these *personal* criteria as to the successful prediction and control of events based upon the model. Kelly assumed that any event is open to as many reconstructions of it as our imaginations allow. He rejects an absolutist view of truth and contrasted his position with that of *'Accumulative Fragmentalism'* ie the notion that knowledge is a growing collection of substantiated facts or *'nuggets of truth'*.

Constructive alternativism provides the basic philosophy within which Kelly elaborated his personal construct theory. Kelly's approach to the development of a person was based upon the metaphor of '*man-the-scientist*'. He invited us to entertain the possibility that looking at people *as if* they were scientists - ie view their 'scientist-like aspects' - might illuminate human behaviour. Kelly believed that a person's theories were the 'push and pull' of their behaviour and therefore if one wants to understand and communicate with another it is essential to know that person's theories or *personal constructs* about the subject under discussion. Kelly was quite clear in his antagonism towards a psychology that sees people as '*reactive*' rather than '*constructivist*'. He rejected the notion of the person as an '*impotent reactor*' whose behaviour was determined by environmental circumstances or genes. He portrayed people as active agents capable of making things happen, able to construct events and invent their theories.

The contrasting positions of constructive alternativism and accumulative fragmentalism are possible implicit theories held by teachers. The latter position is akin to that held by naive realists who ignore the role of invention in theory building. Lakatos (1970), a philosopher of science, stressed the importance of the invention of theories in science and noted an important demarcation between passivist and activist theories of knowledge:

> 'Passivists' hold that true knowledge is Nature's imprint on a perfectly inert mind: mental *activity* can only result in bias and distortion ... 'Activists' hold that we cannot read the book of Nature without mental activity, without interpreting them in the light of our expectations or theories. Now *conservative activists* hold that we are born with our basic expectations; with them we turn the world into 'our world' but must then live forever in the prison of our world ... But *revolutionary activists* believe that conceptual frameworks can be developed and also replaced by new, *better* ones; it is *we* who create our 'prisons' and we can also, critically, demolish them'.

In our work with science educators we have noted that the positivist, empiricist-inductivist conception of science is in sympathy with an absolutist view of truth and knowledge and that if teachers hold to that conception of science then curriculum content and the manner in which students are taught will place little or no emphasis on the student's own conceptions and active participation (Pope and Gilbert, 1983). The model of the learner as 'impotent reactor' is appropriate here as are passivist theories of knowledge. Kelly's philosophy is akin to Lakatos' description of the revolutionary activist's views. Rather than see constructions of reality as prisons', Kelly would see them as dynamic frames that can limit conceptual development *if* an individual chooses not to accept responsibility for the creation of such frames.

One of our concerns with respect to the teaching of psychology student teachers is the extent to which lecturers present such a 'body of knowledge' as conjectural. To what extent are student researchers encouraged to make explicit their implicit personality theories and implicit epistemologies? Indeed to what extent are the epistemologies embedded in many psychological theories made explicit by the lecturers? When, if ever, does the lecturer address his/her own view of knowledge? In our own work we have noted a distinct neglect of such issues within the process of teacher education despite a growing interest in constructivism within psychology and education (Pope 1982, Magoon 1977).

An implication of Kelly's notions for education is that teachers will have a range of epistemologies and perspectives on learning and teaching and individuals will differ as to their aims, aspirations, expectations, and ways of operating in relation to how they construe education. Postman and Weingartner (1977) state:

> 'There can be no significant innovation in education that does not have at its centre the attitudes of teachers and it is an illusion to think otherwise. The beliefs, feelings and assumptions of teachers are the air of a learning environment; they determine the quality of life within it'.

These writers acknowledged the subversive implications for education of relativistic knowledge. They argued that it was a direct challenge, and possible threat, to teachers who saw their role as passing on bodies of substantiated facts or 'absolute truths' without the necessity of presenting these as problematic or conjectural. The idea that 'all facts are subject' leaves the door open for the student to question the knowledge presented by the teacher. If control rather than negotiation, understanding and facilitation is the dominant value of the lecturer and if s/he has an epistemological view in line with Lakatos' categorization of a passivist view of knowledge then the medium of teaching and learning offered to the student teacher is likely to be such that the message may be one which supports an accumulative fragmentalist perspective. The notion of constructive alternativism recognizes the importance of differing views. For an educational system to function adequately there is a need for an understanding of these different perspectives. Problems can arise when an individual or group operates with one set of assumptions (subsystem of constructs) and tries to impose or communicate these assumptions to others without any acknowledgement or understanding of an alternative framework or set of assumptions which the other values. Olson (1980) drew attention to how teacher constructs about innovations can affect the progress of implementation. As he pointed out, attempts at curriculum change were often translated into a "*pale reflection of the original as they are implemented in the classroom*". In Olson's study some important constructs identified were concerned with the teachers conceptions of their role and issues of control and influence. In our own work we have found that problems can arise when there is a lack of negotiation between students and tutors about fundamental epistemological issues, as the following quotation indicates:

> 'What anybody says is gospel in this place. And there's no point trying to vary it because they won't agree with it. You've got to comply with everything. There's no room for voicing your own opinions. They like you to argue, but if you do then there's trouble'. (B.Ed. Year 2)

Student teachers are in the unique position of alternating roles between student and teacher whilst also studying the formal theories of the processes that they themselves are involved in i.e., processes of learning and knowledge development. During their studies they may be faced with a range of views on teaching, learning and underlying epistemologies. Pope and Keen (1981) outlined what they saw as four major themes in the development of Western educational ideology, each of which exhibits a continuity based upon particular assumptions on the nature of knowledge. Some or all of these themes may be presented to student teachers as formal theories. Each student is likely to have a personal epistemological stance which is in part compatible with one or other of such theories but, as Pope and Keen point out, one should be wary about rigid classification as it may do an injustice to the diversity of views which do exist. In our opinion, student teachers should be encouraged to adopt a reflexive stance and consider the inter-relationships between their own views and those inherent in formal statements offered in lectures and educational texts.

It is clear from our research that students do have personal epistemologies some of which can be categorized as falling within the global themes outlined by Pope and Keen. In the next few paragraphs we will present crude thumbnail sketches of each of the four themes identified within Pope and Keen's book in order to provide a backcloth against which some personal statements of student teachers can be gauged.

The traditionalist *cultural transmission* approach is encapsulated in a quote from Hutchins (1936):

> 'Education implies teaching. Teaching implies knowledge. Knowledge is truth. The truth is everywhere the same. Hence education should be everywhere the same'.

Here we have an indication of an educational ideology based upon a particular epistemology – one which corresponds to the basic principles of naive realism i.e. that the world we perceive is not a world we have created or recreated mentally but is the world as it is. True knowledge is knowledge that corresponds to the world as it is and the task of the educator is the direct transmission of a body of knowledge. The learner in this system is seen as essentially passive.

The Romanticists' ideology stresses the importance of a pedagogic environment within which the inner forces of 'good' and 'bad' in each person can be dealt with so that the inner good will unfold and inner bad come under personal control. Repression by the teacher is to be avoided. There are similarities between the views of Romanticists and educators and those inherent in Idealist philosophies which hold that ultimate reality is spiritual in nature rather than physical and mental rather than material.

When considering Progressivists' views on teaching and learning a dominant theme is the provision of a challenging environment so that learners have to struggle and, through their own activity, develop their knowledge. Knowledge acquisition is an act of change in pattern of thinking brought about by experiential problem solving. A basic principle of Pragmatism is that the world is neither dependent on nor independent of a person's idea of it. Reality is the interaction between a person and his/her environment and such transactions produce truth that is a property of experiential knowledge.

The fourth ideology, which Pope and Keen labeled 'de-schooling', can be seen in the view of Illich (1971), Goodman (1972) and others who are concerned with the lack of personal relevance to the learner of much of the school curriculum.

Similarities can be seen between the de-schoolers and existentialist views that knowledge is personal and that reason is informed by passion. Buber (1965) wrote about what he believed to be the tyranny of impersonal knowledge and advocated that the teacher needs to find the subjects s/he teaches personally relevant before s/he can offer this knowledge to students. The responsibility rests with the student to reject or accept the teacher's interpretation and the pedagogic environment should be one of mutual trust within which such questioning can evolve. Here we have a view of knowledge and educational practice that is far removed from that of the transmitter of knowledge model of the teacher.

These sketches can be equated with four major philosophical traditions: realism, idealism, pragmatism, existentialism, and we suggest that each would imply a different educational practice. There have been attempts to develop checklists to measure a teacher's epistemological position on scales based on a conceptual analysis of such philosophical stances (Ross, 1970). However, as Young (1981) suggested, the apparent lack of consistency between epistemological views based upon such checklists, and similarly constructed questionnaires to establish other educational beliefs, could be due to the form of research inquiry itself:

> 'The categories used in the instrument construction are not very useful categories for describing teacher epistemologies, and that there are other differences between teachers about epistemological questions which may be more important than those identified by conceptual analysis of philosophical viewpoints'.

Young reviews some attempts to measure epistemological style and, notes the difficulties that researchers have had with mutual exclusivity of categories used in such instruments. Within our research group Swift (1983) has found Cawthron & Rowell's' questionnaire on epistemology within science education too restrictive.

Denicolo (1983) has decided against a categorization of her data on teachers' views on explanation in science according to a conceptual analysis of philosophers such as Kuhn, Popper, and Bacon. Both of these researchers have found less procrustean techniques more acceptable in collecting and interpreting their data.

Young has pointed out there have been few attempts to study teachers' conceptions of knowledge and of these there is a marked absence of detailed ethnographic work which might illuminate the range of epistemological beliefs held. Our own notions of educational research lead us to place particular value on research strategies that take into account the ways in which persons involved in educational interactions interpret and construct their personal viewpoints. Research can be seen as a formalized version of human inquiry and educational research as a formal instance of people trying to understand people and processes within educational settings. An ethnographic approach experimenting conjointing *with* people rather than on them seem to us appropriate given our interest in personal epistemologies and views on teaching and learning.

From this conceptual approach emerged a case study of students attending four different courses in Primary Education. Two courses (B.Ed. and P.G.C.E.) were for initial teacher training and two (B.Ed. Inset and Dip. Ed.) were for in-service teachers. Participants totaled 51 (see Table I).

Table 1. Number of Participants.

	Initial		In-service	
Course	B.Ed. Yr 2	P.G.C.E.	B.Ed.is	Dip.Ed.
Number of participants	18	16	10	7
Total in Class	45	40		14

The study followed students throughout one academic year: before during and after the first teaching practice in initial training and during the college year for in-service students. Research strategies included semi-structured interviews and observation. Interviews were loosely structured to encourage the student to reflect on his/her own views of knowledge and pedagogic practice focusing on themselves both as teachers and learners. All interviews were recorded and transcribed verbatim. Initial training students were interviewed prior to and after their first block teaching practice. In-service teachers were interviewed towards the beginning and end of their seconded year. Observations were made of selected college classes of all groups. Additional classroom observations were made of some initial students on teaching practice. Informal chats, e.g. over coffee, were encouraged with staff and students. The following quotations represent some of the views expressed during the semi-structured interviews.

The first quotation is from a student whose personal views on teaching and knowledge development embraced many of the tenets of "de-schooling" and progressivism but she acknowledged the existence of a cultural transmission perspective.

> '... I think it is to do with how you perceive teaching because if you view teaching as you have a store of knowledge and its your job to impart the facts A or B or whatever to the children or the students, then you would perceive teaching in that way but if you view teaching as helping children or students to reach their full potential and to acquire the knowledge and make it their own and develop in their own terms, then your role, then teaching you know you're helping them move to acquire this knowledge. You're not. giving them facts and saying 'right you've got to go and memorize them' you're discussing with them and helping them to try and make those facts or that knowledge their own, so therefore they can use what you're trying to show them or teach them'. *(Dip. ed.)*

The following quotations, however, indicate the students' views of the teacher as 'expert transmitter of knowledge', knowledge as subjects or established facts building one upon the other and as an entity legitimated by the teacher.

> '... as a direct imparter of information because I think that some things can only be learnt and taken in by being told or educated by another, by a person who is experienced at that other thing'. *(P.G.C.E.)*

> 'So knowledge could be...'knowing about things that have happened in the past and also things that are going to happen in ... could possibly happen the future ... and within that you could have the various skills of knowledge. The subject disciplines that are needed to separate out from the big idea of knowledge ... like your maths and your physics and your language'. *(P.G.C.E.)*

> 'Well you could say it's (knowledge) established fact and there are established facts even though some people say there are not. There are, there must be'. *(B.Ed. Year 2)*

> 'I think knowledge should be built upon itself, so it sort of builds up like building blocks and it has to have very firm foundations to build upon. So a child has to be taught the basics ... so that knowledge is built up step by step'. *(P.G.C.E.)*

> 'I suppose knowledge is what somebody has deemed as being important, and they consider that it's important that it should be continued to be taught as knowledge, and if you can reproduce it, then I suppose that that's knowledge. Because I definitely think that somebody's got to believe that its knowledge in order for you to think that you know something, because we think that we, when we think something in our own head, which hasn't been taught to us that's not knowledge in a way I think it is formal, and its got to be considered knowledge by the person who's imparting with me (laughter) on the part of the teacher I suppose'. *(B.Ed. Year* 2)

In contrast, the following quotation is from a student who rejects the notion of knowledge as 'facts' and prefers a knowledge as personal experience model.

> '(Knowledge) It's experience. I don't think knowledge is ... a body of facts. I don't think that is true. I think knowledge is when you've experienced ... different things, different things in life, different people, attitudes, values, beliefs, and you accept or, when I say accept or understand those beliefs even though you if don't accept them that is knowledge. When you are aware of other people's views, and do not discard them'. *(B.Ed. Year 2)*

Adopting a Kellyan perspective would suggest that the teacher recognize pupils' personal constructs as having both important epistemological value and high educational status. Encouraging students to articulate <u>their</u> beliefs about knowledge and learning whilst in the role <u>of student teacher</u> seems to us to be equally important as Postman and Weingartner (1971) stressed:

> 'Following the medium is the message or you learn what you do theme, it is obvious that teacher education must have prospective teachers do as students, what they as teachers must help their students, in turn, to do'.

The following quotation from one of the student teachers in our study indicates that this state of affairs does not always exist:

> '... there's one particular module which tends to be more on the traditional, almost chalk and talk type of delivery, in which there isn't, they're not really requiring feedback from us. They're not really requiring our active participation, and we feel you know far too passive and er... they are presenting ideas of teaching involving children and they seem to be you know teaching us in the opposite way to the way they are trying to suggest that we would teach children'. *(B.Ed. In-service)*

Many of the student teachers commented that they had found the process of talking about their epistemology and ideas about teaching and learning during the interviews to be helpful in focusing and clarifying their thoughts:

> 'It's (the interview) been of great value to me actually. I've not really had to sit down and think about things like this as much. I do wrangle about it with other people, but not just sit down and me think and no-one else come back and with counter argument. So it's helpful, very helpful. Perhaps we could do with more of it'. *(P.G.C.E.)*

Similar views were expressed by student teachers in Pope's (1978) study using Repertory Grids. Many indicated that it was the first occasion that they had really thought about their *own* views on teaching as opposed to those presented during their course. We would argue that such discussions should be an integral part of teacher education. The formal concepts presented on college courses need to be transformed and assimilated by the student teacher and their own frame of reference is important in this process.

Our work indicates that student teachers can, if encouraged, become aware of their initial epistemologies. Constructivist teacher educators would seek to continue this process by encouraging the student teacher to challenge his/her initial epistemology and explore the potential of alternative views.

Diamond (in press) describes an approach to in-service teacher education that does seek to build upon and help to challenge student's personal constructs. It is based upon Kelly's Fixed Role therapeutic technique during which clients were encouraged to give a self-characterization sketch of themselves and discussion would take place about the constructs that the client developed. During the course of discussion it might be noted that a particular construct, or way of seeing the world, may be causing ineffective functioning or '*problems of living*'. Kelly would encourage the client to recognize that these constructs are of the client's making. If clients accept the responsibility for their own ways of thinking about the world then they can be encouraged to experiment with a way of construing which is opposed to the construct which may be causing problems ie they can be encouraged to try out an alternative theory. The client would be invited to *put this alternative theory into practice* over the next few days. In this way clients are putting themselves in a situation in which the possibility of the alternative, and perhaps more adaptive, viewpoint can be validated. If alternative theories are not experimented with and one's own theories go unchallenged then one will be a captive within a prison of one's own making. Diamond used Fixed Role Treatment (FRT) as an educational procedure rather than a therapeutic method and has found it useful in helping in-service teachers to review personal construct systems underlying their classroom pedagogies. His study demonstrated that "*teachers would give up a viewpoint, even if it were an integral part of them, providing that they had become aware of the personally meaningful implications of an alternative.*" Rix (1983) argued that to "*develop and stimulate reflective conversation is a most preliminary and important step towards improvement in our educational system*".

Whilst we would not suggest that radical changes in an educational system would ensue, we do suggest that reflective conversation within teacher education will make explicit fundamental perspectives. Teacher educators in colleges and staff developers in educational institutions need to develop strategies, such as FRT; in order that they can understand, for example, what a particular student's epistemology is and, if deemed appropriate, encourage the student to explore alternatives. Student teachers and practising teachers may hold personal views that are highly resistant to change and, as has been found with studies in concept development, simply giving them 'facts' about alternative positions may not be sufficient (Driver, 1973; Donaldson, 1978).

We have been concerned to raise the issue of the educational status of students teachers epistemologies and our preliminary analysis of data supports Young's 1981 contention that: "epistemologies, curriculum codes, assessment approach and approach to pupil control constitute elements of a coherent pedagogical ideology".

However, mere documentation will not improve teacher education. We suggest that a process of articulation and challenge of personal views is a necessary step which the college tutor or staff developer must facilitate if they are to help the student teacher "*further realise their potential as personal scientists and intelligent educators*". (Tomlinson, 1982).

Author Reflection Twenty Years On

The emphasis on student teacher's implicit theories has, if anything, increased since Eileen Scott and I wrote our ISATT paper in 1984. Kelly's epistemology, on which the research was based, has had a pervasive influence in educational research. In a recent book (Pope and Denicolo, 2001) I have provided ample evidence to justify this claim. In my view Kelly's legacy was the notion that education should be about personal meaning. It is this emphasis on personal meaning that links Kelly to contemporary constructivist approaches in education. In particular issues such as: the 'perspective of the personal', his focus on relevance and responsibility within the teaching and learning process, his epistemological stance and his recognition and valuing of alternative perspectives, have all had an impact on education.

Our paper drew attention to the need for teacher educators and researchers to address the students implicit theories. Since then, I and a number of my colleagues in our Centre for Personal Construct Psychology have worked with experienced teachers and explored their epistemology and implicit theories.

Kelly's metaphors of person-as-scientist and, more recently, person-as-story-teller underpinning personal construct psychology have become guiding metaphors for many contemporary educationalists. Constructivist educators, including teacher educators, now pay attention to the learner's current ideas and how they change in addition to the structure and sequence of teaching "received" knowledge. In teacher education there has been a move away from teaching the "ologies" without reference to practice. Psychology as a separate subject has disappeared from the curriculum for initial teacher trainees. Reflection on practice and personal perspectives is becoming more commonplace. Learning is now being portrayed as a personal exploration and student teachers are encouraged to recognize the need for the teacher to come to some understanding of the experiments, lines of enquiry and personal strategies used by the learner.

Implementing such an approach within teacher education poses a challenge, particularly if governmental policies militate against it or such a view runs counter to the teacher's implicit theories of teaching and learning. Teacher educators may wish to encourage conceptual change in the way teaching and learning is promoted. However, the goggles through which others view their worlds might not be easily altered. This is an aspect that has been stressed more frequently in the year's post 1984. The teacher educators wishing to encourage students to articulate their implicit theories need to be aware of the extreme sensitivity of the reflective material that can be evoked. They must be prepared to give time and support during periods of deconstruction and reconstruction when an individual is confronted with an image or action that he or she may wish to change. This is a fundamental requirement within constructivist approaches to professional teacher education.

Clark (1995, p. 124) alluded to the fact that, for many, the phrase 'professional development of teachers' contains "*a great deal of negative baggage: it implies a process done to teachers; that teachers need to be forced into developing; that teachers have deficits in knowledge and skills that can be fixed by training; and that teachers are pretty much alike*". I agree with Clark that such a deficient model constrains development and pays insufficient attention to support mechanisms that may be needed during reappraisal.

I am happy to say that this is now increasingly recognized through research and professional developments that have taken place since my chapter was written in 1984. Nevertheless, I stand by much of what I wrote at that time. Perhaps life in teaching has become more fraught and opportunities for reflection more limited. One of my students found this so in her research with teachers. One of whom said, "*Everyday life is so hectic so there is little time to evaluate what you do, what you have done and if you have achieved the goals you set up*" (Apelgren 2001, 327).

In my own work with teachers I have found similar reactions. Two issues I may have given insufficient attention to in the 1984 chapter were that personal and professional change are inextricably linked and that quality education depends on the extent to which we understand the position of others with whom we interact. This is a central message of Kelly's (1955, 1991) work for those striving for a more democratic educational process in schools and in universities and is one that many teacher educators throughout the world now advocate.

Chapter 11

Teacher Thinking and Teacher Routines: A Bifurcation?

Joost Lowyck

Summary

During the last decennium, a steadily growing number of studies are devoted to the important question of how teachers function in the complex teaching-learning environment. Along with the basic concepts of cognitive psychology, teachers are conceived as information processors, interacting with the environment in a unique and idiosyncratic way. Consequently, the process-product paradigm has been abandoned. Teaching behaviour cannot be understood as the sum of relatively isolated, elementary behaviours.

Within the cognitive approach of teaching, some attention is paid to the role of routines in pre-interactive and interactive phases of teaching. The lack of a conception of what is meant by 'routines', however, is one of the main problems in this line of research. At the same time, it seems difficult to integrate a concept of routine-development into a cognitive approach of teaching. Some tendency to isolate both kinds of teaching activity is discernable.

In this paper, the theoretical background for the study of routines is elaborated, using the work of Miller, Galanter and Pribram (1960), Newell and Simon (1972), van Parreren (1982) and the work of many scholars in the field of skillfulness in teaching.

The problem of the connection between teacher's thinking and routines must be rendered more concrete. Therefore, different research reports are analyzed from some questions, such as: what is the operationalisation of 'routines', and 'teacher thinking', what kind of relationship is reported between them, what research methods are used for studying both kinds of teaching activity and how can the results fit into the theoretical background?

This research analysis will serve as a platform for further exploration of possible ways to bridge the gap between routines and thinking. Although the author is convinced of the problematic character of finding clear solutions, the results could reach a rather heuristic value.

Introduction: Historical Context

During recent years, a steadily growing number of studies have been devoted to the questions about how teachers function within the complex teaching-learning situation. Putting the teacher's intentional activity in the core of research is an almost logical consequence of the recent evolution in research on teaching.

Without any doubt, researchers in the field of teaching always have tried to find out what 'good teaching' represents, what features it includes, how the quality of teaching can be measured and improved. This knowledge about teacher effectiveness can deliver, then, interesting information for the improvement of teacher education and in-service training.

Before we examine more closely the framework for research on teacher thinking and routines, it is useful to review briefly the antecedent traditions in research on teaching.

The teacher's personality paradigm

A first tradition in research on teaching has been characterized as the search for 'good teachers'. Due to the influence of the test-psychology, man is conceived of as a cluster of specific and stable traits. Teaching, consequently, is seen as an activity of the person who possesses the necessary professional traits. It is not as much the activity in itself, but the person as the source of activity who is studied.

The main research question to be answered here is: can one discover in a systematic way the personality traits, characteristic of a good teacher? The instruments used are selected from existing psychological tests (e.g. the Minnesota Multiphasic Personality Inventory, M.M.P.I.) or from tests specific to teachers, such as the Minnesota Teacher Attitude Inventory (M.T.A.I.). The results of these tests subsequently are correlated with some criteria of good teaching gathered mostly by means of ratings or opinions from pupils, colleagues, directors, inspectors, and researchers.

Getzels and Jackson (1963) have reviewed more than 800 studies since 1950, as did Domas and Tiedeman for more than 1,000 studies up to 1950. In their main conclusions, the authors agree with the vagueness and unsuccessfulness of this tradition. Getzels and Jackson (1963, p. 547) note that "despite of the critical importance of the problem and a half century of prodigious research efforts, very little is known for certain about the nature and measurement of teacher personality, or about the relation between teacher personality and teaching effectiveness".

This rather pessimistic comment must certainly not be interpreted as saying that good teaching has nothing to do with the personality of the teacher. It is more likely that we attribute the lack of outcomes to the inadequacy of both the framework and the methodology within this paradigm.

The teaching behaviour paradigm

As a reaction to the lack of information about relevant teacher characteristics, along with a more analytical, behaviouristic view of teaching, research was focused on teaching behaviours in the classroom. The criterion for effectiveness is not sought in stable traits of the teacher as a person, but in the concrete teaching behaviours within the teaching-learning situation: the classroom. In order to gain insight in the classroom behaviours, there has been a proliferation of observation systems. A well-known inventory of observation categories can be found in the work of Simon and Boyer (1970). A more content oriented review is written by Rosenshine and Furst (1973) and by Dunkin and Biddle (1974). As in the former paradigm, the methodology is a correlational one: researchers correlate observed and coded teaching behaviour with several measures of learner outcomes as indicated by differences between pre- and post-tests.

From the meta-analysis of the research results, some promising teaching behaviours have come to light (Rosenshine and Furst, 1971); examples are: clarity, variability, enthusiasm, task-oriented and/or business-like behaviour, student opportunity to learn criterion material, use of student ideas and general indirectness, criticism, use of structuring comments, types of questions, probing and level of difficulty of instruction. This conclusion has been discussed by Potter (1973), Heath and Nielson (1974) and many others. They indicated weaknesses in the methodology of the original studies as well as in the way data were treated.

In our opinion, it was not only the lack of an adequate methodology nor the lack of precise criteria in making a meta-analysis, but also the concept of teaching behaviour itself, which contributed to the rather problematic character of these studies. Teaching cannot in its wholeness be conceived of as the sum of a limited number of isolated effective teaching behaviours often called "skills." Teaching behaviour has to be understood in relation to the intentions of the teacher and to the situational complexity. It is, for example, not the frequency of the questions in the classroom that is so important, but rather what questions about what content is asked, at what moment, to what pupil.

The teaching intentionality and complexity paradigm

One reason why the process-product paradigm has been abandoned by some researchers is because of the too restricted interpretation of teaching behaviour, which excludes the internal processes and the specific situation that influence observable behaviour. This conception of teaching parallels the cognitive approach in psychology. Attention is now paid to the complex processes and activities during pre-active, interactive, and post-interactive teaching.

Frequently, overreaction against former paradigms leads to overemphasis and so we see teaching as a highly rational activity. This stress on rationality is also expressed in a fan of metaphors for the teacher, who is now called "decision-maker" (Shavelson, 1976), "hypothesis-tester" (Coladarci, 1959), "information-processor" (Stolurow, 1965), "problem-solver" (Turner, 1964; Fattu, 1965), "planner" (Clark and Yinger, 1978) just to give a few examples. The tendency to pay an almost exclusive attention to the internal processes can be seen in the name of the research paradigm, namely: "teacher thinking".

From most of the descriptive studies on teacher thinking we have done during the last years, it became clear that the teacher is not exclusively a rational professional. On the contrary, he/she reduces the complexity of the job using many non-reflective behaviours. This insight leads to the modifications of an extreme rational view, particularly by revising the functional relationship between thinking and routines. How we can handle the tension between cognition and action will be the next question.

Toward an Integrative Theory of Teaching

One of the basic questions is the following: how is it possible to reconcile the different and often divergent approaches to the study of teaching? The answer is an optimistic one if we start from a purely theoretical point of view. We have to search for a framework in which the different important teaching variables can be put together: personality traits, observable behaviour, internal processes, and situational characteristics. We have thus to combine intentions with actions, persons with environments, thoughts with routines, anticipation with unpredictability, processes with products. Researchers should leave for a while their pragmatic empirical research and clarify the paradigms, the frameworks, the methods, the concepts. If we could agree with an integrative theory of teaching, we can avoid splitting up our forces in a centrifugal direction.

Seen from a more pragmatic point of view, it will not be so easy to build up an integrative theory of teaching. Each researcher has his own concept of teaching, his preferred methodology, his framework for thinking, and his personal motivation. Therefore, it seems as difficult to communicate within the same professional language or to use equally loaded terms, as it is to make the methodology clear and verifiable or to integrate and not to separate.

Although presently there is no elaborated model on hand for the study of intentional and complex teaching, some features can nevertheless be elucidated.

(1) Teaching is a meaningful activity of a teacher, including the complex, cognitive processing of external and internal input. Observable behaviours, underlying processes and situational variables interact in a subtle way.

(2) Teaching is an activity that shows different characteristics, depending upon the specificity of the teaching phase: planning, execution, and reflection. The quality of the teaching activity as well as the specificity highly depends on the characteristics of the phase. When, for example, preparing a lesson, the teacher will use other information sources than during his interaction with the pupils.

(3) Teaching is an intentional activity. Goal orientation is one of the most striking features of teaching. On the other hand, not all teaching activity can be explained by intentions, because at some times the environment is a very powerful influence. Thus, teaching is both an independent and a dependent variable. The teacher is self-steering as well as steered by the field.

(4) Teaching is a professional task. It seems however impossible to describe extensively and objectively all the features of the complex task: the so-called "task-environment" (Newell and Simon, 1972). This lack of a uniform definition of the task is a direct consequence of the complexity of the situation as well as of the philosophical assumptions of the teaching activity.

(5) Teaching is as an activity not fully defined by specific "teaching behaviours", for a teacher uses many spontaneous, daily behaviours while teaching: answering questions, giving information, telling a story. These behaviours are not professional in their essence. Moreover, some daily behaviours can fulfill other functions when used within the teaching context. Asking questions, for example, is not necessarily intended to gather that information which is relevant to the teacher but it is frequently used to stimulate the pupils themselves to remember, to think, to formulate.

At last, there is a number of very specific teaching activities, such as the diagnosis of the pupil's state of learning, the organization of suitable learning activities, remedial teaching.

(6) Teaching is conceived of as an activity on two tracks: the teacher and the pupil. When a teacher only behaves from his own standpoint, he has no contact with the pupils. If he is operating exclusively on the track of the pupils, the intentionality of teaching comes into danger.

(7) Teaching is in essence a molar activity (Miller, Galanter and Pribram, 1960). This view on teaching enables the researcher to study teaching from the most significant point of view. In this way, he avoids reducing teaching behaviour to small pieces of non-significant behavioral elements. Broader activities (strategies) as well as smaller behavioural units (tactics) have to be studied at the same time. This conception of teaching reminds us of the choices we have to make for relevant units of analysis in research on teaching.

The possible bifurcation between teacher thinking and teaching routines has thus to do with how we conceive teaching. It is our opinion that it must be possible to investigate several aspects of teaching activities by bridging the gap between cognition and action, observable behaviour and cognitive processes, behavioural elements and molar activities, self and environment. On the other side, we are aware of the difficulties in reconciling precise, valid and reliable research aims with thick descriptions of the teaching reality.

Teacher Thinking and Teacher Routines

We define teaching as "the intentional activity of the teacher which allows the pupils to reach the educational goals by means of a lesson content in as optimal a way as possible within the context of the teaching-learning situation; this activity concerns planning as well as execution and evaluation of teaching and learning activities and processes" (Lowyck, 1980, p. 4-5). In this definition, the priority of the intentionality of teaching is made clear, but this does not mean that there are no "routines" in the teaching activity. It depends highly on how we define "thinking" versus "routines."

Although one can emphasize the "unity" of behaviour, including conscious and unconscious aspects, the research literature shows choices being made for one or another aspect of teaching behaviour. On the one side, attention is almost exclusively paid to the "rational" side of human behaviour, often called "activity" and, on the other, more emphasis is laid on spontaneous, unconscious behaviours, often called "skills" or "routines". How can we reconcile this bifurcation?

Teaching as a "rational activity"

In the more rational oriented conception, teaching is conceived of as an intentional, conscious, reflective activity. Dewey (1910, p. 9), for example, sees reflection as the means of escaping from impulsive and routinized behavioural elements. He defines reflection as "active, persistent and careful consideration of any belief or supposed form of knowledge in light of the grounds that support it and the further consequences to which it leads." Even in the literature about skills, more attention is paid to the integration of behavioural elements and to the underlying mental processes. Broadbent (1964, p. 295) refers to the cognitive processes when defining skills: "The picture of skilled performance built up by modern researchers is one of a complex interaction between man and environment. Continuously, the skilled man must select the correct cues from the environment, take decisions upon them which may possibly involve prediction of the future, and initiate sequences of responses whose progress is controlled by feed-back, either through the original decision-making mechanism, or through lower-order loops."

Within this more cognitive oriented approach, different theoretical models have been used to interpret teacher thinking. Smith and Geoffrey (1968, p. 96) pointed out "teaching must be seen as an intellectual, cognitive process. What goes on in the head of the teacher is a critical antecedent of what he does." This way of defining teaching as a complex and intentional activity has generated new ways for research. Shavelson (1973), for example, has studied teaching using the decision-making metaphor. Many researchers have been inspired by his original approach or have used alternative models. Recent overviews of research on teacher thinking by Shavelson & Stern (1981) or by Clark and Peterson (1984) are very instructive ones. Consequently, it seems unnecessary to summarize their findings here, taking the expert position of the readers into account.

The preference for more rational aspects of teaching behaviour, however, holds a possible restriction for the understanding of teaching in its complexity. In the research tradition of teacher thinking, much work has been done on the planning activities of teachers, with some neglect of the relation between planning and interactive teaching. When studying cognitive processes in the classroom, much attention has been paid to the description of explicit rational processes, using the decision-making model, the problem-solving metaphor, or the information-processing approach. Along with others, we like to search for a more integrative approach on teaching, in which teaching in its complexity is studied from several entry-points, often called "triangulation". The latter term has been described by Denzin (1978, p. 295), who says, "it is convenient to conceive of triangulation as involving varieties of data, investigators and theories as well as methodologies."

With Jackson (1968, p. 151) we conclude that, "sometimes teaching is described as a highly rational affair. Such descriptions often emphasize the decision-making function of the teacher, or liken his views with elementary teachers raise serious doubts about these ways of looking at the teaching process."

As happens regularly in the history of research, the overemphasis of one or another aspect provokes an analogous over accentuation of some opposite aspect. As we will see, it remains difficult to hold the meaningful teaching activity in mind, when making decisions about research.

'Teaching as a "routine"

The term "routine appears rarely in the current literature on teaching. It has been used by Marie Hugues (1967, p. 21) in contrast with decision-making and connotes a "stereotyped" response. Good (1973, p. 8) defines routines as "activities engaged in by a teacher and children which are necessary for efficient classroom management, so conducted as to provide a classroom conducive to group learning and wholesome personality development, for example health inspection, collection of lunch money, recording of attendance etc.".

Such definitions are not specific enough to point out the important features of routines. We will now examine some recent descriptions of routines.

Definitions of "routine"

Because each theory on teaching rests on the validity of the concepts, it is an important task to analyse seriously the meaning of the terms. One of the main problems of putting research results together in a consistent theoretical framework is the divergent use of concepts; different terms are used for the same phenomenon or the same concepts grasp a very different reality. Heath and Nielson (1974, p. 475) express strong criticism against the lack of precision in the operationalization of teaching behaviours, as reviewed in some meta-analysis reviews on process-product studies. They conclude, "continued research on such sterile definitions of teaching seems unlikely to provide a basis for training teachers." It is an important task for researchers in a developing field of study, to handle very precisely the meaning of the used concepts. If this is not done, many investigation results will be inadequate in their contribution to theory building. What different meanings are found in the literature about "routines"?

(1) Some authors refer to routines as to recurrent units of behaviour. Unwin and McAleese (1978, p. 683) define a routine as "a regular course of procedure, a more or less mechanical or unvarying performance of certain acts or duties." Creemers and Westerhof (1982, p. 24) see routines as standardized behaviours which occur frequently during planning and execution of teaching and which are initiated automatically by the teacher.

(2) Other researchers define routines in contrast with conscious activities. A routine, in this case, is all the teacher does when not thinking. An example of this conception is given by Morine (1973), when she describes the planning as a routinized activity of writing down what pages of the textbook will be covered.

(3) Especially within the context of psychological research on skills, routine has a connotation of "automatism." The criteria for defining a routine are here: speed, precision, adequacy, ease, minimal use of time and energy.

(4) Other authors speak about routines as established procedures whose main function is to control and coordinate specific sequences of behaviour (Yinger, 1979). This interpretation is close to the cognitive interpretation of routines. Welford (1968), Whiting (1975), Bartlett (1964) and Fitts (1965) have pointed out that in most of skilled activities, cognitive aspects are very important, such as complex functioning, goal orientation, organization and co-ordination, learning processes, adequate selection of input, anticipation, decision-making, use of feedback. In their approach to complex skills, the authors come very close to the more "rational" or cognitive interpretation of behaviour. Schriffin & Schneider (1977), who have studied information processing in human beings, make a distinction between conscious processes and "automatic processes." The latter are processes where (a) the sequence of nodes from the long-term memory always becomes active in response to a particular input configuration and (b) the sequence is activated automatically without the necessity of active control or attention by the subject.

(5) An important element for the definition of a routine is dependent upon the moment at which an activity did become a routine. Tillema (1983) refers to the "expert" quality of a routine, which supposes a high level of professionalism in teaching after a serious time of learning. Bromme (1982) also draws attention to the complex problem of studying expert versus novice problem solving. This transition from a very conscious to an almost automatic activity is a central issue in the study of routines and teacher thinking. Welford (1968, p. 194) looks

at the evolution from novice to expert as at the mastering of larger and larger units in which the smaller units occur more automatically. Whether one operates on the more cognitive or the more automatic level, is highly dependent upon the situation in which the activity occurs. Where conditions require or permit virtually exact repetition of a behavioural unit many times, the performance tends to become more stereotyped. If the demands of the situation are more open, or the performance is in an early stage of practice, the activity is more consciously steered.

In the literature about teaching routines, it seems to us that authors often have used only a "term" without examining the complex dimensions of it. It makes a difference, when we call teaching routines all these teaching activities which occur frequently (asking questions, organize group work...) or when we define routines as the result of a long professional learning process, with a gradual "abbreviation" of control points in the complex teaching activity, The latter definition brings us closer to bridging the gap between thinking and routines.

Teaching routines in pre-active and interactive teaching

In the literature concerning teaching routines, a distinction is often made between planning routines and the interactive ones. During planning the external input consists of accessible materials, like handbooks, curriculum materials, information about pupils, notes. The teacher has at his/her disposal internal input: expectations, experiences, reflections, habits, and feelings. There is time to reflect upon eventual problems and the teacher is not hindered by complex reality. The interactive phase, on the contrary, is characterized by simultaneity, multidimensionality, unpredictability, time pressure, and so on. At the same time, a teacher must take the position of the pupils into account: their reactions, their knowledge, and their feelings. It seems also that the “double-track” position of the interactive teaching is one of the most salient features which needs more attention in research.

Pre-active routines

Jackson (1968, p. 151) suggests that the teacher often seems to be engaged in a type of intellectual activity that has many of the formal properties of a problem-solving procedure, when he works before and after the face-to-face encounter with students. Pre-active teaching is then interpreted as a rational process of deliberate actions. Yinger (1979) has pointed to some routinized activity during the planning of instruction. From the results of his well-known research, two types of routines have been identified: the routinization of planning itself on the one hand, and making decisions about interactive routines on the other.

The routinization of planning activities can be seen as a kind of “abbreviation” or “shortening.” The teacher gradually condenses the lesson content, because the control points are reduced. He also anticipates in a cumulative way the possible reactions of the pupils (Lowyck, 1980). Yinger (1979) suggests that the routinization of planning activities ensures flexibility. Because of our study, it seems that the abbreviation or reduction of control points does not guarantee the quality of the reduction. The question is whether the teacher focuses on the essential or critical points. We have revealed in our study that the teacher can loose sight of some critical features of the content when preparing a lesson.

Inter-active routines

Shavelson and Stern (1981) have summarized the results of a number of descriptive studies that focus on routines during interactive teaching. They point out that teacher's interactive teaching may be characterized as carrying out well-established routines. The teacher reduces his complex information processing to some cues, such as the participation of the students. This monitoring is automatic, as long as the perceived cues are within an acceptable tolerance. If not, he/she searches for a suitable routine. If no routine is available, the teacher reacts spontaneously and she continues the teaching activity. Again, we encounter the problem of the quality or relevance of the controlled cues. How can we conclude that we are dealing with a trial and error behaviour, with spontaneous reactions or with routines in a more technical sense of the word? Here we meet an analogous problem in the study of skills. Whiting (1975, p. 59) noticed that the term "skill" is a theoretical construct, so "that we do not observe the skill directly, but infer its presence by the behaviour of the person and that a concept of this nature is known in the psychological literature as *intervening variable*."

In order to know whether the teacher is using a routine, two types of knowledge are necessary. First, we need a description of how teachers select the space for the teaching activity (Newell and Simon, 1972). What does a teacher perceive in the complex task environment, what in the complex situation can be handled with routines? Second, we have to know if and/or how routines come about as a kind of abbreviation of originally cognitive activities. In other words: how are routines learned? Both questions must be studied thoroughly in future research.

Some Concluding Results

The problem of a possible bifurcation in the study on teacher thinking and teacher routines cannot be solved without a clear definition of what is meant by both terms. Moreover, there is a need for a framework in which the complex activities of the teacher can be studied in an integrative way. Researchers must pay attention to the danger of splitting up reality by inadequate use of words. Unless we have insight into the complex relation between cognition and action as well as into the genesis of expert teaching behaviour, important conclusions will not be reached. Miller, Galanter and Pribram (1960, p. 11) refer to it, saying, "Our point is that many psychologists, including the present authors, have been disturbed by a theoretical vacuum between cognition and action". Many researchers have tried to understand something more about the thought processes of teachers. They use different models, mainly oriented to the "inner side" of the teaching activity during planning, execution, and reflection. Due to the historical evolution of research on teaching, they tend now to neglect the observable behaviour of teachers in the classroom.

Maybe it is time to focus on the complex teaching activity, including thought processes and spontaneous, automatic activities. The latter as well as the former are worthy of in depth studies. Before researchers will be able to understand the complexity, they must understand fully the way in which teachers behave. It seems unwise to over-emphasize some types of activities.

Teacher thoughts and teacher routines could be integrated if we conceive of teaching as the fulfillment of a task, in which different modes of behaviour occur which depend on the complex situation as well as on the degree of expertise of the teacher.

More attention has to be paid to the description of the task environment of teaching, which raises the question of what constitutes teaching as a situation and as a process. The problem however is that a pure description of what constitutes teaching is hardly possible: there are always more philosophical principles underlying our view on what teaching is. It is the lasting tension between what teaching is and what teaching ought to be.

Beside the focus on task-environment as an object of research, there is a need for studies that are oriented to the gradual reduction of cues, called "abbreviation." The transition from an initial teaching behaviour with many conscious moments into expert behaviour is a challenging question for further research. Without any doubt, the cross section approach has to be complemented by longitudinal studies that can deliver interesting information about how a beginning teacher becomes a professional. However, here too we cannot evade the definition of the criteria by which teaching is called "expert" teaching.

Within an integrative theory of teaching, cognition and action, thought processes and routines may be put together. Of course, at some time, we can choose for one or another aspect of teaching behaviour, on the condition that we guarantee a scientific communication, including shared meanings of terms, frameworks and results.

Chapter 12

An Attempt to Study the Process of Learning To Teach From an Integrative Viewpoint

Jan Broeckmans

Summary

Research on teacher thinking and other recent approaches in research on teaching offer interesting perspectives for studying the process of learning to teach. Research on teacher education, however, would be furthered most by a conception of teaching behavior that integrates complementary approaches. The action-oriented interpretation of teaching behavior offers a heuristic model of such an integration. A description of a current study on student teaching exemplifies methodological implications of this model. This study makes use of observations of overt teaching behaviors, as well as of intro and retrospective techniques for gathering data on the pre-active, interactive, and post-interactive phases of teaching. The method of analysis is also described. This method is distinguished by the simultaneous study of the different kinds of data in their mutual relations.

Introduction

Research on teacher training, as research on teaching, has been dominated by behaviorist approaches. Studies on teacher training have mainly focused on changes in overt teaching behaviors between pre- and post-tests. Few studies have actually looked at the process of learning to teach as it unfolds over a period of time (Fuller and Bown, 1975; Popkewitz, Tabachnik and Zeichner, 1979). Furthermore, attempts to clarify the processes and variables underlying changes in overt teaching behavior are scarce. This is a rather serious lack, as we know that any learning process involves changes in processes and variables, underlying overt behavior.

In research on teaching though, approaches have emerged or revived, which try to look beyond overt teaching behavior. Research on teacher thinking focuses on action and relevant cognitions (Clark, 1978-79; Shavelson and Stern, 1981). Research on the meaning and functions of teaching behavior in its complex environment accentuates the importance of contextual factors in shaping teaching behavior (Doyle, 1980).

These approaches offer interesting perspectives for meeting the research needs that are mentioned in the area of teacher education.

Complementary Approaches in Research on Teaching

When applying these approaches to the study of the process of learning to teach, however, it should be acknowledged that they complement each other. Each approach points to aspects of teaching behavior and – implicitly – of the process of learning to teach, which others neglect.

(1) First, research on teaching attempted to identify characteristics of teachers that are related to students' achievement and attitudes (Gage, 1963). Consequently, teacher education had to focus on the development of these characteristics.

(2) Next, research focused on the overt acts of teachers during interactive teaching, i.e. the instructional processes linking teachers' characteristics to students' achievement (Gage, 1963). The emergence of training methods like microteaching made apparent that the acquisition of a repertoire of specific, concrete, overt behavioral elements was considered as essential to the process of learning to teach.

(3) Research on teacher thinking has the aim to clarify the cognitive processes and variables underlying overt teaching behavior that was neglected by the preceding 'process-product-approach'. Researchers focus then on diverse processes and variables, such as teachers' judgments, decisions, problem solving, 'thoughts', routines, implicit theories, belief and knowledge structures etc. These cognitions are studied both during planning and interactive teaching. The different approaches in research on teacher thinking seem to accentuate different aspects of an information-processing view on teaching behavior, on which they rely implicitly or explicitly. According to this view, overt teaching behavior is organized and directed by basic psychological processes that are thought to occur in the teacher's mind both prior to and during interactive teaching. These processes do not operate in a vacuum. They are embedded in a psychological context (teachers' implicit theories, values and beliefs about teaching and learning) and in an ecological and social context (Clark, 1978-79; Clark and Yinger, 1980). This conception implies that changes in overt teaching behavior rely on changes in the teacher's cognitive information processing and/or in its psychological context.

(4) Studies, focusing on the meanings and functions of teaching behavior in its complex context, elucidate another aspect of teaching, which behaviorist approaches had overlooked. This kind of research is not new (e.g. Jackson, 1968; Smith and Geoffrey, 1968). Since about half a decade, it revived in studies that are mainly ethnographic or phenomenological. In these approaches, teaching behavior, including thoughts, is seen as a functional response to environmental demands (Doyle, 1980). The situation influences behavior with its objective features, as well as with the actor's representations of it (Wilson, 1977). This implies that learning to teach requires the acquisition of 'classroom knowledge', i.e., a network of semantic representations, or a 'schema', that reflects the event structure of the classroom, and

of cognitive and behavioral skills congruent with the environmental demands of that setting (Doyle, 1980).

The Need for Descriptive Research on an Integrative Viewpoint

Each of the forenamed approaches points to important aspects of teaching behavior, but reduces the complexity of teaching from a different perspective. To gain a thorough and fundamental understanding of teaching, it must be known how these different aspects interact. The different approaches must be integrated. Teaching behavior should be studied as the result of an interaction between teachers' characteristics (including implicit theories, knowledge structures, etc.) and the teaching context. Research should acknowledge that this interaction could take many different forms. Teaching behavior can be more or less consciously guided and/or steered by different kinds of mental processes, but it can also run off automatically, once begun, as well as it can be 'provoked' by the situation. An important question is how the different mental processes of teachers are related to each other and to overt teaching behavior, i.e. the much-discussed gap between cognitions and overt behavior should be bridged. Furthermore, planning, interactive teaching and the less studied reflections of teachers on their professional behavior should be investigated in their mutual relations.

Concerning research on teacher training, an integrative perspective is important because it may reveal the influence of the acquisition of one aspect of teaching behavior on the acquisition of an other.

A full, integrative description of teaching behavior is a prerequisite for research, focusing on specific aspects (De Corte, 1982; Lowyck, 1982). This is not to say that research on teaching has to start from scratch. Results so far certainly may contribute to a full, integrative description. In research on the process of learning to teach, however, only a few attempts have been made to apply cognitive or environmental approaches.

The Action-Oriented Model

A basic outline of an integrative conception of teaching behavior is offered by the action-oriented interpretation, which Lowyck (1982) and Peters et al. (1982) have described. This interpretation is based on a conception of 'behavior', coming from Russian psychology and introduced into Western psychology mainly by Van Parreren (1979). It is also related to the TOTE-concept of Miller, Galanter and Pribram, (1960).

Central to the action-oriented conception of behavior is the concept of 'action'. With 'action' is meant: every intentional activity performed upon objects, whether overt activity upon material objects or internal activity upon cognitive representations of objects (Van Parreren, 1979, p. 5). Application of this concept to teaching behavior points to some important features.

(1) Teaching is a very complex activity. It is composed of many actions, which differ in complexity. These 'sub-actions' of teaching are subordinated and/or juxtaposed to each other, so that they form a hierarchical structure. On top of this structure are the most complex sub-actions planning, interactive teaching and post-interactive behavior (i.e. teachers' evaluations and reflections). These are also made up of sub-actions, which are still rather complex and which can be analyzed further into less complex sub-actions and so on.

(2) Being related to the context in a meaningful way, the sub-actions of teaching are 'molar' units of behavior, rather than 'molecular' elements.
(3) Teaching as well as its sub-actions are intentional, i.e. the teacher has a representation of the goal to reach. The intention of a super-ordinated action ties its sub-actions together.
(4) With respect to the intention of the super-ordinated action, the sub-actions can have different functions towards each other. Van Parreren (1979) uses the distinction the Russian psychologist Gal'perin makes between the functions of orientation, execution and control. As to teaching, planning, interactive teaching and post-interactive behavior seem to have functions of, respectively, orientation, execution and control and/or orientation. The sub-actions on the lower level of complexity can also have these functions towards each other.
(5) Mental processes and overt behaviors are equally considered as actions of teaching, which presumably have different functions.
(6) Finally, it is assumed that the sub-actions of teaching can be very different in nature. The intentionality of teaching does not imply that all the sub-actions are consciously steered by a rational interaction with the context. Some sub-actions are more or less routinized and more or less consciously controlled by the teacher. Other sub-actions may be, 'spontaneous' reactions to the situation or may be steered by the context without any conscious control by the teacher.

According to Van Parreren (1979, p. 5), a thorough understanding of an action can be gained by describing the nature of its sub-actions, their mutual relations, their functions towards each other and towards the whole action, the relevant situational factors and the way in which these influence the action. As to teaching behavior, such a description should be given for each of the sub-actions on all the different level of complexity. Such a multi-leveled description would reveal what will be labeled here as the 'psychological structure' of teaching behavior.

This conception of teaching behavior implies that the process of learning to teach has to be seen as a process of qualitative changes in the psychological structure of teaching behavior. These changes rely on developments in the potential for teaching actions, available to the (student-) teacher. These developments can result from all kinds of actions of the (student-) teacher, e.g. studying the content of a training course, observing a model-lesson, interpreting pupils' behavior while teaching, reflecting on own teaching behavior, etc.

Research Goals and Design

The mentioned study uses the action-oriented interpretation as a heuristic model and is an attempt to develop it further on an empirical base. More specific aims of the study are: describing (a) the psychological structure of teaching behavior of student teachers and (b) development in this psychological structure as they occur over a short period. (See note 1).

The study consists of two parts. In both, participants were volunteering first-year students of three training colleges for primary school teachers. Their teaching behavior was studied during the practice lessons that are part of the weekly schedule in this type of teacher education. All lessons studied were in grades 3 or 4.

In the first part of the study (1981-82), 12 student teachers were followed for two successive practice lessons. In the second part (1982-83) six student teachers were followed for four consecutive practice lessons.

Methods for Gathering Data

Methods for gathering data were chosen from an inventory containing different kinds of methods in research on teaching (Cf. Lowyck, 1980; Shavelson and Stern, 1981; Huber and Mandl, 1982).

Principles for choosing methods

Methods were chosen according to a number of principles. (1) The first principle was to use external observations as well as self-reporting techniques because the action-oriented interpretation requires studying overt as well as internal aspects of teaching behavior. Overt aspects are accessible by means of external observations, internal aspects by means of self-reports (Van Parreren, 1981). (2) The second principle concerned the time perspective from which verbal reports are gathered. Huber and Mandl (1982) have made a distinction between pre-, peri- and post-actional forms of self-reporting, in which the data are reported respectively before, during and after the studied activity. Pre-actional self-reports seem less suited for the goals of this study, because it also concerns teaching behavior in situations that were not anticipated. As to peri- and post-actional self-reports, it is not clear, which of both forms gives most valid information about internal aspects of behavior (De Corte, 1982; Huber and Mandl, 1982). Therefore, it was decided to use both peri-actional and post-actional self-reports wherever possible. This decision implied the assumption that both kinds of data would be complementary and that they would allow a validation by confronting them with each other. (3) The third principle concerned the range of aspects of teaching behavior on which self-reporting techniques focus. This focus can vary on a number of dimensions, for instance: range of teaching situations, 'psychological nature' of aspects of teaching behavior (e.g., only rational decisions or all 'thoughts'), momentary processes versus underlying cognitive variables. On each dimension, the focus can be more or less limited. In order to keep assumptions about the nature of teaching behavior at a minimum, it was decided to keep the focus as broad as possible on all dimensions except one. Rather than allowing student teachers to report on processes usually underlying teaching behavior in similar situations, it was tried to focus the verbal reports on the processes occurring in the specific situation reported about. This limitation of focus is necessary when trying to study changes in teaching behavior over a short period of time. (4) The fourth principle applies to the structure of self-reporting techniques, i.e. the degree to which it is tried to determine the quality and quantity of the data reported (cf. Huber and Mandl, 1982, p. 16). It was decided to keep structure at a moderate level, in order to reconcile two conflicting demands. On the one hand, it seems preferable to keep structure low, in order to keep assumptions about the nature of teaching behavior at a minimum. On the other hand, a high degree of structure is a means of avoiding the student teachers filtering out relevant data because of their possibly biased interpretations of instructions and questions.

The last two principles also apply to the choice of methods of external observation. Furthermore, they have implications for the concrete design of methods.

The chosen methods
Table 1. Presents a survey of the chosen methods. Due to space limitations, neither argument for the specific choices made, nor a more detailed description of the used methods can be given here.

Table 1: Survey of the used methods.

	METHODS USED		
PHASE OF TEACHING	External Observation	Peri-actional Self-report	Post-actional Self-report
Pre-active	° lesson plan(s) ° (provisional) notes, 'cribs' ° documents		
Interactive	- videotaping - narrative descriptions		
Post-interactive a) immediately after the lesson	0	Written thought listing	Retrospection of thought listing
b) further reflections	°° written lesson critiques °° occasional notes	°° written thought listings	Retrospection of logbook

° = joined in a 'logbook on planning'
°° = joined in a 'logbook on further reflections'

Coping with problems in the concrete design of methods
When using self-reporting techniques, one is faced with many problems and limitations regarding validity (Ericsson and Simon, 1980; Huber and Mandl, 1982), and with the need to develop a 'hard qualitative methodology' (De Corte, 1982, p. 89).

A number of recommendations for the concrete design of the self-reporting techniques in this study were deduced from the Ericsson and Simon (1980) self-reporting model. This model predicts how different features of self-reporting techniques will influence the processes which respondents use to report verbally on cognitive, task-related processes. This enables one to predict how these features will affect the internal and ecological validity of self-reported data.

It cannot be discussed here how the recommendations, stemming from the Ericsson and Simon model were applied in the concrete design of methods. It must be noted, however, that it was impossible to follow strictly all these recommendations, partly due to pragmatic reasons.

The Procedure of Simultaneous Analysis
At the end of the first part of the study, data of four student teachers were analyzed provisionally, in order to develop a procedure of analysis. Two principles regarding analysis were deduced from the action-oriented interpretation. (1) The different kinds of data available should be integrated. (2) The data should be rearranged into a

chronological description. Arrangements of data, according to a number of themes (e.g., for planning: choice of lesson content and choice of teaching method), largely reduce the possibility of exploring mutual relations between activities concerning different themes. These mutual relations are important aspects of the psychological structure of teaching behavior.

Initially, the different kinds of data were analyzed independently and were integrated afterwards. This procedure was abandoned, because one kind of data couldn't be interpreted nor be rearranged into chronological order without consulting the other kinds of data. Furthermore, the separate analyses proved very incomplete and even faulty when taking all kinds of data into account.

Therefore, it was decided to develop a procedure of simultaneous analysis, in which the different kinds of data are studied at the same time and in their mutual relations.

Arranging data for simultaneous analysis

For such a simultaneous analysis, a number of conditions have to be fulfilled (a) For each phase of teaching per lesson, the different kinds of data must be arranged on a common timeline. (b) The time-units used for this purpose must not determine the later 'unit of analysis'. (c) Each time-unit can contain a large amount of information. To present this information in an orderly fashion, it must be divided into 'units of information'. This unit has to be uniform for all kinds of data. (d) This unit must not define the unit of analysis either. (e) Because data are rather different in nature, it seems preferable to translate them in standard formulations. (f) These formulations must be true representations of the original data and must express clearly whether the unit of information concerns activities of the student teacher, contextual factors or characteristics of the student teacher.

Taking into account these conditions, a procedure was developed for arranging data for simultaneous analysis. By way of illustration, the procedure for arranging data on planning behavior will be described here. This procedure consists of six steps.

(1) A common time-line is defined by the data in the logbook on planning, because the retrospection is based on it. First, all data in the logbook (see Fig. 1.) are divided into time-units and rearranged into chronological order. The used time-unit is: each new entry in the thought listings or – for the lesson plans and notes – each lesson phase described

(2) Next, the data in the logbook are transcribed on a protocol in their chronological order. (See protocol I b.). In these protocols, each time-unit is divided into units of information. Such a unit contains all the data that can be presented by one main clause. The units of information are translated into standard formulations. Text arrangements indicate whether the unit of information is about (a) an activity of the student-teacher, (b) a common activity of the student-teacher and others, (c) an activity of others or events or factors in the immediate context or (d) features of the broader context or personal characteristics of the student-teacher e.g. former experiences, beliefs, knowledge, etc.). For activities of the student teacher, the first person singular of the simple present tense is used. The documents, notes, and lesson plans in the logbook are referred to as 'appendices', which are added to the protocol. Text arrangements indicate whether the appendix is a document that was consulted or a note the student teacher wrote.

(3) In the next step, the verbal protocol of the retrospection on planning is divided into similar unites of information. (See protocol 2 a.).
(4) Then, each retrospective unit of information receives two codes: one to place it on the common time-line and one to indicate its relation to other retrospective units of information.
(5) By means of these two codes, the retrospective data are rearranged on a so-called standard protocol. This gives a description of planning which is synchronized with the description, based on the logbook data. (Protocols 1 b. and 2 b. are examples of such synchronized descriptions). In these standard protocols, the retrospective data are likewise translated into standard formulations.
(6) Finally, data on planning, coming from other sources (e.g., the retrospection of the lesson), are arranged on the common time-line into a third synchronized protocol.

Further analysis

Because the further analysis has begun only recently, only a general outline of the procedure of simultaneous analysis of the different kinds of data can be given here.

In order to describe the psychological structure of teaching behaviour a cyclical analysis (Lowyck, 1980) that follows a top-down approach will be used. First, we try to identify the most complex sub-actions and to describe their nature, their functions and relations to the context and to each other, etc. This is done independently for each lesson. By means of recurrent comparisons between lessons, an attempt is made to mark out the unit of analysis and to identify and label differences in the psychological structure on this highest level of complexity. Next, we try to describe, for each lesson, the psychological structure of the complex sub-actions that were already marked out. New recurrent comparisons must mark out the unit of analysis and identify and label differences in psychological structure on this lower level of analysis. This procedure is repeated several times on ever-lower levels of complexity.

This should enable us to describe the psychological structure of teaching behaviour in the 48 lessons studied, as well as developments in this structure. As to the latter though, a problem will be to differentiate between developments, resulting from changes in the potential for teaching behaviour of the student teacher, and differences, due to fluctuations in situational factors.

Notes

1. The research goals and design, as well as the heuristic model, also concern the actions of student-teachers on account of supervisory sessions and the role of these actions in the process of learning to teach.

Chapter 13

Describing Teachers' Conceptions of Their Professional World

Staffan Larsson

Summary

An approach to the qualitative description of "teacher thinking" is presented. It is an approach that recently has produced a number of studies in Sweden. These studies are focusing on the description of teachers' assumptions about phenomena in their professional world. It is the teachers' conceptions of those phenomena that are described and consequently the approach is a "phenomenographical" one. Two examples from the work of the author are presented. The second is about certain paradoxical situations that teachers might experience. These situations concern teachers' and students' influence in the classroom.

Introduction

Teachers' thinking on their teaching has been researched using various approaches. This pluralistic situation reflects the dynamics of this research field, different approaches elucidating different aspects of teachers' thinking. In this way different approaches can be complementary to each other and are not always contradictory. One set of approaches with a common theme is that which concentrates on descriptions of the way teachers experience or conceive their working life at a fundamental level. This group contains studies deriving from Kelly's personal construct tradition, studies with an ethnographic approach and studies drawing on attribution-theory.

In this article I wish to present yet another approach to the investigation of teachers' thinking which has already resulted in several studies within the research group of the University of Göteborg of which I am a member. This approach was initiated in the study of learning and skill (Svensson, 1976; Saljö, 1982). Its most fundamental characteristic is that learning should be described in terms of its content, i.e. that learning should be described as qualitative changes in the conception of a phenomenon (Marton, 1981). Piaget's early work contains examples of this kind of description in which children's development is described as changes in conceptions of certain phenomena (Piaget, 1967 (1927)). Another side of

Piaget's work was his interest in the structural aspects of development, a focus that has now been severely criticized (Donaldson, 1979; Hundeide, 1977). Our interest is not in the structural aspects but rather in man's conception of the world around him - in the content of man's thinking.

Interest in the description of learning and study skill has been widened to other fields. One prominent field of study has been that of teachers' conceptions of their professional world, i.e. their way of understanding pedagogical phenomena. From the first study of this kind on teachers' conceptions of grading (Larsson, 1979) a rapid development has occurred: Hasselgren completed a doctoral dissertation on preschool teacher students' perception of free play (Hasselgren, 1981) and Larsson conducted a series of studies on teachers in adult education. The latter studies resulted in descriptions on teachers' conceptions of a number of phenomena and notions in adult education, e.g. the notion of experience (Larsson, 1980, 1983a), teaching skill (Larsson, 1981a), the psychological differences between young and adult students (Larsson, 1981b), the essence of teaching and the restrictions on teachers' freedom of action (Larsson, 1982b, 1983b), cognitive interest and view of knowledge in subjects taught (Larsson, 1998 2c). In 1983 Andersson and Lawenius presented a doctoral dissertation on teachers' conceptions of teaching based on teachers from comprehensive school (includes age-groups from 7 to 16). Aside from the research outlined above at the Department of Education at the University of Göteborg, there is a project at Uppsala, using the same approach (Lindblad and Hasselgren, 1983; Hasselgren, 1984).

Our interest lies in teachers' conceptions, not in what is actually happening. We believe that studies into teachers' conceptions of the world are complementary to the study of the world as it is. In describing this, Marton (1981) makes the distinction between a first-order and a-second order perspective. In educational psychology questions are, for example, frequently asked about why some children succeed better than others in school. Any answer to this question is a statement about reality. An alternative kind of question would be: "What do people think about why some children succeed better than others in school?" Any answer to this second kind of question is a statement about people's conception of reality. These two ways of formulating questions represent two different perspectives. Marton calls the former a first-order and the latter a second-order perspective.

Table 1. Conceiving Educational Change.

Aspect Conception	Basis of Teacher Behaviour	Role of Research	Innovation Activity
Rationalist	Compliance with policy directives	Find ways of engineering compliance	Disseminate policy directives - couple elements of system
Ecological	Controlled by nature of context	Discover causal relations between context and behaviour	Alter nature of context
Reflexive	Intentions of teacher	Discover what teacher intentions really are. Compare intention to practice	Engage teaches in critical analysis of practice

The Qualitative Analysis

My analysis aims at finding qualitatively different categories and describing them as accurately as possible. The focus is on description, not on explaining the genesis of conceptions or finding correlates to them. Our unit of description is the conception. Marton specifies this term by stating that it often relates to what is taken for granted, but it can also be explicit, its most prominent characteristic being that it constitutes the basis of our reasoning. The term does not differentiate between scientific ideas and common sense. In many instances, scientific ideas and common sense reasoning can be described the same level of description - as the same conception of a certain phenomenon. Scientific ideas often become common sense (Marton and Svenson, 1978).

My aim is not to describe the most prominent or most important conception, but to describe *the variation.* The qualitatively different conceptions of a certain phenomenon or notion can be detected. One example of this kind of study into teachers' conceptions of their professional world is Larsson's study (1983a) of their conceptions of how *students' experience* could be used in the classroom setting in relation to the content of instruction. Five different conceptions were described.

Conception A: Experiences that can be used in teaching are experiences that one or a few students have had and which can be used as information to the rest of the class.
This conception focuses on the fact that pupils are sometimes experts on something, for instance, an electrician has practical knowledge of electricity that can be used in physics, or a shop steward has an acquaintance of bargaining that may be used in social science.
Conception B: Experiences that can be used in teaching are experiences that most students in the class have had and which can be used to direct the students' attention to the relevant context Here the focus is on the fact that adults are familiar with certain phenomena like, for instance, taxes. It is this shared, common knowledge that is encompassed by conception B.
Conception C: Job experience can develop practical knowledge that the students may use in the educational context.
Conception D: The student brings into the classroom an outlook on the world that is in conflict with the view of the subject taught. This is a conception built on the conviction of a conflict between everyday knowledge and school knowledge.
Conception E: Experiences from life give a certain capacity for empathy that makes it possible for the adult student to understand different perspectives as serious alternatives. The combination of teachers' conceptions of experience and frame factors like the allocation of time gives the teachers reasons for not using students' experience to any great extent.

It is quite reasonable to wonder about the use of this kind of description. Olson (1983) describes three different ways of conceiving educational change both in practice and research. He calls them the rationalist, ecological and reflexive conceptions of change, respectively.

The first aspect is the way in which teachers' behaviour is looked upon. My interest is in the way teachers conceive phenomena around them, but this is not to say that their conceptions are the only background for the understanding of their acts, since, for example, conceptions can be both intentions and rationalizations. However, a conception describes the way teachers conceive of some phenomenon, i.e. it is the unit of description to use when you want to characterize how things appear to teachers. As a special cognitive interest, however, the description of *variation* can be used by teachers to reflect upon. Olson's "reflection"-category encompasses the innovative activity as the engagement of teachers in critical analysis of practice. In contrast, my analysis can be used by teachers for reflection on the conceptual possibilities (the different conceptions) that exist when they think about teaching in practice. In this way it is not an analysis of practice but of alternative ways of conceptualizing practice-relevant phenomena. This idea of reflection as a source of change can probably be compared to the "emancipatory cognitive interest"-category of Habermas (1972), especially in Olson's version. If we return to the idea of conceptions as fundamentally descriptive units, it is worth indicating the possibility of using this in different ways. Conceptions have been used in the description of learning to describe the impact of a certain curriculum on pre- and post-test-design (Dahlgren, 1978; Alexandersson, 1981; Hasselgren, 1981). I will present one elaboration of the pure description of conceptions below. In this research investigation I combine conceptions into a description of a situation experienced by teachers.

The study concerns teachers' conceptions of restrictions on their freedom to act according to their convictions and the consequences of these restrictions. These consequences are described as certain relations between students and teachers. This work is published in another version in Instructional Science (Larsson, 1983b).

The material for the analysis consists of interviews with 29 teachers in adult education, who were teaching at least one of the following subjects: Mathematics, physics, chemistry, history and social sciences. The interview was a semi-structured one concerning a number of phenomena like learning, knowledge, etc. The purpose of the interviews was to get as good an insight as possible into the convictions of the teachers. The interviews were tape-recorded and typewritten.

Teachers' Conceptions of Restrictions

From the interviews, it is evident that teachers reckon with restrictions that they consider significant to them. When they plan, their way of conceiving restrictions can be important and can also affect their evaluation of their own teaching.

The interview data shows that a crucial restriction is the time limit experienced by teachers. The reason for this experience is the difference between the amount of time allocated to a certain content in adult education in comparison with compulsory secondary school (age groups 16-19). In fact, the time in adult education is considerably shorter for specific courses. One could say that secondary school sets the background – the norm for what is normal – from which the "abnormalities" of adult education arise.

The other prominent restriction conceived is the students' conceptions of teaching – what students think is a meaningful way of arranging the instructional process.

The conceptions of teaching which teachers express can be differentiated into two conceptions, i.e. two qualitatively different ways of conceptualizing what teaching ought to be. The first can be described thus:

Conception A: The essence of teaching is that content should be presented and structured for the students. This means that the content should be prepared so that the student can learn without too much interpretation. This conception relies on the view, taken-for-granted or not, that the teacher must do some of the interpretative

and/or structuring work for the students. In some cases, teachers believe that the student can interpret and structure himself, but only as a complementary task to the fulfillment of conception A. The second conception is characterized in the following way:
Conception B: Teaching ought to involve the students in interpretative and structuring work. If they are not involved, real changes will not occur or they will not develop real knowledge. This conception has two parts, one concerning the essence of teaching and one that, explicitly or implicitly, is a criticism of conception A. The criticism is in all cases founded on the idea that the learning effect of such teaching is weak. In some cases the teachers refer to their own experiences as students. In some cases they exclude mathematics as a subject impossible to teach from a B conception basis.

Out of 24[6] Cases 8 teachers were found to have an A-conception and 16 a B-conception. Teachers always describe the students' view of teaching in the same way – that the dominating view among the students is always an A-conception. It was this student-conception that teachers referred to as a restriction. Not every teacher reported the students' view of teaching, but those who expressed a view on the matter reached the same conclusion. Fifteen out of sixteen teachers with conception B and two out of eight with conception A reported the students' conception as describable in terms of conception A.

If we return to the restrictions, it is clear that the experience of the time-restriction dominates among teachers with conception A. It is also notable among teachers with conception B, but in the latter case the other restriction competes in importance: the students' conceptions of teaching. There are some indications that the latter restriction has the power to change the teachers' main form of teaching, while the former encourages minor changes.

By reconstruction, we get a picture of the teachers' situation as it is conceived. What teachers think teaching ought to be comes into conflict with what they experience as restrictions and as a result some of them experience major discrepancies between what they ought to do and what they actually do. If we tabulate this as a description of their conception of teaching and their actual teaching, we find the following pattern:

Table 2: Conceptions of Teaching.

		Conceptions of teaching	
		A	B
Teachers' report of their actual control	Strong	8	10
	Weak	0	6

From the teachers' report, actual teaching can be divided into one category of strong and one of weak teacher control over the communication in the classroom. The first category includes expressions like lecturing, discussion under surveillance of the teacher or going through the content. The standard

[6] Five were not possible to identify clearly, either because they did not express themselves clearly or because they expressed conceptions that were mixed forms of A and B.

view is that the teacher has control over the forms of communication and the messages sent, and is in fact also the dominating sender. The second category refers to what teachers call problem-orientation, group-work or explorative classes. Here, the general opinion is that the teacher exerts a certain lack of control over the forms of communication, especially of the messages sent. The distinction has much in common with the one Olson (1981) calls "high influence" and "low influence" teaching.

On scrutinizing the findings, a pattern appears. All teachers with conception A actually exert strong control over the communication process when they teach, i.e. there is no contradiction between their view of what teaching ought to be and what they actually do. In relation to teachers with a B-conception, the picture is divided; six of them exert weak control over the communication process, i.e. act as they ought to do according to their convictions. Ten out of sixteen exert strong control over the communication process and thus act in contradiction to their fundamental view of teaching. This divided picture shows that teachers decide whether they should let restrictions be a barrier or whether they should break this barrier: the students' view of teaching. The dilemma is illustrated in the following quotation:

> What chance do you have of doing the teaching as you wish?
>
> Yes, in fact the opportunities are very great, we can say that the external frames are given then I think that I can control my working situation quite a lot, at least in theory I can, and if I just want to have group-work for the whole year, I can do that. If I just want to have educational visits, I can do that. If I want to skip a textbook, I can do that. There are schools where the students work exclusively with newspaper articles in our subject. So, the form of teaching I can obviously choose just as I like, either alone or in cooperation with the class. In practice, I do not think that the freedom is that wide since the students, as I feel it, want a textbook and become very disturbed when you read things that cannot be found in the textbook and want to know where it can be read and what they should prepare. We have talked about grades now. It is said from certain quarters that most students do not want grades in school; I have never believed in that. We carried out a questionnaire here at school, and it revealed that 21 out of 23 classes clearly expressed that they wanted ... they wanted to see a result from their work. And that leads us to, I suppose, having, after every time, some sort of report, if not, they feel that they have not learnt anything; they want to show that they have prepared themselves. They want to see a result; they want to be rewarded at once. And that means that then one must work in a certain way, so that it is possible to have a test, so to speak. Then you cannot exclusively have group-work, because it becomes very difficult to measure, but instead you must have some sort of reading … the same text for everyone.

In this case the forms of teaching are chosen on the students' premises and students immediately oppose deviations from their ideal. They in fact force the teacher to exert strong control over the communication process. This can be looked upon as a paradox: Students use their control to compel the teacher to exert control over them.

Other teachers choose to enter into conflict with their students and force them to accept their intentions, regardless of resistance. As in the above quotation, there is a conflict but here the teacher is the "winner":

> This about the students' conceptions of what the subject should be about, as you say about the government, the parliament, and just memorizing facts. How do you handle such situations, where *you* have another view than *they* have?
>
> There *can* be problems in certain classes to enforce such a form of teaching. I think in fact that I have succeeded in most cases. I do not think that I have been confronted with resistance to the extent that I have had to leave the idea and go back to one hundred per cent lecturing. But neither do I have problem-oriented teaching during all teaching time, nor do I have to do a summary of certain central points in social sciences.

A teacher in physics described his desire to do teaching in line with his intentions, i.e. a B-conception, thus:

> And I shall probably start a new course in physics now in the spring and then I want to build further on what I have worked with during this period. And this will be a very difficult start, difficult to start and difficult to know how to structure it and how you should make them accept a different way, because there is a restriction there too ... one could call it a conservative perspective on what teaching in physics means.

In the above cases, we can see the students' perspectives as a restriction on the teachers' freedom of choice. But here the teachers enforce their view on the students. In the first quotation, this is described as a rather uncomplicated task, while in the second it is a difficult challenge. I maintain that in this case too there is a paradox - but of another kind. Here teachers with B-conceptions force the students to accept the teachers' weak control of the communication process. I wish to stress that the use of teachers' authority or force is contradictory to their intention: the weak control of the communication process. Thus, students are supposed to accept greater responsibility for the teaching process contrary to their view that this could best be controlled by the teacher.

So far I have described two forms of relations concerning control over the communication process: these were based on the conflict between the teachers' perspective on teaching a B-conception and the students' view, an A-conception. This paradoxical character stems from the constellation of perspectives, but the variation in forms is related to the "solution" of the conflict. The ways in which conflicts are "solved" are paradoxical.

Finally, let us look at the teachers with the A-conception and their students. In this case there were only two teachers who describe the students' view. Presumably, the reason for this was the lack of conflict. If we suppose the students' conception of teaching to be the same among the "silent" teachers as among those who describe the students' view, it is logical for them not to talk about the students' conception of teaching as a restriction, precisely because in their case it is not a restriction. On the contrary, the views of students and teachers are in harmony here. If they act according to their perspectives they will take complementary poles - the teacher controlling and the students accepting control. In this case control of communication is explicitly in the hands of the teacher.

What is seen as a dilemma can at a deeper level be seen as a paradoxical situation. This can also be interpreted within the conceptual framework of the "communication theory" of Bateson (1973) and Haley (1963) presented in the article previously referred to (Larsson, 1983b).

This presentation of my findings illustrates our research group's thinking - descriptive studies where the level of description is characteristic. We are attempting to examine the teachers' fundamental convictions and describe them in detail. These conceptions can be seen as basic elements in the understanding of teachers' way of looking at their work and in certain cases also a key to the understanding of the facts of teachers themselves. Whether the conceptions have any counterpart in acts or not, they constitute teachers' professional lore.

Author Reflection Twenty Years On

To have the opportunity to comment on an old article is a second chance to say something better. What I now consider as the main contribution in the article is the latter part of the article where I present my study on "Teachers' conceptions of restrictions". I still have not had such a strong feeling of discovering something new as in that case. I spotted early that the teachers told me something interesting about the power plays in the classroom, but I had great difficulties to portray it properly. Eventually I came to think of the theoretical constructions that were developed by Bateson and his collaborators in their research to understand schizophrenia – double-bind is probably the most famous term that came from this work, where paradoxical communication is a key focus. One of the team, Jay Haley, had developed of a system of communicative power-relations and I found out that this made sense as a tool to understand the pattern that I was struggling to describe. I consider it as the hit of my research career. In English it was published in Instructional Science with the title "Paradoxes of teaching". In that text the theoretical level is more highlighted than in the text above. If I had the opportunity to rewrite the article I would have stressed the theoretical interpretation, since it would have made it more interesting. In ethnographic studies of power plays in teacher education and in folk high school teaching the theory has been useful as an interpretational tool. As far as I know, few have used this thought tradition in relation to educational phenomena, in spite of an increased interest in power relations. Engeström has tried to incorporate Bateson's learning theory into activity theory, but Haley's theory is focusing on something else within the general frame of thinking. I think the potential of analyzing paradoxical communication in educational settings has not been realized by the research community.

The second comment is on the approach that is also presented in the article, often called ”phenomenography”. I have personally taken other routes the last decades, so I am not representing this approach. However, the perspective is fully alive and has crossed borders to a great extent. Anyone who wants to know more can browse on the web for ”phenomenography”.

Chapter 14

What is 'Personal' in Studies of the Personal

D. Jean Clandinin and F. Michael Connelly

Summary

The chapter presents an analysis and interpretation of studies of 'teacher's theories and beliefs,' which focus on individual teachers' thoughts and actions and which are called by us 'studies of the personal.' The analysis is presented in terms of 'what is asserted,' i.e., the key terms, stipulated definitions and origin of ideas and 'what is done,' i.e., the problem, method and outcome of each of the studies. While there is wide diversity in key terms and stipulated definitions, there is an underlying commonality in problems undertaken and in resulting outcomes. The second part of the article interprets the analytic descriptions of the studies by examining two relationships of importance in understanding the personal: the relationship of personal evidence to personal experience *and* the relationship of the personal to practical action. These relationships are based on a conception of the composition of teacher thought composed of prior experience (biography), ongoing action (practice) and thinking in isolation of these two. The interpretation suggests that most of the studies *assume,* but do not inquire into, an explanatory relationship between prior experience and the way teachers think. However, several assume a dialectical relationship and one assumes a problematic relationship between knowledge and action. Most studies operationally define teacher thinking independent of prior experience or action although some studies use complex combinations in their collection of data and in their resulting conceptions of thinking. This context of possible relations is the basis for a discussion of a variety of generic possibilities for research.

Introduction

Recent reviews (Feiman-Nemser and Floden, 1984; Clark and Peterson, 1984) of teacher thinking and teacher knowledge have drawn attention to a small set of interesting studies named by the reviewers 'teacher theories and beliefs.' What is especially interesting about these studies is that, one way or another, they purport to study the personal, that is, the what, why and wherefore of individual pedagogical action. As Clark and Peterson (1984) say, the purpose of this cluster of studies is 'to make explicit and visible the frames of reference through which individual teachers perceive and process information' (p. 19); and Feiman-Nemser and Floden (1984) say the intent of these studies is to 'get inside teachers' heads to describe their knowledge, attitudes, beliefs and values' (p. 4-5). Thus, in contrast to studies focused on group action, and to others focused on generalized patterns of behavior in populations, this research, if the reviewers' assessment is correct, is focused on individuals' thought.

But what is the personal and how is it related to pedagogical action? The first clue to this question is given by the reviewers' descriptive language. Their key terms are 'frames of reference,' 'knowledge,' 'attitudes,' 'beliefs,' and 'values,' terms presumably derived from the substance of the studies. The conceptual scope implied by these terms is considerable and we must turn to the studies themselves to assess the meaning specific researchers make of the idea of 'studies of the personal.'

Here, the scope remains wide with a somewhat bewildering array of terms naming the research. For studies noted by Clark and Peterson, there are 'teachers' conceptions' (Duffy, 1977), 'teacher perspectives' (Janesick, 1982), 'teachers' understandings' (Bussis, Chittenden and Amarel, 1976), 'teacher constructs' (Olson, 1981), 'teacher principles of practice' (Marland, 1977), 'teacher beliefs and principles' (Munby, 1983) and 'teacher practical knowledge' (Elbaz, 1981). When we examine the Feiman-Nemser and Floden review, and add relevant research reported at the ISATT meeting (Halkes and Olson, 1984), we have the personal additionally defined in terms of 'teacher conceptions' (Larsson, 1984), 'teachers' thinking criteria' (Halkes and Deijkers, 1984), 'personal constructs' (Pope and Scott, 1984), 'personal knowledge' (Lampert, 1985) and 'teachers' personal practical knowledge' (Connelly and Clandinin, 1984).

To the extent that key terms are telling of how the personal is conceptualized in inquiry, this list suggests little commonality among researchers. Matters often thought to be different in principle are treated by different researchers, for example, 'conceptions,' 'perspectives,' 'criteria,' 'beliefs,' and 'knowledge.' Still, it is possible that because this small sub-field is relatively new, these various terms, for example, 'personal knowledge,' 'personal constructs' and 'conceptions,' are simply different words naming the same thing.

The task of resolving these questions and of assessing the implied concept of the personal at work in specific inquiries consists, for our purposes, of two different tasks, the analysis and description of the set of inquiries and, subsequently, an interpretation of the description in terms appropriate to the personal. We show that there is a degree of commonality in how researchers conceive of the personal in inquiry and we set forth a possible conceptual structure of inquiry in this field based on a commonplace notion of thinking and on different conceptions of teacher thinking evident in the research.

Describing the Studies of the Personal

Method of Analysis

The first task results in a description of what is asserted (Table 1: Part 1) and of what is done (Table 1: Part 2) in inquiry. Our underlying assumption is an operational one to the effect that what one does in inquiry defines the phenomena and resulting knowledge claims. Thus, what a researcher does reveals his/her meaning. Perhaps because this set of studies on the 'personal' has only a small tradition on which to draw, there is a fairly rich content of conceptually asserted material in each paper as each author explains his/her work. We have already noted that each author specifies a key term by which the work is identified and, presumably, which names the phenomena under inquiry. These terms are conceptualized and defined in each of the papers and conceptual origins are noted. For instance, Halkes and Deijkers' (1984) key term is 'teachers' teaching criteria' which they define as 'personal subjective values a person tries to pursue or keep constant while teaching.' Conceptually their work originates, in part, in the curriculum implementation literature. These matters are summarized in Table 1: Description of Studies of the Personal – Part 1: "What is Asserted" in three columns titled, 'key term,' 'stipulated definition' and origin of ideas.'

Table 1 – Part 2: "what is done" presents a description of the 'pattern of inquiry' used in each study, structured according to 'problem,' 'method' and 'outcome.' Halkes and Deijkers (1984), for example, define their problem as one of identifying teacher criteria used in solving classroom disturbances. Their method is to analyze the relevant literature; identify seven criteria; convert them to sixty-five operational statements presented in a Likert-scaled survey instrument; and have teachers respond. Their claims, relative to the problem of determining teacher criteria, constitute a set of summary statistics, and discussion of them, based on the survey.

Commonalities among the Inquiries

1.1 *What is Asserted*: With respect to what the authors assert they are doing (Table 1: Part 1), we have already noted in our opening remarks that a wide diversity of key terms, and definitions of them, is offered. There is somewhat less diversity in the asserted origins of the work because three of the studies claim to use Kelly's (1955) construct theory. (It is interesting to notice the differences in language that emerge from Kelly's (1955) work for these three authors. In a relatively new and fluid field such as this, even elementary terminology is apparently subject to re-naming.) Overall, it is apparent that a flood of different theoretical resources are entering this field from general curriculum studies, psychology, philosophy of various persuasions, and empirically based typologies.

1.2 *What is Done*: Turning from what is asserted to what is done we find that similarity begins to overshadow difference. The patterns of inquiry for the studies in question (Table 1: Part 2) reveal more commonality in this research than is implied both in the reviews (Clark and Peterson, 1984; Feiman-Nemser and Floden, 1984) and in our observations of 'what is asserted' by the researchers.

Table 1: Description of Studies of the Personal.

Part 1: What is asserted

Authors	Key Term	Stipulated Definition	Origin Of Ideas
Halkes & Deijkers	Teachers' teaching criteria	Personal subjective values a person tries to pursue or keep constant while teaching.	Literature on innovation and curriculum implementation; teacher thought, judgements, decisions and behavior; and attitudes to education
Marland	Principles of practice	Principles that guide a teacher's interactive teaching behavior and that can be used to explain teacher interactive behavior.	Not stated
Pope & Scott	Personal constructs/theories/ epistemologies/	Teachers' view of knowledge and of pedagogic practice.	Kelly's personal construct theory
Munby	Beliefs and principles (implicit theory)	Coherent structures that underlie a teacher's practices.	Kelly's personal construct theory
Bussis/ Chittenden/ Amarel	Teachers' understanding	Teachers' beliefs about curriculum and students in terms of classroom activities (surface content) and teachers' learning priorities for children (organizing content) and the connections between the two.	Phenomenological inquiry
Janesick	Perspective	A reflective, socially-derived interpretation of experience that serves as a basis for subsequent action. Combines beliefs, intentions, interpretations, and behavior. A frame of reference within which the teacher makes sense of and interprets experience rationally.	Symbolic interaction

(table continues)

Authors	Key Term	Stipulated Definition	Origin Of Ideas
Larsson	Teachers' conceptions	A conception describes the way teachers conceive of some phenomena. The conceptions are basic elements in the understanding of teachers' ways of looking at their work and in some cases, of understanding their acts.	Phenomenography
Duffy	Teachers' conceptions of reading	A conception of reading is the approach to reading, which the teachers believe is most like him or her.	Research-based typology of approaches to teaching reading
Lampert	Personal knowledge	Personal knowledge is knowledge of who a teacher is and what he/she cares about *and* knowledge of students beyond paper and pencil tests that is used by a teacher in accomplishing what he/she cares about, what students want and what curriculum requires.	Not supplied
Elbaz	Practical knowledge	A complex practically-oriented set of understandings which teachers actively use to shape and direct the work of teaching.	Phenomenology (particularly Schutz and Luckmann)
Connelly/ Clandinin	Personal practical knowledge	Knowledge, which is experiential, embodied, and reconstructed out of the narratives of a teacher's life.	Experiential philosophy

Table 1: Description of Studies of the Personal

Part 2: What is done

Authors	Problem	Method	Outcome
Halkes & Deijkers	To identify teacher criteria used in solving class disturbances.	Analyze literature to develop seven categories of teaching criteria; convert to sixty-five operational statements presented in a Likert-scaled instrument; have teachers respond.	A set of summary statistics of the teaching criteria used by teachers.
Marland	To identify teacher thoughts that guide teaching behavior and can explain teaching behavior.	Videotape teaching events; conduct stimulated recall interview describing 'thoughts' while teaching; sort individual statements from interview into categories of statements called principles.	Five principles of practice that guide teaching behavior.
Pope & Scott	To explore pre- and in-service teachers' epistemologies; their views on teaching and learning and what interactions, if any, there are among these interactions with implications for teacher education.	Identified four theoretically-derived epistemologies; conducted interviews and observations of pre- and in-service teachers in which teachers reflected on own views of knowledge and practice; researchers derived teacher's personal theory or epistemology. Administered REP test; personal constructs derived from teacher's response to Repertory Grid test; sorted data from REP tests and interviews and observations from individual teachers into four theoretically-derived categories.	Concludes that student teachers can become aware of their initial epistemologies.
Olson	Problem of implementing a new science curriculum; how teachers deal with new curriculum.	Identified different science teaching methods; prepared and presented statements of 20 science teaching events to 8 teachers; teachers sorted events; discussed and located basis of grouping; labeled basis of grouping.	Identified underlying constructs in implicit theories of teaching; identified main feature of new science curriculum; identified ways teachers changed curriculum project to make it compatible with personal constructs.

(table continues)

Authors	Problem	Method	Outcome
Munby	Problem of explaining how and why a nominally common curriculum is interpreted and implemented differently by each teacher.	14 teachers each generated 20 descriptive statements (elements) of what a visitor to his/her class would see; teachers sorted elements and discussed basis of grouping; terms and phrases used became constructs within teacher's implicit theory; constructs further analyzed through interviews which led to labels for groups and for their relationships; produced statements called teachers' beliefs and principles.	Illustrated wide individual differences in beliefs and principles of teachers working at same school and within same subject matter specialization.
Bussis/ Chittenden/ Amarel	To investigate the understandings and constructs of teachers implementing open or informal teaching.	Clinical interviews with teachers implementing open or informal teaching. Analyzed data to identify orientations for aspects of teachers' belief systems.	Identified four contrasting orientations for each of four aspects of teachers' belief systems; curriculum priorities; role of children's needs and feelings; children's interests and freedom of choice; importance of social interaction among children.
Janesick	To describe and explain the classroom perspective of a teacher.	Participant observer with teacher; interviews with teacher, other school staff and students; analyzed database to offer interpretation of teacher's perspective.	Identified the teacher's classroom perspective and offered an account of how the teacher gave meaning to the day-to-day events in classroom and how he constructed curriculum.
Larson	To describe teachers' assumptions about phenomena in their professional world.	Semi-structured interviews with 29 adult educators on phenomena of learning and knowledge in order to gain insight into conceptions of teachers. Analyzed data in order to identify conceptions and restrictions of teaching.	Identified two restrictions and two conceptions of teaching (two qualitatively different ways of conceptualizing what teaching out to be).

(table continues)

Authors	Problem	Method	Outcome
Duffy	To describe the distribution of conceptions of the teaching of reading among teachers.	Identified five conceptions or contrasting reading approaches from the literature; added a sixth conception (confused/frustrated); six prepositional statements derived from each approach.	Summary statistics on reading approaches used by teachers.
Lampert	To examine teachers' thinking about the problems of practice.	Teacher-researchers gathered weekly over a two-year period to discuss everyday work dilemmas; during these discussions they were observed and interviewed. Conversations/ discussions considered as 'text' to be interpreted and understood as an expression of the way they think about work dilemmas. 'Text' analyzed to produce comparisons.	Noted comparisons among theories of teaching constructed by scholars, theories constructed by teachers themselves and teachers' concrete reflections on practical problems that arise daily in their classrooms.
Elbaz	To offer a conceptualization of the kind of knowledge teachers hold and use.	A series of semi-structured interviews with one teacher; analysis of transcript database; researcher interpretation of teacher's practical knowledge.	Identified five content areas of practical knowledge; five orientations of practical knowledge; three ways in which practical knowledge is held.
Connelly & Clandinin	To understand the kind of knowledge teachers hold and use; how to talk about personal knowledge.	Participant observation of practices of several teachers over several years; semi-structured interviews with participants; researcher participant mutual construction of teacher's narratives; development of theory of personal practical knowledge.	A theory of personal practical knowledge as made up of participants' rhythms, images, personal philosophy and as based on narrative unities in participants' experience.

Problem: To begin, there is an underlying commonality in problems undertaken. Several focus directly on teacher thought by 'identifying' (Halkes and Deijkers, 1984; Marland, 1977), 'exploring' (Pope and Scott, 1984; Bussis, Chittenden and Amarel, 1976), 'describing' (Larsson, 1984; Duffy, 1977), and 'examining' (Lampert, 1985) some particular content aspect of knowledge. Two others (Olson, 1982; Munby, 1983) specify their problem as one of accounting for action in terms of teacher thought, specifically, of determining why curricula are differentially implemented by teachers. In inquiry, however, the difference in problem between Munby and Olson, and the others, mostly disappears when their method is reviewed. Both adopt a version of Kelly construct theory methodology and have teachers generate constructs (terms and statements) that, in effect, constitute the content of teachers' thinking, something that marks the nature of the problem for the others. Implementation *per se* is not subjected to inquiry in either Olson's or Munby's work. The three remaining studies (Janesick, 1981; Elbaz, 1981; Connelly and Clandinin, 1985) also focus directly on teachers' thoughts. The difference between the problem in these studies and in others is that these three studies are only indirectly concerned with the content of teacher thought. Their problem is one of re-imagining its form and composition. Elbaz is interested in the structure of teacher thought; Janesick is interested in the overarching features, which shape the specific bits, and pieces of a teacher's knowledge; and Connelly and Clandinin are concerned with the language for thinking about teachers' knowledge. Accordingly, a review of the problems in the twelve studies shows that nine are aimed at discovering the content of teacher thought, and three, while also concerned with content, are primarily focused on the form of and/or the language of discourse about, personal knowledge.

Method: As might be expected, there is somewhat more diversity in method than there is in problem. But, because different methods are often used in aid of a common kind of problem, diversity in method is not the best indicator of the degree to which studies differ.

Two used formal teacher response instruments (Halkes and Deijkers, 1984; Duffy, 1977); one used stimulated recall (Marland, 1977); three used versions of Kelly construct-based statement sorting techniques (Pope and Scott, 1984; Olson, 1982; Munby, 1983); four used versions of interview techniques (Bussis, Chittenden and Amarel, 1976; Larsson, 1984; Lampert, 1985; Elbaz, 1981) and two used participant-observation (Janesick, 1981; Connelly and Clandinin, 1985). There are, of course, variations in detail within each of these methods, especially the interview method for which there are clinical, semi-structured and conversational approaches in the listed studies.

Outcome: The resulting interpretations using these various methods are formed by the more common underlying problems. For instance, Halkes' and Deijkers' instrumentation leads to particular statements about the content of teachers' thought just as does Munby's more elaborate teacher statement-sorting process. In the end, the question of what is in the teacher's mind is answered by several of the researchers with quite similar kinds of statements. From the point of view of what it is we learn of teachers' personal knowledge, 'principles,' 'constructs,' 'criteria,' and 'conceptions' seem to name the same kind of thing. The terms and methodology differ but the problem and conclusions arising from the inquiry are remarkably similar in kind. None of the five remaining studies uses specific words (e.g., to name a construct), phrases (e.g., to name a belief), or short descriptions (e.g., to describe assumptions) to account for teacher thought. Their interpretations consist of longer, more complex, less precise, and more context dependent accounts. Accordingly, while there is some similarity in the problem, both method and outcome are different. There is a sense, then, that these five studies yield understandings different in kind from the other seven.

Interpreting Descriptions of the Studies of the Personal

The second task of assessing the implied concept of the personal at work in specific inquiries consists of an interpretation (Table 2) of the studies described in Table 1. This interpretation has been structured by a commonplace conception of what it is that composes teacher thought. This commonplace conception is outlined below as a preamble to the interpretation.

The Composition of Teacher Thought

Table 2 is designed on the assumption that one useful way of thinking about teachers' practical thought is to conceive of it as composed of ongoing action and prior personal experience. To state this is virtually to state a commonplace. When we teach, we think about our teaching and this thinking is composed in part of all the various things we do in our teaching. Schön (1983) names this aspect of practical thought 'reflection in action.' Furthermore, it is impossible, when a person thinks, to do so without reference to themselves as people. All the things one knows, has studied, and has done are in one way or another present in one's thoughts. This aspect of thinking is explicitly addressed in biographical and autobiographical studies. Thus, both action and prior experience compose our practical thinking.

Now, while the distinction between ongoing pedagogical action and past experience as elements composing thought may be commonplace, there are uncommon consequences of this view for our understanding of what is done, and of what might be done, in inquiry into teacher thinking. The functional separation in this body of literature between thought and action (Clark and Peterson, 1984) is, clearly, one of the possible directions for inquiry given this commonplace conception. When thought and action are imagined to be separate, each may be studied in isolation. Classroom interaction studies, especially popular a few years ago, are illustrative of the latter. Currently, as our interpretation below shows, the former is occurring, where thought is studied directly outside the context of a classroom or other practice setting. Those who study either thought or action might, of course, argue that the other was assumed within the work. For example, those who pursue classroom interaction studies might claim that they assume that people are thinking when they interact; and those pursuing thinking studies might claim that they assume that actions flow from thought. The point we are making, however, is that in these studies the two are not simultaneously subjected to inquiry. Rather, the study of one *assumes* something about the other. There is a difference between inquiries, which assume a relationship and those where the relationship is subjected to inquiry. Accordingly, the commonplace conception permits the identification of studies which define different relations of thought and action in inquiry.

Furthermore, the assumptions at work encourage different ways of imagining the uses of research results. Specifically, when thought is assumed to direct action, descriptions of thought stand to be taken as more or less abstract shorthand accounts of the rules governing practice. Given these rules, methods of intervention, which the philosopher McKeon has labeled 'logistic,' may be imagined. 'Logistic' methods of relating theory and practice are ones where practice is treated as being under the control of theory via strategic means. However, the commonplace conception of thinking makes it easy to imagine still other thought-practice relations; ones, for example, where practice directs thought or, possibly, interact with thought. In McKeon's terms, studies embodying such relationships might be called 'problematic' or 'dialectical' in their working relations of theory and practice.

If we extrapolate these commonplace notions of thought and action to encompass prior experience, as well as future action, a complex array of possibilities for inquiry are imaginable.

Interpretations

Returning to the studies, the most telling point of departure for an interpretation, in terms of prior experience and ongoing action, is the particular act and/or kind of thought which serves as evidential of the personal in inquiry. Whatever it is that passes as fact may be said to be as experientially close as one will get in the inquiry to touching whatever it is that is 'personal' in the phenomena of the inquiry. For instance, for Halkes and Deijkers, what is personally evidential is the teacher's self ratings on the Likert scales eventually summarized in tabular format; for Olson, what is personally evidential is individual teacher names for teacher-sorted categories of teaching events; and for Elbaz, what is personally evidential is informal teacher interview statements describing and reflecting upon the teacher's experience.

Following our remarks on the composition of teacher thought, whatever is taken to be personally evidential in the research potentially exhibits two relationships of importance in understanding the personal: the relationship of personal evidence to personal experience and the relationship of the personal to practical action. Table 2: Interpreting Descriptions of Studies of the Personal is, accordingly, presented in three columns, 'what is evidential of the personal,' 'relationship of personal evidence to past experience' and 'relationship of knowledge to action.'

1.1 *What is Evidential*: In our discussion of Table 1, we noted that most of the studies reviewed focus on the content of teacher thought. These studies use as evidence of content, specific teacher statements about themselves (Table 2, column 1). For instance, Duffy's teachers rate themselves on a prepared instrument, Munby's teachers effectively generate their own instrument by the use of describing and categorizing procedures, and there are variations in-between such as Olson's, where teachers are given a list and asked to sub-divide it into categories and to name each. The principle difference in the evidential basis within this set of studies is that, in some, the descriptive terminology is supplied by the researchers (e.g., Halkes and Deijkers, 1984; Duffy, 1977) and, in others; it is supplied by the participating teachers (e.g., Marland, 1977; Pope and Scott, 1984). However, in both kinds of study the teacher eventually lays claim to the content description of his/her thought.

Three of these studies (Janesick, 1981; Elbaz, 1981; Connelly and Clandinin, 1985) while sharing the teacher-determined evidential base of the other eight studies use qualitatively different evidence. Two of the studies (Janesick and Connelly and Clandinin) use teaching practice, as well as teachers' statement about themselves, evidentially. Given the interpretive direction in these studies, practice precedes teacher assertions and is ultimately considered more telling in the accounts of teacher knowledge offered. Given this dual evidential base and relative significance of each basis in understanding thinking, these studies effectively define personal knowledge in terms of both of 'thought' and action. Their claims about teacher thinking are, equally, claims about action and cognition. Elbaz and Connelly and Clandinin go one step further and include biographical teacher statements as part of their evidential base.

Table 2: Interpreting Descriptions of Studies of the Personal.

Authors	What is evidential of the personal	Relationship of Personal evidence to Past experience *	Relationship of Knowledge To action **
Halkes & Deijkers	A teacher's self-rating on prepared statements	By implication: Assumption that teachers' rating of statements	Logistic
Marland	A teacher's statement of guides to his/her interactive teaching behavior collected in stimulated recall interview	By implication: An implied explanatory relationship between principle given and teaching acts	Logistic
Olson	A teacher's grouping and labels for predescribed teaching events (How a teacher sorted 20 statements and names the sort)	By implication: Assumption that description of pre-set teaching events corresponds to teaching practice	Logistic
Munby	A teacher's generation of statements describing his/her class; grouping and labels for groupings of statements; researcher selected labels and phrases from teacher's discussion of basis of groupings	By implication: Assumption that a teacher-generated description corresponds to a teacher's practice. Implied relationship between term and experience through descriptions of teaching and explanatory reason for grouping	Logistic
Janesick	A teacher's practices and interview responses	Practices and interview responses based on teacher's interpretation of his own experience	Dialectic
Larsson	A teacher's responses in a semi-structured interview	By implication: Interview responses assumed to be based upon teacher's experience	Logistic

(table continues)

Authors	What is evidential of the personal	Relationship of Personal evidence to Past experience *	Relationship of Knowledge To action **
Pope & Scott	In-service and pre-service teachers' descriptions of their practices, and researcher observation of teachers' practices	By implication: Interview responses and descriptions assumed to be based on teacher's experience	Logistic
Bussis/ Chittenden/ Amarel	A teacher's descriptions of his/her planning and teaching	By implication: Interview responses assumed to be based on teacher's experience	Logistic
Duffy	A teacher's self-rating on prepared statements	By implication: Assumption that rating of statements of teaching approaches corresponds to teaching practice	Logistic
Lampert	A teacher's interview description of practice and conversations between a group of teachers	By implication: Assumption that teacher interview responses and contributions to conversations based on teacher's experience	Problematic
Elbaz	A teacher's description of her practice and of how she feels about it	Teacher interview responses based responses based on teacher's professional experience in school where she worked	Dialectic
Connelly/ Clandinin	Biographical detail, emotionality, morality, aesthetic, based on narrative of experience	Narrative accounts based on mutual reconstruction of experience as expressed in practices in interview and in classroom	Dialectic

* By implied, we mean situations where experience is used as a way of explaining the personal evidence but where the relationship is not under study in the research.

** Following McKeon, as described in text.

1.2 *Knowledge to Action*: From these interpretations of the evidential basis of each study, it is clear that different relations of thinking to practice are operationally at work in the research. Those studies (Janesick, 1981; Connelly and Clandinin, 1985) that use action evidentially adopt a reflexive 'dialectical' relationship between thought and action. In Janesick's study, for instance, particular classroom actions bring forth, and are seen to be part of, the teacher's perspective. Elbaz also appears to hold an assumption of the dialectical relationship of thought and action although this relationship is not subjected to inquiry in her work. Lampert appears to imagine the relationship as 'problematic' – one in which teachers face, and work through, certain classroom dilemmas. In the remaining studies, thought as described in the studies is assumed to direct action. This 'logistic' assumption is most clearly in evidence in the Munby and Olson studies, where it is assumed that the descriptions of different knowledge contents for different teachers accounts for differential implementation of curriculum materials. Accordingly, such studies tend to have a more practical, useful flavor about them.

1.3 *Personal Evidence to Prior Experience*: With respect to prior experience, most studies appear to assume that prior experience 'explains' why teachers think the way each study describes they do. This assumption is mostly of little influence on the actual conduct of various inquiries although it presents research possibilities in some. Munby, for example, assumes his claims about teachers' knowledge will be more valid if teachers generate descriptions of teaching based on their experience. Contrast this use of prior experience, for example, with Olson's, whose descriptions of teaching are prepared from the research literature. Elbaz goes further and focuses her interviews on her participating teacher's teaching history. In both Munby and Elbaz's case, however, this experiential history is eventually submerged in the way results are presented, rather than being treated as components of teacher thought.

1.4 Cognitive and Affective Evidence and Knowledge: Our examination of what is personally evidential in these studies is revealing in yet another important respect. Most of the studies admit as evidence material, which, for our purposes might be called 'cognitive' aspects of teacher thinking. Marland, for example, has statements of principle: the three Kelly construct studies use statements, which may be called 'constructs;' and so forth. Ultimately, this research leads to an account of the cognitive functioning of teachers on such matters as values and beliefs held, principles adopted, constructs formed, and knowledge of subject matter, classrooms and teaching. Four studies (Janesick, 1981; Lampert, 1985; Elbaz, 1981; Connelly and Clandinin, 1985) additionally admit as evidence 'affective' states of mind of the teacher. Elbaz's teacher, Sarah, for example, describes how she was unhappy in certain teaching situations and of how she created others in which she felt better about herself and her teaching. In the end, for Elbaz, this affective evidence tended to be utilized to produce claims on the cognitive state of the teacher's mind, for example, that practical knowledge is oriented to social situations. This use also characterizes Lampert's work. Janesick and Connelly and Clandinin, however, create statements of teacher knowledge, which combine affective and cognitive evidence. Janesick's teacher's group 'perspective', for example, is one laden with value. This use of value is not only cognitive, as it would be if one were to study the teacher and the teacher were to assert (or the researcher to assert on the teacher's behalf) that one of the teacher's values was group action. Instead, Janesick's claim is that the teacher holds a perspective and that this perspective is, in part, composed of a structure of, and belief in, group action *and* also a feeling of a certain intensity, altered by circumstances, of the worth of particular group teaching situations. Her term 'perspective,' and Connelly and Clandinin's term 'image,' are terms which refer to the emotional quality of personal knowledge.

Conclusions

The analysis presented in this paper may be summarized in three points. The first refers to the theoretical and linguistic diversity in the field and the remaining two provide answers to the question of the paper, "What is personal about studies of the personal?"

First, there is more commonality in the patterns of inquiry into the personal than is apparent in the authors' language and stated intentions. People using different terms such as 'construct,' 'criteria,' 'beliefs and principles,' appear in fact to often mean much the same thing. To the extent that this commonality holds, we may imagine some positive evolution in our understanding of the nature of the personal as it applies to teacher thinking in the years to come. Furthermore, the variety of theoretical resources should enrich inquiry in the field. We do think, however, that the theoretically borrowed language, and the various theoretical origins of the asserted ideas, may also have a tendency to inhibit understanding among researchers. The differences in theoretical origin and corresponding differences in theoretical language tend, we think, to divide the field, making researchers skeptical of using and cross-referencing one another's ideas. Diversity of thought is a mixed blessing.

The second point is that there are significant differences in how we may imagine the composition of teacher thought. The three general components suggested in this review are practical actions, biographical history, and thoughts in isolation of these two. Most research reviewed above focuses on the third. Various combinations not now treated as part of the literature in reviews of this topic are possible as seen, for example, in biographical studies (Pinar, 1981; Grumet, 1978; Darroch-Lozowski, 1982); in experientially based studies of thinking (Hunt, 1976); and in narratively oriented studies of all three components in combination (Connelly and Clandinin).

The third point, also a matter of the composition of teacher thought, is that most studies conceive of teacher thought in cognitive terms. But it is possible, following experiential epistemologists, such as Lakoff and Johnson (1980), and experiential ethicists, such as McIntyre (1981), to imagine thinking simultaneously in cognitive and affective terms. To claim that knowing something means having aesthetic, moral and emotional states of mind about that thing has a limited tradition in the pages of philosophy and in education, where affect is often treated as an impurity in reason. Our view is that a cognitive and affective understanding of the personal practical knowledge of teachers will help produce more living, viable understandings of what it means to educate and to be educated (Clandinin, 1985; Connelly and Clandinin, 1985, in press).

Finally, we should like to note that the latter two points create a variety of generic possibilities for research. A cognitive approach to biography, for example, might yield a kind of positivistic study of personal biographic origins much, in the same way, as a psychoanalyst might discover certain features in one's background that explain in some causal sense one's current state of mind. A narratively oriented study emphasizing cognitive and affective aspects of action, thought and biography would, of course, require different methods and would yield different outcomes. A narrative approach to thought and biography, for example, might yield stories linking, as it does in Schafer's (1981) psychotherapy, thought and biography, not as cause and effect, but as one among several possible explanatory narratives. Accordingly, one might undertake research, which combined a narrative interpretation of biography, thought, and action in experiential terms and which joined cognitive and affective matters in the narrative. Readers will be able to imagine other possibilities of interest.

Author Reflection Twenty Years On

When we wrote this paper in the mid 1980's teachers' theories and beliefs was one of the preoccupations of research on teaching. This topic provided the main contextual framework for positioning studies of the personal. Though there are still studies on teacher's theories and beliefs, the main contextual framework in the current literature is teachers' knowledge. By and large the key terms identified in the 1986 chapter still hold and are relevant from the point of view of teachers' theories and beliefs. While there have been additional studies the terms have not changed. However, from the shifted conceptual framework of teacher knowledge this paper would now be written quite differently. Fenstermacher (1994) did an analysis that pretty much stands up today. He identified three programs of research in the area of teacher knowledge, comparing and contrasting them primarily along epistemological lines. Our own work was one of the three, the other two being lines of work associated with Shulman's (1987) pedagogical content knowledge and Schön's (1983) reflective practice. Since our chapter was written, our work has shifted from the personal to the personal in context. Our work is now positioned at the interface of the personal with the social. A brief account of this shift is seen in our retrospective introduction to Chapter 15.

Chapter 15

Personal Practical Knowledge At Bay Street School: Ritual, Personal Philosophy and Image

F. Michael Connelly and D. Jean Clandinin

Summary

This chapter presents an overview of research on teaching aimed at developing an understanding of teachers' personal practical knowledge. The purpose is to develop a language for understanding how teachers know their classroom situations. The work uses anthropological method and its theory development is done inductively from participant observation, interview, and text data. In this chapter one school event, a professional development day, is used to illustrate the method. One of the terms in the language under development, teachers' "personal philosophy" is discussed and its relationship to two other terms, "ritual" and "image", is indicated. The chapter, therefore, serves to introduce both the method and meaning contained in the notion of the study of teachers' personal practical knowledge.

Introduction[7]

The research reported in this paper is part of a long-term study of personal practical knowledge in school settings.[8] To date, research results are presented in a series of

[7] This section, and the two following, follows closely the accounts given in the working papers, articles and chapters written on this work.

[8] The work is supported by grants from the National Institute of Education, grants for Research on Knowledge Use and School Improvement (grant # NIE-G-81-0020) and the Social Science and Humanities Research Council of Canada (grant #410-80-0688-XI).

working papers[9], and in a forthcoming Yearbook chapter (Connelly and Clandinin, 1984). The work is focused on the development of a person-centered language and perspective to account for school practices and for the actions of school practitioners. We adopt the position that actions, such as teaching acts, are knowing actions. They are both the expression and origin of the personal knowledge of the actor. Thus, we imbue action with knowledge and we imbue knowledge with passion. Action and knowledge are united in the actor and our account of both is of an actor. Hence, we use the term personal practical knowledge.

We view the teacher in an active, semi-autonomous role in the classroom and we focus on teachers' practical experiential knowledge. Teachers are, of course, affected by their origins, their teacher training, the instructions they receive, and the mores of the school. They are, however, far more than this. They are persons with a kind of school knowledge possessed by no one else and this kind of knowledge affects their perception, their interpretation and their response to classroom and school events. In short, they are, in important respects, intellectually autonomous. Further, we acknowledge the existence of a form of teacher knowledge that is at once practical, experiential, and shaped by a teacher's purposes and values (Elbaz, 1983).

Our intent is, then, to give an account of teachers' practical knowledge from the side of the persons who hold and use it. It is through coming to understand this knowledge as it is embodied in the teachers and principal *and* expressed in their practices that we will be able to both view and talk about teachers and their practices in schools with new concepts and a new language. Our work is, on the one hand, a detailed account of practice from the perspective of specific school practitioners. It is, thus, an account full of details, events and people. In providing such an account of practice, we also offer a new language and new theoretical constructs that help make sense of school practices.

We view the practices of the practitioners with whom we work as minded (Jennings, 1982).[10] Their practices, i.e. their verbal communication about their work, their actions and their products, are seen as expressions of their personal practical knowledge. We see a mind at work when we see an action. Even actions that might be seen as error or idiosyncratic personal response are assumed by us to be minded. Thus, when the principal turns his back on a gymnasium audience and straightens a welcoming display, we assume that, in his mind, "things" and their proper order are somehow a part of his image of how a school should look. Thus, an action that might be dismissed as a nervous tic is treated by us as a minded practice. The action becomes a sign that the mind is at work in the school environment and we begin to trace its features.

Our inquiry is shaped in such a way that we search for experiential, biographical, and historical interpretive constructs to account for school practices. Our interpretive movement is from practitioners' personal experience and history to

---

[9] Those directly related to this chapter are Clandinin, D.J. and Connelly, F.M. Personal Practical Knowledge and Narrative Unity (1983) and Teachers' Personal Practical Knowledge: Calendars, cycles, habits, and rhythms (1984); Connelly, F.M. and Clandinin, D.J., Personal Practical Knowledge at Bay Street School: image, Ritual and Personal Philosophy (1983).

[10] Our use of the term was triggered by Jennings' notion of "minded action" in his discussion of religious ritual.

personal knowledge rather than from established theory and associated analytic constructs and clue structures.

The Notion of Personal Practical Knowledge[11]

Virtually everything we have to say is based upon the notion of personal practical knowledge. This phrase is both so commonplace and so complex that it will be misunderstood without a clarification of it at the start. We begin this clarification with two negatives. We do not mean by "personal" something apart from the society and circumstances in which an individual lives. On the contrary, we are merely emphasizing, as against Marxism and some other views, that in addition to the long sweep of tradition and the embodying structure of a culture and society, there is an individual local factor that helps to constitute the character, the past, and the future of any individual. Nor by "personal" do we define the knowledge in question as being privately owned, possessed and secret to the individual in question. In many respects, as we shall try to make clear, it is private but it need not remain so. It is knowledge that can be discovered in both the actions of the person and under some circumstances by discourse or conversation. What we do mean by "personal" as defining knowledge is that the knowledge so defined participates in, is imbued with, all that goes to make up a person. That is, it is not such "impersonal" knowledge, as two times two equals four. It is, rather, a knowledge which has arisen from circumstances, actions, and undergoings which themselves had affective content for the person in question. The word "knowledge" also has two possible misunderstandings. The phrase "personal knowledge" could be taken to refer to intellectual possessions of immediate access to the person expressible in language of one kind or another, i.e. common language or some technical language involving oral or written symbols. Personal knowledge does have this sense of an accessible possession since we do refer to person's speech about oneself. But we mean much more. We mean as well other intellectual events and bodily actions. Action, and the possession that characterizes it, is part of what we mean by personal knowledge. Indeed action, broadly conceived to include intellectual acts and self-exploration, is

---

[11] Our own forerunners on the study of practical knowledge sprang from Schwab's curricular writings on the practical, and the philosophical writings of Dewey, Aristotle, McKeon and others. The work has progressed through studies on deliberation and choice, decision making and now on personal practical knowledge. See, for instance, Connelly, F. Michael, The functions of curriculum development. *Interchange,* Special Issue on school Innovation, 1972, 3, 161-175; Connelly, F. Michael and Diennes, Barbara, Teacher decision-making and teacher choice in curriculum planning: A case study of teachers' uses of theory. In K. Leithwood (W.), *Studies in curriculum decision making*. Toronto: OISE Press, 1982. Freema Elbaz's dissertation, *The Teacher's 'Practical Knowledge': A Case Study,* (University of Toronto, 1980) is a highlight in this respect. Clandinin has completed a dissertation focussed on teachers' personal imagery entitled A *Conceptualization of Image as a Component of Teacher Personal Practical Knowledge,* (University of Toronto, 1983). Several recent dissertations at the University of Toronto and the Ontario Institute for Studies in Education expand on the notion of personal practical knowledge; Siaka Kroma, *Personal Practical Knowledge of Language in Teaching: An Ethnographic Study* (1983), Nevat Ephraty, *The Relationship Between an Educational Technology Program Innovation in a Teacher-Education Institute and Teachers' Personal Practical Knowledge* (1983), and Pita Deverell, *Arts Policy, Society and Children: Towards Guidelines for the Inclusion of the Arts for Children in Arts Policy in English Canada,* (1984).

all we have to go on as researchers. When we see an action we see personal knowledge at work. Second, the phrase "personal knowledge" could be taken to refer to truths possessed by the person in question. This too is not necessarily the case in the light of some parts of one's past (whether intimately personal or social or traditional); later parts of one's life may well be interpreted in such fashion as to be quite literally untrue. This is to say that people have convictions arising from experience that differ markedly from convictions of other persons undergoing similar events and leading in both cases to actions which are sometimes failures and sometimes successes for the purposes which the actor had in mind. What we do mean by knowledge is that body of convictions and meanings, conscious or unconscious, which have arisen from experience, intimate, social and traditional, and which are expressed in a person's actions. We use the term "expression" to refer to a quality of knowledge rather than to its more common usage as an application or translation of knowledge.

In this chapter we illustrate our method by giving an account of one aspect of the school principal's personal practical knowledge as we came to understand him in a case study using a methodology we refer to as Interactive Interpretive Inquiry: The Study of Minded Practices.[12] The methodology uses participant observation and interview data collected over a three-year period with selected practitioners in Bay Street elementary school. Bay Street is an inner city school in a large urban city. Bay Street's principal is Phil Bingham, the key actor in this paper.

The Research Site: Bay Street School[13]

Bay Street School is a Junior Kindergarten to Grade 8 core Inner City school in a large metropolitan board in Ontario, Canada. The school has 47 staff members and 750 students of ethnically diverse backgrounds. Approximately one-third of the students are oriental, one-third Portuguese, and one-fifth Black. The remaining students are of other, miscellaneous ethnic origins. The school receives special inner city funding and is one of five schools designated by the board as a Language Project School.

Among the many Board policies impinging on the school is its most high profile policy, the Race Relations Policy. Again, Bay Street School is an experimental school under this policy as defined by the Human Rights Leadership Project.

The school is developmentally alive in many other ways. Its principal is relatively new to the school and 20 teachers joined the staff in September 1981. The new teachers were replacing teachers who had asked to leave rather than become part of the three-year Language Project. As part of the Language Project mandate, the school is engaged in establishing shared decision-making mechanisms. It is trying to realize a working philosophy consistent with the principal's child and community oriented views and with the language orientation of the Language Project.

Teachers are expected to participate in the committee structure of the school; to justify a detailed curriculum plan for purposes of an observation based

[12] The methodology is explored in depth in a forthcoming chapter of the same title in Forms of Curriculum Inquiry, ed., E. Short.

[13] To ensue the privacy of participants in this study, pseudonyms are used throughout the paper.

evaluation; to become actively involved in the community through home visits and through participation in community events; and to always be on the alert for racial incidents both interpersonal and in print. On top of these expectations, there is an almost constant demand for teachers to be involved with in-school professional development.

Methodology

Negotiation of Entry, Method, and Interpretations

Methodologically, our work is cast in terms of a particular notion of the relationship between theory and practice. These notions are more fully developed and conceptualized in Clandinin (1983). In brief, we adopt what the philosopher McKeon (1952) refers to as a "dialectical" relationship between theory and practice and between researcher and practitioner. For us, this notion commits us to a methodology of negotiation. We negotiate (a) entry, (b) methods of research work within the school, and (c) our interpretations of practice and practitioners.

(a) We began negotiation of entry with the Toronto Board of Education in October 1980 when we discussed the study with central office personnel our work was approved at the board level in November by the Research Review Committee and negotiation with Bay Street School began in January 1981. We had our first meeting with interested staff in March 1981; we met the school cabinet in April 1981 and gained school approval to participate two weeks later. Participant observation began in April 1981 and continued on an intensive basis through June 1982. We continued to be involved in the school until June 1983 when this paper was being finalized and we plan to maintain this involvement in coming years. The 1982/1983 year contained a shift in emphasis from classroom participant observation to the discussion of ideas, concepts, and working papers described below in our negotiation of interpretive concepts.

(b) The negotiation of research working methods within the school began with the principal and school cabinet and resulted in the position that we would pursue our research interests provided we were seen to be of some use to the school. Both parties were to benefit. The principal invited teachers to volunteer to work with us with the result that Clandinin was invited to work in a Grade 1 teacher's classroom and Connelly to work in the library. It was understood that Clandinin would work three days per week and Connelly would participate on a less regular but weekly basis. In addition to this, we were invited to participate in a full range of school activities and to become involved in the major school committees and their main decision making body, the cabinet.

Because of the special dialectical relationship between researcher and practitioner, our negotiation of research methods within the school included a process that we call "the negotiation of minded practices." For instance, Clandinin's negotiations with the Grade 1 teacher developed in such a way that Clandinin became, variously, a teacher aide and co-teacher of the class. She was actively involved in working with individual students and in discussing programs for individuals and for the class as a whole. Both researcher and teacher had their own ways of doing things. Their practices would differ on, for example, the teaching of reading. Behind this, of course, were two different minds at work.

In short, the minded practices of the researcher and teacher were different. Negotiation of these practices was inevitable and occupied both planning and reflective sessions when researcher and teacher evaluated lessons and planned future ones. In the process, the negotiation of minded practices for the classroom occurred.

(c) The third feature of negotiation in our work consists of the sharing of all written material on our interpretive constructs with school staff. This process, from the point of view of the staff, is seen as a professional development activity. We have been invited to privately discuss papers with teachers and we have been invited to lead noon hour colloquia on our interpretations. From our point of view the purpose of these sessions is to obtain a "reality check" on our observations and to engage in reflective discussion on the constructs developed by way of interpretation of minded practice. For instance, this chapter was discussed with Phil Bingham, the school principal, and through these discussions we gained insight into community aspects of his personal philosophy noted in subsequent sections of this chapter.

Our active involvement in school activities precluded, for the most part, note taking during our school visits. Short notes were kept during meetings. After leaving the school, and usually the same day, the events of the day were reconstructed using Dictaphone. The dictated notes were then entered into a DEC-10 computer through either a word processor or a computer text-editing system.

Unstructured, open-ended interviews were held with the school principal and other school staff. While some interviews were tape recorded, most were recorded in field notes. All material was entered into the computer.

Epistemological Order Vs. Methodological Order

The researcher, through tracing biographical and historical elements in the speaker's experience, can partially reconstruct a telling understanding of personal practical knowledge. Methodologically, the relationship between the two appears somewhat different than it does epistemologically since *we* reconstruct personal practical knowledge through the links between its expression and personal practical knowledge that are provided by accounts of experience. We move from expressions of personal practical knowledge to personal practical knowledge through experience when, for the agent, experience suffuses and creates both. Thus, the order in which we present our account may leave the reader with a false impression of the relationship among the elements we wish to portray. We have begun with various minded expressions in the school as represented by the Professional Development Day and our first visit to the school and we made a transition to personal practical knowledge. This is not the epistemological order which we wish to reconstruct; it is merely our methodological order.

With the description of the school and our research methodology completed, we now turn to our account of the principal's life in our research school.

A Theoretical Expression of Personal Philosophy: An Illustration of the Method

The remainder of this chapter illustrates our method of work by recounting our interpretation of a specific school event, a professional development day. We give some sense of one of our key terms, teachers' "personal philosophy."

As we wrote the first draft of this paper, the school had just survived a crisis that threatened the job of the principal, challenged the tenets of the "developing philosophy," and caused each teacher to whom we talked to question their professional life.

Now consider the following scene, on February 19, 1982, two days after the meeting which brought the crisis to a head and which resulted in a reduction of tension. The staff is away from the school in relaxed surroundings at the city's Harbourfront for a full Professional Development Day. The teachers are seated on chairs and on, what the principal views as, comfortable floor cushions arranged in a semi-circle. Charts depicting the aims and goals of the project school are displayed on room dividers around the roan. Phil Bingham, the principal, is standing at the open end of the semi-circle beside an overhead projector and a screen as the meeting breaks up and teachers move to get their morning coffee. Phil had just completed one and a half hours of one-way presentation, labeled by a teacher as "jug to mug", summarizing the six points of his personal philosophy and how it relates to his expectations of staff working in his Project School.

The Professional Development Day came at the end of two weeks of meetings and consultations with school staff, administrative staff, and community representatives all of which culminated in the crisis meeting already noted. This had been a particularly stressful time and we see the reiteration of Phil's philosophy as one in which he both *re-confirms* his commitment and *invites* a similar re-confirmation for established staff. The expression of his philosophy with its attendant re-confirmation and invitation functions constitutes a kind of ritual. If one thinks of the ritualistic expression of the philosophy, in a physically structured environment, performing specific gestures of body and speech, as minded bodily knowledge; of the ritual as expressing and re-confirming the personal knowledge of the speaker; and of the ritual as one way a community can participate in a common practice, then the expression of the philosophy in a professional development day takes on new meaning and points to a kind of practical knowledge obtained and shared in joint action.

A reader may recognize our meaning by reflecting on his/her own experience of what we call minded bodily knowledge. For example, in one's participation in a religious ritual, the body movements in rising, kneeling and praying are expressions of religious meaning. Often the body anticipates the actions to be taken. Or, consider a teacher's bodily attitudes containing a myriad of body movements, such as head nodding, eye movement and hand gestures as he/she stands in front of his/her class. These constitute minded bodily knowledge of teaching in both, the religion and teaching examples; the expressions are personal and may exhibit individual features.

The Philosophy as a Ritual in a Ritualistic Context

The Philosophy

The theoretical expression of Phil's personal knowledge in the form of a "philosophy" is easily presented due to its brevity and directness. The setting in which the philosophy is presented is considered by Phil to be contextually important. Thus, the professional development day introduced above is significant because it represents Phil's notion of an appropriate setting. Likewise, our rendition of this philosophy to our readers is best presented in context and we have, accordingly, chosen to make its presentation in the following paragraphs as it appeared to one of us in our field notes on the morning of the presentation. We have also presented a paragraph's follow-up to the philosophy in order that the reader obtain a sense of the significance Phil attaches to the philosophy. Our field notes show that:

> Phil started the session by saying "thank you". He had put up the overhead that said "Bay Street Today For Our Children Tomorrow". With that overhead on the screen, he drew our attention to the automobiles passing on the Gardiner Expressway. They gave the illusion of being on the roof of the building. He said something about bringing children down and getting their impressions of what they thought it looked like.
>
> He then said that he hoped that they would be able to relax today. He made a comment about them belonging together, working together to fulfill the mandates of the Board of Education and the Ministry of Education. He said that the Bay Street staff was "here today for our children tomorrow".
>
> He said that he had gone through what he planned to do that morning with about 300 people at Open House but that the staff had missed it.
>
> He started out by saying that he tried to date most of his planning and thinking on a set of guidelines. He made reference to them being "a set of fundamentals" that he took from E. C. Kelly in his book "In Defence of Youth". He said he found the book at a used bookstore when he was working 142 with Hill Street (an alternative school) kids in the sixties. He started out with a blank overhead and wrote the following points on it as he discussed them. He said that the following were his fundamentals for survival:
>
> 1) The importance of other people;
> 2) Communication;
> 3) In a loving relationship or atmosphere (he said they could translate loving for caring);
> 4) A workable concept of self;
> 5) Having the freedom to function (which Phil says stands for responsible action and interaction);
> 6) Creativity (having the ability and the right). He said this is what we, as teachers, are trying to do with kids.
>
> (Each point in the philosophy is repeated as each new point is introduced so that by the time the sixth point is raised a long sentence containing the previous five points is stated.)
>
> He said "these fundamentals are in my mind when I work with you." He said it doesn't always work but that he does get satisfaction from trying to make them work. He then made an apology to those who have already heard it. (He was going to use the same overheads that he had used down in the cafeteria at the Open House).

> He said that what he was going to present today is what 'I've been working on for the past 3 years at Bay Street School" and that he "felt that things were really moving in that direction.' (Field Notes, Feb. 19, 1982)

The philosophy could easily be ignored following a barebones reading of the six points. In the context presented here, however, one is inclined to query more deeply its meaning since it was obviously presented with sincerity, with a sense of personal prescription directing Phil's relationship to the staff, and it was used to orient teachers to a full day of professional development. The philosophy is often used on those occasions when Phil is groping to give a central account of himself. He used the philosophy to introduce himself to us and us to the school; he used it when he introduced himself and the Board of Education's Language Project to the teachers; and he used it again when he did the same thing for a community presentation. It is not simply that the philosophy is trotted out to fill in time but it is used in situations where a central understanding of the man, and the school, whose "ethos" he believes he is structuring, is called for.

The Ritualized Context

The meaning conveyed upon, and by, this ritualized philosophy is a function of its assertion-context and its concrete grounding in existing and intended school settings. These are treated in order below.

1. *The Assertion-Context.* At this point we shall content ourselves by giving an account of the meaning conferred on Phil's statement of his philosophy by its assertion context. We view this as being composed of two elements within the situation: its significance as an occasion for calling forth a central understanding of the man and its physical environment. These contextual elements operate intellectually and are no mere window dressing in our effort to grasp the meaning of the philosophy.

1.1 Significance of the Occasion for Calling Forth a Central Understanding of the Man

The philosophy is always stated, in our observation, in situations where Phil is giving an account of himself. We have never observed it in use as the basis for defending a plan or action. It does not appear in situations of dialogue or debate. Rather, it appears when Phil is before an audience. And it appears on those occasions when Phil deems it important that the audience understand him and what he stands for. Such a situation may be called forth by the audience, as it was in our own April 15th, 1981 interview with Phil and later in his presentation to the Community; or it may appear when Phil decides that it is pedagogically important for his audience, as in his presentation at the professional development day noted above and in staff meetings. The significance of these settings is easily recognized by members of the audience, and the intention to convey an important personalized conceptual message is obvious. The stage is set, therefore, for members of the audience to wonder what is "in behind" the statement and, as we have done over our time at Bay Street School, to try to fill in the meanings left unstated. The significance of the setting, therefore, leads to, and contributes to, the personal authority of the speaker.

1.2 Physical Environment for the Theoretical Assertion of the Philosophy

Phil is careful to present his ritualized statement of the philosophy in particular surroundings. Furthermore, we shall later see that the character of the physical environment is a crucial expression of Phil's personal practical knowledge. He strives to create a certain kind of physical environment for the school as a whole, for visitations to the school, and for crisis community meetings.

We had little insight into the significance of the physical environment on our first visit to Bay Street. There was a kind of shabbiness and dourness about the halls and walls, mixed with spots 144 of brightness. The dourness has almost completely been transformed in our period of stay in the school. Returning to our field notes, the first clue to the significance of environment for Phil was evident in our April 15th, 1981 meeting in which he had said that he wanted an occasion to discuss school philosophy with the whole staff in the coming September we had offered meeting rooms at The Ontario Institute for Studies in Education and he had declined in favour of the more informal Harbourfront setting. According to our field notes:

> We had a discussion about Phil's philosophy. He recited 4 or 5 points, Kelly's philosophy – one had to do with being closer to the earth. Somewhere in this discussion Mick asked if teachers were in early in the fall. Phil said he did not know but would like to have time together. Mick offered O.I.S.E. facilities but Phil said they use pillow rooms at Harbourfront. They want to set up their own pillow room. Phil sketched out how he wanted to have a weekend retreat so people could discuss school philosophy. Grace has a copy of the school philosophy document. We are welcome to see. (Field Notes, April 15, 1981)

2. *Concrete Grounding of "My Philosophy" in Existing and Intended School Settings.* A detailed account of the physical environment and its connection to the philosophy would constitute a paper in itself we wish only to point out that the ritualized expression of the philosophy in a particular physical setting contributes to the state of mind in which it is to be received and it reflects the kind of school environment compatible with the personal knowledge behind the statement. In a highly symbolic gesture in our first visit to the school Phil first took us to Cynthia who was supervising the construction of a large mural.

This mural was later transferred to a prominent spot in the entrance hall and is now a focal point in the hallways. It is no accident, in our opinion, that this is the first location within the school shown to us by Phil. The construction of the mural and its intended placement on the wall was a clear-cut physical expression of his personal philosophy and it was one of the most direct ways for him to express this philosophy to us. From these considerations we see how Phil's personal philosophy is related to the physical environment of the school and to its ritualistic mode of expression.

Phil's Personal Practical Knowledge

For purposes of this chapter Phil's actions during the Professional Development Day illustrate what we mean by the notion of minded practice. In this case the practice is "minded" by Phil's personal philosophy. Unlike the usual connotations attached to the term "philosophy" the expression of Phil's personal philosophy exhibits ritualistic features akin to those in religious studies. The ritualistic expression of the personal philosophy in particular school environments gives particular meaning to school events such as the Professional Development Day. It is through understandings such as these that schooling may better be understood as the product of participant's personal practical knowledge.

When Phil refers to "my philosophy" and on other occasions, presents it as a long six-point sentence, the two are not synonymous. The six points constitute a theoretical expression of practical meaning and, as with theoretical statements which express theoretical meaning, personal theoretical statements are torn from the reality to which they refer and only partially and selectively account for it. Polanyi (1958) in writing about the articulation of personal meaning draws attention to the ultimate ineffability of personal knowledge. Polanyi gives examples, drawn from his earlier medical career, to show that a novice diagnostician will, even with comprehensive physiological knowledge combined with diagnostic rules, be unable to make sense of a set of symptoms. With repeated tutelage he gains the necessary diagnostic expertise, but he still cannot account for it in a scientific way that would permit him to convey the skill to other novices. Polanyi utilizes the notion of "subsidiary awareness" as a concept to account for the submersion of telling instance and detail below the level of consciousness in the interests of the focal attention aimed at some end such as a diagnosis. If, in fact, one were to pay attention to information in subsidiary awareness, focal attention would be lost, a feature frequently noticed in novice doctoral students in their efforts to define a problem. Phil's ritualized expression of his philosophy is, therefore, a sign pointing to a rich content of personal knowledge, most of which is unavailable to Phil for purposes of verbal expression. It is what he refers to as "my philosophy" and, in its complete professional expression, the school "ethos". This "hidden" content of personal knowledge is, however, partially amenable to interpretive analysis using data drawn from the narrative of the actor's life. This point is illustrated in other writings noted above on this study.

Conclusion

The burden of this chapter has been to briefly describe our studies on teacher's personal practical knowledge and to illustrate our method with an account of one school event. Coincident with this purpose is our more general purpose of developing a language for understanding the origin, character and uses of personal practical knowledge.

Practical knowledge includes what we call personal practical knowledge and its expression in the minded practices of practitioners. Personal practical knowledge is composed of such experiential matters as personal philosophy, ritual, image[14], narrative unity[15] and rhythm.[16] Personal practical knowledge is expressed

---

[14] Image is here conceptualized as a kind of knowledge embodied in a person and connected with the individual's past, present and future. Image draws both the past and the future into a personally meaningful nexus of experience focused on the immediate situation which called it forth. It reaches into the past gathering up experiential threads meaningfully connected to the present. And it reaches intentionally into the future and creates new meaningfully connected threads as situations are experienced and new situations anticipated from the perspective of the image. Image emerges from the imaginative processes by which meaningful and useful patterns are generated in minded practice. Minded practice involves the emergence of images out of experience so images are then available as guides for the person to make sense of future situations (Clandinin 1983). Many images have a metaphoric quality and as such are close to Lakoff and Johnson's (1980) "experiential metaphors".

in such minded practices as practitioners' verbal explanations of self and actions including both doing and making.

This chapter, using anthropological style data collected during a Professional Development Day at Bay Street School, has allowed us to elaborate the distinction between personal practical knowledge and minded practice, and to reflect upon their developmental and expressive relations. In particular, the chapter revolves around the principal's expressed philosophy, his "personal philosophy". We account for this philosophy in terms of "rituals" and "makings" found throughout the school as expressions of the philosophy. The idea of the principal's "personal philosophy" provides the basis for subsequent exploration of the notion of image as a personal knowledge construct operating in the life of the principal. In sum, then, this chapter gives an introduction to how we view the modes by which teacher practitioners know their classrooms.

[15] Narrative unity is a continuum within a person's experience which renders life experiences meaningful through the unity they achieve for the person. What we mean by unity is the union in a particular person in a particular time and place of all that he has been and undergone in the past and in the past of the tradition which helped to shape him. We are informed on this notion by McIntyre (1981). (See footnote No.3 for reference.)

[16] We use the term rhythm in the commonplace sense of something recurring repetitively, perhaps cyclically, and which has an aesthetic quality as performed. Rhythms as part of personal practical knowledge serve to modulate an individual teacher's imagery and the narrative unities of which the imagery is a part (See footnote No. 3 for reference.)

Author Reflection Twenty Years On

This was the first paper in a two decade long research program. There are three program threads all foreshadowed in this paper: the personal, the personal in context, and narrative inquiry as a way of thinking about these matters.

The research program took many twists and turns as our own research program developed in different parts of the country and as graduate students joined the program and wrote on a wide array of topics. The first major publication was *Teachers as Curriculum Planners: Narratives of Experience* in which we focused on curriculum planning as a personal narrative act of teaching. Research on the place of the personal in context followed, our major publications being *Teachers' Professional Knowledge Landscapes* and *Shaping a Professional Identity*. The former advanced a landscape concept of the personal in context and the latter interwove the personal with the contextual by developing a narrative notion of professional identity. The narrative thread is marked by three key publications, a 1990 article on *Stories of Experience and Narrative Inquiry,* a 1994 article on Personal Experience Methods, and our 2000 book *Narrative Inquiry: Experience and Story in Qualitative Research.*

Chapter 16

Self-Evaluation: A Critical Component in the Developing Teaching Profession

Miriam Ben-Peretz and Lya Kremer-Hayon

Summary

The study relates to several questions concerning teacher thinking in the process of self-evaluation. Teachers' goals and criteria for self-evaluation, their perceptions of the evaluation process, their main concerns and personal styles of thinking about this issue are investigated. Findings are based on questionnaires responded to by two groups of ninety teachers and in-depth interviews with nine teachers. Teachers' goals of self-evaluation cover a wide range of topics including students' needs and teachers' needs – professional and personal. A progressive view of education was found to underlie these needs. All traditional means of self-evaluation were perceived to be used much more than desired, probably because they are relatively more observable and operational. As far as the style of self-evaluation is concerned, several categories emerged: Professionalism; cognitive/affective orientation; structure; externality/internality; criticism; self vs. other emphasis; negative/positive image; difficulties. Findings may serve as a basis for reconsidering teacher education and staff development toward a higher degree of professionalism.

Perspective

Teachers' self-evaluation is one aspect of the overall process of teachers' thinking. Studies of teachers' thinking represent an attempt at revealing covert, tacit processes conceived of as affective teaching acts. Research findings indicate that teachers' thinking in the process of planning instruction plays an important role in bringing about student cognitive and affective achievements (Marx, 1978; Clark et al., 1978). While a good deal of work has been done on teachers' planning, there have been very few investigations of teachers' self-evaluation. The investigation of teachers' planning must, however, be followed by studies on self-evaluation as a form of feedback, which is considered a necessary condition for the enhancement of learning and change. Psychologists may differ on the principles of learning they suggest, but there is agreement on the importance of feedback. Since the professional development of teachers is a process of constant learning on the part of the teacher, feedback becomes an indisputable need (Moustakas, 1972).

There are several sources from which teachers may derive feedback; supervisors, colleagues, principals, students and parents. Supervision of instruction has been extensively studied as a tool for providing feedback. Within this context, various strategies of supervision have been developed and put to use, such as clinical supervision (Cogan, 1973), conferencing techniques (Hunter, 1980) and interaction analysis (Blumber, 1980).

Teachers themselves as a possible 'self-feedback system' are a relatively neglected topic. Techniques and strategies for teacher – evaluation that have been developed, investigated and proved to be effective in improving instruction, are based on criteria, that are derived from educationists' and supervisors' views. All these factors are located outside teachers. A few examples will suffice to illustrate the point: rating forms (Ryans, 1960), micro-teaching (Allen, 1966; Fuller, 1973), interaction analysis (Flanders, 1970). The anthology of instruments "Mirrors for Behaviour" (Simon and Boyer, 1967) suggests a variety of such instruments. The fact that self-evaluation based upon teachers' own criteria has not been thoroughly studied, is surprising in view of the extensive demands for professional growth and for teachers' autonomy in planning and in processes of instructional decision making. Fostering teacher self-evaluation is perceived as deriving from the need for constant feedback, from trends toward professionalization and the development of an autonomous teacher.

The present study attempts to answer questions related to teacher thinking in the process of self-evaluation.

Specifically, the objectives of the study were:

- To determine teachers' goals in the process of self-evaluation;
- To identify specific topics of concern of teachers in the context of self-evaluation;
- To identify main criteria used by teachers for self-evaluation purposes;
- To disclose sources of evaluation teachers use in order to evaluate their professional work;
- To disclose the extent to which teachers' self-reported evaluative activities are congruent with a their stated preferences;
- To explore whether teachers perceive their training as contributing to the process of self-evaluation;
- To detect personal styles of teachers in the manner in which teachers discuss the process of self-evaluation.

Method

Two investigative procedures were used in the study – a questionnaire and interviews.

A questionnaire "Teacher Self-Evaluation" (TSE) was developed based on answers given by teachers in an open-ended interview regarding goals, criteria, and sources of self-evaluation. Analysis of answers yielded a long list of items. After the removal of duplication, the remaining items went through a process of item analysis. The questionnaire in its final version consisted of 21 items cast into a five-point interval scale ranging from 1 (low) to 5 (high). Respondents were asked to related to each item three items as follows: (a) to what extent do they actually act as stated in the items? (b) To what extent would they prefer to act accordingly? (c) To what extent were they trained to act as specified in the items? Background data were included in the questionnaire and analyzed with regard to the questions raised and was processed by the aid of SPSS programs, yielding frequency distributions, means, standard deviations, and t-tests.

A group of 90 teachers randomly selected from the northern district of Israel from various school levels, with differing seniority and training background served as subjects for questionnaire administration. (Partial findings were reported elsewhere, Kremer and Ben-Peretz, 1984). A series of in-depth interviews were carried out with nine elementary school teachers, ranging from 10-12 years of seniority.

The interviewers were a group of graduate students in education, who had been especially trained for this purpose. Following the literature on interviewing, the interview focused on one open-ended question, stated as follows: "What can you tell about yourself as an evaluator of your own teaching?" Interviewers were instructed not to interrupt the flow of talk unless they were sure that the participants concluded their responses. In this case, probing questions could be raised and some elaboration could be asked for, if needed.

Protocols of the interviews were analyzed by the investigators using the process of analysis described by Miles and Huberman (1984).

Findings and Discussion

Analysis of teachers' goals of self-evaluation points to a wide variety of interests, starting with perceived students' needs and followed by varying professional and personal needs. In descending order, teachers indicated their goals of self-evaluation as follows: "Improving students' achievement" ($x = 3.75$, $sd = .72$), 'improving the teaching profession" ($x = 2.83$, $sd = 1.03$), "career promotion" ($x = 2.20$, $sd = 1.09$).

With regard to various criteria suggested for diagnosing developmental stages of teachers' concerns, the listed goals of teachers' self-evaluation may be classified relatively high (Fuller, 1969). Thus, teacher educators may derive some satisfaction from the fact that 'improving student achievement' achieved the highest score and that the next three listed goals pertain to professional development, because these interests may serve as a sound basis for improvement. Since the highest rated goal of self-evaluation focuses on students, a progressive orientation may be implied. This implication is supported by additional findings regarding differences between actual and preferred criteria of self-evaluation; for instance: traditional-oriented criteria are perceived by teachers to be used more than is desirable, whereas progressive-oriented criteria are used less than is perceived to be desirable. This direction of discrepancies may be related to the nature of the traditional characteristics of teaching that are observable and relatively operational as compared with progressive teaching characteristics (Hofman and Kremer, 1980), and hence more easily evaluated. However, the possibility of social desirability of progressive orientations should also be kept in mind.

Except for 'tests,' the traditional means of evaluating students' achievements, all sources of self-evaluation are considered to be used less than desired.

The general declared lack of training for self-evaluation is not surprising in view of this relatively neglected topic in teacher education programs. However, some additional interpretations of this phenomenon may be suggested. Even if attempts at training for self-evaluation are included in the processes of pre- and in-service education, teachers may not be sensitive to their implications for the daily practice of teaching. These findings indicate the need for developing programs that integrate and foster self-evaluation in the process of teacher education.

In-depth interviews with teachers provided more specific insights into their main concerns.

Eight out of nine teachers interviewed in the study mentioned *pupil achievement* as their main concern of evaluation, a clear tendency to classify the achievements into information, intellectual hierarchy. Increase in information was perceived to be the easiest to assess and evaluate and value education was perceived as the most difficult area of evaluation. Teachers expressed their frustration, because of the high importance they attached to value education on the one hand, and lack of professional knowledge to evaluate their works in this area. Another topic of concern was *teaching methods.* The eight teachers who mentioned this topic related it to pupil achievement, wondering if they chose their teaching methods wisely, and whether they performed well. *Discipline* was an additional topic of concern mentioned, however only by four teachers. Interestingly, no connection was found between seniority and discipline concerns.

While topics of concern pertain mainly to the content of self-evaluation, personal styles pertain to the form and manner in which these concerns were expressed by teachers.

The following categories for defining styles of self-evaluation emerged from our analysis;

1) *Professional approach*, namely, the use of professional educational terms in discussing the process of self-evaluation. Such terms as, for instance, "taxonomy of objectives," "affective domain," "learning for mastery," are used by teachers with different frequencies ranging from 12% - 56% of statements. These varying frequencies, as well as the nature of the terms used, form part of the personal style of teachers commenting on self-evaluation processes.
2) *Cognitive and/or affective approach to self-evaluation.* Examples of statements reflecting a cognitive approach are: "How does a teacher use the educational potential of a given content ... what does he do to ensure its being learned ... how does he transmit its cognitive messages." Examples reflecting an affective approach are: "Self-evaluation is accompanied by negative side effects ... One has to remember that the professional status of teachers is low in comparison with other free professions."
3) *Structure* is another characteristic of teachers' personal style in thinking about self-evaluation. Some teachers exhibit a quite rigid structure which is reflected in the fixed sequence of their description of self-evaluative attempts. Examples of this style relate to the necessity of defining objectives which will serve as guidelines for assessing one's teaching efforts.
4) *Reliance on external sources*: A surprisingly high percentage of statements, ranging in frequency from 3% - 39%, reflected an expressed need for external sources as aids in teachers' self-evaluation. Thus we find teachers who say: "I need the feedback of colleagues in order to assess my teaching," or: "when one sees

how complicated the issues are, it is best to turn to external experts."

5) *Emphasis on difficulties*: Several teachers who participated in this study emphasized the difficulties associated with self-evaluation. The percentage of such statements ranges from 3% - 20%. Among the difficulties mentioned were the following examples: "It is not so simple to evaluate oneself, the classroom climate may suffer as a result," or: "there is the danger of not being able to accept criticism." The last three categories deal with teachers' perception of self in the process of evaluating their work.
6) *Self-criticism*: Relatively few statements expressed teachers' readiness to view their professional activities critically. The frequency of such attempts ranged from 3% - 16%. Statements in this category were, for example: "If you believe that things are not going well, you have to examine yourself and not blame the students" or "I am able to judge what I am doing, I strive to be satisfied with my professional activities."
7) *Emphasis on self*: This is the style of teachers who put themselves in the center of the stage and tend to emphasize their own personal satisfaction in the process of self-evaluation. The percentage of such statements ranges from 3% - 25%. Statements reflecting this style are: "I am reinforced when I perceive how hard I work," or: "The question is what do you expect of yourself …"
8) *Negative image*: Sometimes the personal style of teachers discussing self-evaluation reflects a negative image of themselves. The following statements exemplify this style: "I was stunned, I understood that something is wrong with me." The percentage of such statements ranged from 1% - 14%.

Conclusions

It is interesting to note that much of teachers' concern in evaluation focuses on their pupils' achievements. This finding may be interpreted as reflecting a high developmental level in their profession (Fuller, 1969).

It seems clear that teachers exhibit different personal styles of relating to self-evaluation. Some are highly cognitive in their approach while others are predominantly affective. Some rely heavily on external support for self-evaluation while others seem to be oblivious to such possible assistance. The use made of professional terms varies greatly. Most interesting, some teachers are burdened with the difficulties of self-evaluation, while others neglect this aspect completely.

Can we detect any regularities in the personal profiles of teachers as they relate to self-evaluation? Several tendencies emerge. There seems to be a consistent positive relationship between a structured approach reflected by teachers and their use of professional terms. A possible interpretation is that awareness of professional terms allows for a more structured, systematic way of handling self-evaluation.

In almost all cases reliance on external sources is linked to a low level of self-criticism. This seems logical if one relies on others to assess one's professional activities there is no need to be too self-critical. Some relationships are inconsistent. Thus, the perception of difficulties is sometimes related to a great emphasis on self, while in other cases the opposite phenomenon emerges – the more awareness of difficulties the less emphasis on oneself.

This study is an explanatory attempt to reveal some of the personal styles of thinking about self-evaluation. It is contended that the categories for defining personal styles of teacher thinking about self-evaluation can be used in further research.

Many questions call for further investigations:

- Are teachers' personal styles in self-evaluation reflected in other evaluative contexts as well. For instance, if teachers reveal an affective approach to their own evaluation, do they use the same approach toward peers or students? If they tend to focus on negative views about their teaching, do they do the same when evaluating their students or peers?
- What is the nature of relationship between personal styles in self-evaluation and personal background or situational variables? Is there any connection between self-assessment style and professional achievement?

Insights into ways in which teachers think about evaluating their own work may provide important knowledge about professional modes of teaching. The more we learn about teachers' self-evaluation the closer we may come to the enhancement of professionalization of teaching.

Chapter 17

Teachers' Beliefs and Classroom Behaviours: Some Teacher Responses to Inconsistency

B. Robert Tabachnick

and Kenneth M. Zeichner

Summary

This paper utilizes data from a study of two beginning teachers in the United States and analyzes the strategies employed by the teachers to reduce contradictions between their expressed beliefs about teaching (in four specific areas) and their classroom behaviour. The individual and contextual factors related to the choice of a particular strategy and to its eventual successes or failures are discussed. One of the teachers sought to change her *behaviour* to create a closer correspondence between belief and action, while the other teacher changed her *beliefs* to justify behaviours that were inconsistent with her expressed beliefs.

The Problem

This report of research examines consistency and contradiction in teacher beliefs and draws upon data from a two-year longitudinal study of four beginning teachers in the United States (Tabachnick and Zeichner, 1985; Zeichner and Tabachnick, 1985) and will analyze: (1) patterns of relationship between teacher beliefs and classroom behaviours; (2) strategies employed by teachers in an attempt to bring about greater consistency between beliefs and behaviours; (3) the individual and contextual factors that influenced the relationships between teacher beliefs and classroom behaviours.

In that two-year study, we aimed to explore the range of diversity of individuals' responses to the student teaching semester and, following that, to the first year of teaching. Our point of emphasis was to discover what perspectives toward teaching were developed by individual students during student teaching and how these perspectives were influenced by the interplay of the intentions and capabilities of individuals with the characteristics of the institutions of which they became a part, first as student teachers and later as teachers. The paper will be limited to an analysis of the relationships between teacher beliefs and classroom behaviours during the second phase of our study – the study of the first year of teaching.

The construct of perspectives has its theoretical roots in the work of G. H. Mead and his concept of the 'act' (Mead, 1983). *Teaching* perspectives were defined in our study as 'a coordinated set of ideas and actions which a person uses in dealing with some problematic situation.' This view of perspective is derived from Becker et al. (1961). According to this view, perspectives differ from attitudes since they include actions and not merely dispositions to act. In addition, unlike values, perspectives are defined in relation to specific situations and do not necessarily represent general beliefs or teaching ideologies.

Teaching perspectives were defined in relation to four specific domains: *(1) knowledge and curriculum, (2) the teacher's role, (3) teacher-pupil relationships, and (4) student diversity*. Each of these four categories was further defined in terms of several specific *dilemmas* of teaching that had emerged in the analysis of our data from the study of student teaching (e.g., public knowledge vs. personal knowledge; knowledge as product vs. knowledge as process). Altogether 18 dilemmas of teaching were identified within the four categories of perspectives, and it was these dilemmas that gave direction to our data collection efforts during the second phase of our study.

A key assumption underlying the use of teaching perspectives as the organizing construct for our study is that teacher behaviour and thought are inseparable and part of the same event. We assume that the meaning of teacher thinking cannot be understood in the absence of analyses of behaviour engaged in by the actors to complete the ideas, to 'express' them. Thinking and beliefs are, of course, not directly observable. We assume that classroom behaviour expresses teacher beliefs in a way similar to the use of language to answer the question, "What are you thinking?" or "What were you thinking when you did that?" It may be that classroom behaviour is a way of thinking about teaching analogous to the craftsperson or artist who 'thinks with his (or her) hands.' The interest here moves beyond a concern with either teacher thinking or teacher behaviours represent active expressions of thought and the ways in which teacher behaviours represent apparent contradictions of expressed beliefs. We are interested in knowing if teacher behaviour and beliefs move toward some kind of internal consistency over time. What appear to be contradictions between behaviour and belief are often revealed as more consistent, from the teacher's point of view, when behaviour is thought about as a statement of belief.

We utilize the data from our earlier study to probe instances of contradiction and consistency between what teachers say they believed (e.g., about the role of teacher, knowledge and curriculum, etc.), their expressions of intent for particular classroom activities, and their beliefs as expressed in their classroom behaviour. After identifying the strategies employed by the two teachers in an attempt to bring about greater consistency between belief and action, we discuss the various individual and contextual factors in each case that influenced the relationships between teacher beliefs and behaviours.

Much of the research that has been conducted to date on the relationships between teacher beliefs and classroom behaviour has established that there are fairly close relationships between teacher thought and behaviours (e.g., see Shavelson and Stern, 1981). However, (1) most studies have relied almost exclusively on teacher self-reports of their behaviours and not on analyses of observed teaching; (2) few studies have explicated the processes by which behaviours and/or beliefs are modified by teachers in an attempt to move toward greater internal consistency. The paper addresses both of these issues.

Methodology

The subjects for this study are two female first-year teachers who were employed in different school districts in the United States during the 1981-82 academic year. These individuals were selected from a representative group of 13 individuals who had been studied intensively during their student teaching experience at a large midwestern university the previous spring (Tabachnick and Zeichner, 1985). Both of the teachers taught at the eighth-grade level.

Between August 1981, and June 1982, we spent three one-week periods observing and interviewing each teacher. A specific research plan was followed during each of the three weeks of data collection. During four days of each week, an observer constructed narrative descriptions of events in each classroom using the four categories of perspectives and related dilemmas as an orienting framework. Each teacher was interviewed several times each day regarding her plans for instruction (e.g., purposes and rationales for particular activities) and her reactions to what had occurred. One day each week, an observer constructed a narrative description of classroom events with a particular focus on six pupils in each classroom who had been selected to represent the range of student diversity in each classroom.

In addition to the daily interviews with each teacher that focused on particular events that had been observed, a minimum of two in-depth interviews were conducted with each teacher during each of the three data collection periods. These interviews sought to explore teachers' views regarding their own professional development in relation to the four orienting categories of perspectives and also addressed additional dimensions of perspective unique to each teacher that had emerged during the year.

Additionally, we sought to investigate the influence of several institutional elements of school life on the development of teacher perspectives (e.g., school ethos and tradition, teacher culture, administrative expectations about the teacher's role). During each of the in-depth interviews, we also asked the teachers about their perceptions of the constraints and encouragements that existed in their schools and about how they learned what was and was not appropriate behaviour for teachers in their schools. We also interviewed each principal at least once and interviewed two other teachers in each school concerning their views of the degree to which each beginning teacher was free to employ independent judgement in her work. Finally, we also collected many kinds of formal documents in each school, such as curriculum guides and teacher handbooks.

Through the classroom observations and teacher and administrator interviews we sought to monitor the continuing development of teaching perspectives and to construct in-depth portraits of life in each of the classrooms. Tape-recorded interviews and classroom observations were transcribed to facilitate a content analysis of the data. Several analyses of these data led to the construction of case studies that describe the development of each teacher and the individual and social influences on their development from the beginning of student teaching to the end of their first year of teaching. The paper will draw upon the induction year portions of these case studies to examine the relationships between teacher beliefs and classroom behaviours.

Beth: Thinking About Teaching in a Closely Controlled School Environment

Beth was a student teacher in a middle-sized city (about 200,000) in a self-contained fifth-grade classroom, in an elementary school with kindergarten to fifth grade. In that community, this meant that her cooperating teacher was responsible for instruction in all subjects except art, music, and physical education. The prevailing style of teaching in her classroom was characterized by warm personal relationships, and some judicious sharing of curriculum decisions with pupils. Though most of the teaching was fairly routine (reading to answer questions about the text, drill in arithmetic), there was a genuine effort to encourage pupils' creative thinking and problem solving. Beth was encouraged to invent activities that would further these more diffuse goals as well as to further routine classroom learning activities with more precisely targeted goals. Students were from a mixed socioeconomic background, mostly middle class, but some from economically poorer homes. The principal supported an 'active' curriculum which challenged and displayed the results of pupils' creative efforts. The teachers and principal believed that they had firm community and parent support for such an approach.

As a first-year teacher, Beth taught eighth graders in a middle school enrolling pupils in grades six through eight. The school served a middle-class community suburb to a moderately large city. Very few homes could be characterized as near poverty level. The school's organization was quite different from Beth's school during her previous (student teaching) year. Groups of 75 to 100 children were taught all the subjects by teams of three or four teachers. Art, music, and physical education were taught by specialists, and other specialists were available for advice on teaching reading and language and for help in working with poorly achieving or psychologically disturbed children.

Beth and her two co-teachers together taught approximately 80 pupils. Their teaching was directed by lists of 'Performances' in each subject. The curriculum was referred to by the teachers and the principal as Performance Based Education (PBE) with pupil achievement being judged on the basis of Criterion Referenced Tests (CRT's). The lists of performances and the CRT's had been developed some years before by committees of teachers. Bureaucratic difficulty discouraged teachers from changing or adding topics. A CRT identified student inabilities, mainly in reading, language, and mathematics skills and in social studies and science information. Beth and her colleagues decided which of the teachers would be responsible for different groups of students in each subject, for the timing of instruction, and the scheduling of tests. Deviation from these time plans was discouraged. For example, taking longer to explore a topic or 'going off on a tangent' (adding topics not specified in the PBE lists) might force one's colleagues to wait and waste time, since all pupils had to be tested at the same time. The school was built to an architecturally open plan so teachers could easily keep track of what was happening in other areas of the 'pod.' The principal frequently walked around the school and did not hesitate to discipline students or to point out to teachers deviations from established school procedures, either on the spot or in a later conference.

At the beginning of her first year as a regular teacher, Beth refers appreciatively to her student teaching when, from time to time, she decided on a topic to be taught, researched its content, and invented teaching strategies. Beth says she believes an 'open and easy' approach to teaching is valuable because it stimulates pupils to think. In some interview statements, she refers to the routine, or at least 'follow-the-present-pattern' nature of her teaching. In other statements she says she selects some of the topics for study, aims at stimulating pupils to 'sit down and think about things,' tries to think of ways to present the content that will capture the interest of pupils. However, she is observed to teach in a very controlled style. Her planning at the beginning of the year is almost entirely limited to deciding which textbook pages to use in working with groups of 10 to 25 pupils; which and how many math solutions to demonstrate, whether to repeat teaching on a topic or go on to the next item on the PBE list. (Beth is under considerable strain at first, finding her way into the system. This is noticed by the principal who tries to get her to relax, boosts her self-confidence).

One instance is recorded, in five consecutive days of observation, of the 'open and easy' style of teaching; the most capable math group is encouraged to find alternative solutions to problems. The pupils respond eagerly and Beth smiles and says to the observer, "I love this!" But the bulk of her teaching behaviour follows from (1) earlier decisions about how many questions to ask or problems to explain; (2) on-the-spot reactions to time remaining, to student actions (redirecting misbehaviour, answering questions, correcting errors with on-the-spot explanations); and (3) the existence of available materials (booklets, film strips) with previously developed worksheets or test questions. Post-teaching behaviour is mainly correcting tests and selecting the next day's questions, worksheets, and drill practice.

At mid-year, five days of observation reveal no equivalent to the exciting math lesson. All the observed teaching is guided by getting through the PBE lists of objectives. Beth says the main influences on what happens in her classroom are:

> '... the school curriculum in that they say what should be taught ... us pod teachers in deciding who teaches what ... and then me, myself as a teacher, as in how I'm going to teach it'.

Selecting or identifying goals is *not* an important effort. She says her goals:

> '... [are] real sketchy ... I really don't have any big ones set out ... I'd like them to understand what I'm talking about, sure ... and to retain some of the things that I've taught, definitely. But that would be it for goals'.

Beth says she is satisfied with the amount of freedom she has to control what happens in her class, "It sets out things you should be doing, which is nice," she says, "because you know what's expected." She comments that she can generally teach the kind of curriculum that she thinks is important, "as long as it includes what has been set out for me to teach."

At the same time, Beth says she thinks her talents are under-utilized. She says,

> 'School isn't just the place for basic learning, you know; the teacher talks and you learn or absorb it. [It should be] more of an interesting kind of place ... but it's just not coming through anymore. I guess I just don't take the time to sit down and think about it like I used to. Or I don't have the time to design some of the things that I designed that were really neat'.

Asked about prep time, Beth says she has enough.

At the end of the school year, Beth's teaching is observed to have changed little from the mid-year description, except that she is more self-confident and practiced in implementing the PBE curriculum. With the end-of-year tests to face, all of her classroom behaviour is focused on getting her pupils to perform well. Observations describe days filled with assigning drill and practice, giving information, and testing recall.

Beth's statements about her thinking during planning, during teaching, following teaching, have changed in that they no longer contain references to selecting topics, aiming to stimulate pupil creative thinking and pupil reflection, as had appeared in earlier statements of that type. She begins with the PBE lists of objectives, uses materials for which there are information and recall exercises (reading, social studies, science) or decides which and how many math solutions to present, choosing items from the textbook to illustrate. Decisions are often made on the spot regarding what to say about a math problem or what questions to ask, for example, about a story or a section of a science booklet. Consideration for team decisions about time schedules are the strongest determinant for whether to extend or abbreviate teaching, give more time to slower learners or not.

What have also changed are Beth's statements about her perspective toward teaching and about what she thinks she *should* be doing. Her earlier statements of belief placed high value on planning for active ('hands on') learning by pupils, with teacher research into content in order to invent activities that will challenge pupil thinking and stimulate pupil interest. Her statements of belief now indicate that she has learned that she can be successful as a teacher without doing much detailed planning and without the need to do much (or any) research on the topics she intends to teach. Presumably, she finds enough in the Teachers Guides and the pupil materials to support the explanations and presentations she gives.

Beth's thinking about teacher classroom behaviour has also changed in that she no longer sees much value in open discussions and 'hands on' pupil activities. She intends to move more quickly the following year, to spend less time explaining the work, leaving out discussions of topics which are not 'on the test' and 'covering more areas,' especially areas that are tested.

Hannah: Thinking About Teaching in a Loosely Managed School

Hannah was a student teacher in a small village located near a middle-sized city (about 200,000) and worked as part of one of two fifth/sixth-grade teams in a grade 4-6 middle school enrolling about 500 children. There were four teaching teams in this school, each one of which was responsible for the instruction of approximately 120 children in all subject areas except art, music, and physical education. Hannah worked on a team with four certified teachers and had her own classes of around 30 pupils for each subject. During a typical week, she taught almost all of the 120 pupils on her team, since the instructional program was very departmentalized. The school community included few minorities and had a mix of parents ranging from a few who were very poor to some who were highly paid professionals. The majority of the parents were moderately well off financially.

Hannah was expected to follow very closely the highly structure curriculum of the school in all subject areas. She was provided with lists of specific objectives in each subject, which she was expected to cover and with all of the materials and tests that she was expected to use. She was also expected to cover this curriculum within specified blocks of time and had very little choice about when subjects would be taught and for how long. Because of the open architectural design of the school where no walls separated the classrooms, all of Hannah's activities were totally visible to the other members of her team. She was told that very little noise and pupil movement would be tolerated so that the classes would not disturb one another. Hannah was generally provided by her colleagues with models of very formal and distant pupil-teacher relations.

Throughout the semester Hannah questioned the departmentalized school structure, the rationalized curricular form, and the distant and formal relations between teachers and pupils which were a part of the taken-for-granted reality of her school and felt she was being asked to fit into a teacher role that she did not like. Despite isolated efforts, which continued throughout the semester, to implement what she felt was a more varied and lively curriculum and to relate to her pupils in a more personal way than was common in her school, Hannah for the most part outwardly complied with the accepted practices in her school and did not act in a manner consistent with her expressed beliefs. At the end of the semester, despite the lack of confirmation from her experience as a student teacher, Hannah was more convinced than ever ('having learned a lot of things of what not to do') that warm and close relations between pupils and teachers, getting kids excited about learning and feeling good about themselves as people (e.g., by integrating their personal knowledge into the curriculum) were the keys to good teaching.

Hannah's first year in a regular teaching position was spent as the only eighth-grade teacher in a nine-classroom K-8 public school enrolling about 190 pupils. The school was located in a rural farm community a few miles outside of a small city with a population of 9,000. Hannah taught all subjects except civics to her eighth-grade class and taught science to the seventh-grade class. The parents of the children in her class were very diverse socioeconomically, ranging from those who were farm owners and professionals to those who were farm workers. All of the teachers lived in the immediate area with the exception of Hannah and one other teacher who commuted from a city 45 minutes away. Hannah was the youngest and the only first-year teacher in the school and the only one who had not completed a teacher education program at one of the relatively small state teacher's colleges, which were now part of the state university system.

The culture, tradition, and organization of this school were quite different from the school in which Hannah completed her student teaching. On the one hand, there was a very strong tradition of individualism in the school that sanctioned each teacher's right to things in his or her own way, and there was very little cooperation or coordination among the staff. All of the classrooms with the exception of Hannah's and the seventh-grade class were very self-contained, and each teacher was responsible for all of the instruction for a group of around 25 students. The principal of the school was also a full-time teacher and did not observe or confer with teachers except during weekly staff meetings.

Consistent with the individualistic tradition of the school, very few overt controls were exerted on teachers with respect to the planning and teaching of the curriculum. Teachers were given curriculum guides and textbooks for each subject area and were permitted to cover the content specified in the guides in whatever order, at whatever pace, and with whatever methods they thought were most appropriate. Teachers were also free to supplement the texts with any other materials and to go beyond what was listed in the curriculum guides as long as the curriculum was covered by the end of the year.

The only explicit controls that were placed on teachers' handling of the curriculum were in the areas of grading and testing. All of the teachers were expected to give each child 30 'marks' for each subject during each of three report periods and to grade pupils' work according to a standard grading scale. A great deal of emphasis was also placed upon pupil performance on a national standardized test given each spring.

Alongside the tradition of individualism in the school, there was also a very strong and mostly unspoken agreement amongst all but Hannah and one colleague in the seventh-grade class about the ways in which teachers should relate to their pupils. This approach was characterized by one teacher as "the old school method … you can't have someone here who is too soft with the kids." Hannah became aware of this consensus on teacher-pupil relations ("In this school it's the teacher's role to be the disciplinarian.") through observations of other teachers, through her pupils' comments, and indirectly through the school 'grapevine.' Other teachers would rarely confront Hannah directly with criticisms of her more informal style of relating to pupils. On several occasions, however, teachers complained to the principal, who in turn passed the word to Hannah that she had violated the preferred formality between teachers and pupils. All of the classrooms with the exception of the seventh and eighth grades were very tightly controlled by teachers, and this strong, informal agreement among the staff initially made Hannah feel isolated and alone.

> 'You begin to try new things; everything is not out of the textbooks or worksheet oriented. They look down on that. But they don't constrain you and say you can't do things. They would never say you can't do things. They'll do it in a roundabout way … when it comes back to you; you feel that everyone else is against you'.

The community was characterized by Hannah and several other teachers as extremely conservative, suspicious of new ideas, and as holding expectations for teachers to maintain very tight controls over pupils. Hannah initially felt more pressure from the parents than from her colleagues to conform to the unspoken tradition regarding the teacher's role and was initially reluctant to act on her intuitions because she felt that she was perceived as an outsider. From the beginning of the year, Hannah made many efforts to win the trust and confidence of the parents and to learn more about the ways and mores of the community.

At the beginning of the year, despite the lack of close supervision and formal controls, Hannah relied heavily on the textbooks in planning her curriculum; however, she also made efforts from the very beginning to establish warm and close relationships with her pupils in violation of the school's tradition. Hannah continued to describe her basic orientation to teaching as 'humanistic' and emphasized the affective and interpersonal dimensions of her work. She felt strongly that a positive self-concept is the key to learning and wanted to find ways to make school enjoyable for herself and her pupils. Hannah tried very hard to present herself to her pupils as a 'human being' by openly admitting her mistakes, her ignorance with regard to content, and by freely sharing aspects of her personal life with her pupils. She also made many efforts to understand the personal lives of each child in her class and to gain her pupils' trust and confidence.

Initially, Hannah's pupils were very suspicious of her efforts to break down the conventional barriers between teachers and students, and there was a lack of support from her colleagues. Hannah became confused and uncertain in the fall about the direction she should take, and established several classroom practices and rules that violated her own vision of 'humanistic' teaching. Despite these isolated instances where Hannah flirted with more conventional methods of controlling her pupils, for the most part she exerted relatively little direct control over pupil behaviours, and the pupils gradually began to respond to her efforts.

Despite her efforts to establish warm and personal relations with her pupils, who were gradually becoming increasingly successful, Hannah was frustrated with her heavy reliance on textbooks and with her inability to establish a more varied and lively instructional program. While she was very sure of herself in dealing with children in interpersonal matters, she felt that she did not have a clear idea of how to implement her expressed preference for a more integrated curriculum which incorporated children's personal experiences, which gave pupils concrete experiences in relation to ideas, and which elicited their enthusiasm and excitement about solving problems in relation to the world around them. "I just feel like I'm spoon feeding them and opening their heads and pushing the knowledge in."

Knowing that her pupils had been taught 'right out of the textbook' in the past and that they would probably be taught so in the future, and not confident that she was able to explain to others how particular methods were meeting specific academic goals, Hannah worried a lot about handicapping her students and about not giving them what they were 'supposed to learn.' By December, Hannah was so frustrated that she considered quitting teaching and accepting another job outside of education.

As the year progressed, Hannah became more and more satisfied with her classroom program, and her actions began more and more to reflect her expressed beliefs about teaching. She continued to rely mainly on the texts in planning her lessons, but she gradually made increasingly independent decisions, which resulted in a greater emphasis on providing concrete experiences for children and on incorporating their personal lives into the curriculum.

By April, Hannah felt confident enough to drop the basal readers, to have her pupils read novels, and to let two pupils teach a unit on engines to the class that drew upon their experiences in repairing farm vehicles. Throughout the year Hannah continued to expose all of her pupils to the same curricular content and stayed fairly close to the texts in some subjects (e.g., math), but her work in language, reading, and science reflected more and more of the active pupil involvement and problematic approach to knowledge that she had hoped to create since the beginning of her student teaching. By the end of the year, Hannah felt that she had come closer to her ideal where pupils are thinking critically and constantly and where they are always asking questions and trying to apply their in-class learnings to everyday life.

There were several reasons why Hannah was able to move from a point in December where she considered quitting, to a feeling of satisfied accomplishment at the end of the year. Among these were: (1) the support she received from her one teacher ally, the seventh-grade teacher; (2) her ability to mobilize parent support for her classroom program; (3) the pupils' traditions of mutual peer support and the warm acceptance of Hannah as a 'teacher-friend'; and (4) her pupils' success on the national standardized test (scoring the highest of all of the eighth grades in the district). Because of this support from the pupils, parents, and the seventh-grade teacher, and because of Hannah's determination, her skills in dealing with people, and her sensitivity to the political dimensions of schooling, she was able to significantly redefine aspects of her school in relation to her own class and to modify her behaviour to create more consistency between her beliefs and actions.

Hannah maintained her beliefs regarding the importance of 'humanistic' teaching throughout her student teaching and her first-year of teaching with little or no formal support from her schools and gradually, as her pupils and their parents began to respond positively to her approach, Hannah was able to find ways, by acting on her intuitions and through trial and error, of modifying her behaviour to bring it into closer agreement with her beliefs about teaching.

Conclusion

Our conception of 'perspectives toward teaching' is similar to what Clark and Peterson (1986) refer to as 'teacher beliefs and implicit theories.' There is some difference, since we treat classroom behaviour as an expression of a teacher's beliefs or implicit theories about teaching and learning. The teachers we studied were also often able to articulate explicit theories of teaching; they often were aware of their beliefs and were ready to explain and justify them.

At the beginning of her first year as a teacher, Beth made statement of belief about teaching that contradicted or were inconsistent with each other. Her teaching behaviour was inconsistent with those statements of belief that referred to the need for active learning and creative problem solving. The teaching behaviour was consistent with a belief in the value of a curriculum that encouraged pupils to learn prespecified information and skills. As the year passed, Beth's statement of belief contained fewer and fewer of the statements about the value of pupils' creative problem solving. Beth's beliefs changed until they were characterized by statements that affirmed and justified her teaching behaviour; while her teaching behaviour remains essentially the same throughout the year, it is more completely expressive of her statements of belief by the end of the school year.

Hannah also created closer agreement, as the year progressed, between her verbal and her behavioural 'statements' of belief about teaching. She monitored her classroom behaviour, modifying it to bring it into agreement with her beliefs about teaching. Her early lack of success led her to toy with the possibility of abandoning her beliefs (and abandoning teaching altogether), but by the end of the year, she had reaffirmed her earlier commitments to an activity-oriented curriculum that encouraged pupil independence, initiative, and creative problem solving. At no time did her ideas or her behaviour waver in revealing her belief that it was necessary to know children as people – and to be known by them as a person – in order to teach them successfully as pupils.

Both teachers reduced the inconsistencies in their statements of belief but used quite different strategies to do so. Partly, that was a result of their personal characteristics and history, their capabilities, their willingness to risk, their strength of commitment to a particular professional position. Hannah was both intuitively and consciously skilled in managing the political and social context of her classroom, her school, and her school's community. She was also willing to make the effort. Beth avoided 'political entanglements' and was content to affirm principles of action that she seemed to reject early in the year, but whose affirmation created solidarity between her and her co-teachers and the principal.

The schools offered very different opportunities to exercise professional judgement. Edwards' (1979) analysis of methods of control of a workplace is helpful in recognizing differences in the two schools. Hannah's teaching principal had little opportunity to control teaching behaviour. In addition, efforts at control would have violated that school's informal cultural norms of independence (at least for adults). Beth's principal was able and willing to exercise control over what happened in the school. Bureaucratic control through the social arrangement of teaching teams was powerful in Beth's school but weak in Hannah's school, in which teachers could operate more independently behind their closed classroom doors. Control by technical elements – the physical structure of an open architectural plan that made it easy to monitor teacher behaviour, the specificity of a PBE curriculum – was present for Beth but absent for Hannah. Indeed, under the conditions of strict control that characterized her student teaching school, Hannah suppressed the expression of her ideas as behaviour, while reaffirming them verbally. Edwards' theory of control does not account for the presence of the informal school cultures in both schools which either encouraged conformity or else encouraged independent teacher action.

Teacher thinking as described in our study was not merely the result of an individual's personal history and psychological state. However, apparently highly context specific, thinking was not merely shaped by the sociopolitical conditions in the school. Rather, we discovered that in both cases the move to great consistency between belief and behaviour was the result of a negotiated and interactive process between individuals and organizational constraints and encouragements.

Chapter 18

Role Over Person: Legitimacy and Authenticity in Teaching

Margaret Buchmann [17]

Summary

This paper discusses competing norms for justifying teacher decisions, their effects on productivity and legitimacy in teaching, and the teaching profession as a moral and learning community. Drawing on philosophical analyses and studies of elementary and secondary schools, teacher preparation, staff development, and the adoption of innovations, it argues that personal orientations (centering on personal habits, interests, and opinions) remove teacher decisions from the realm of criteria for judging appropriateness. Personal reasons have explanatory value; they carry less weight when justifying professional action. Role orientation involves references to larger, organized contexts, including the disciplines of knowledge, group purposes, and societal issues.

[17] This chapter has been adapted from "Role Over Person: Morality and Authenticity in Teaching" which originally appeared in *Teachers College Record*, Vol. 87, No. 4, Summer 1986. Reprinted by permission of Teachers College, Columbia University. The author gratefully acknowledges Robert Floden and John Schwille who made valuable comments on various drafts of this paper. This work is sponsored by the Institute for Research on Teaching, College of Education, and Michigan State University. The Institute for Research on Teaching is funded primarily by the Program for Teaching and Instruction of the National Institute of Education, United States Department of Education. The opinions expressed in this article do not necessarily reflect the position, policy, or endorsement of the National Institute of Education. (Contract No. 400-81-0014)

Choice in Teaching

What teachers do is neither natural nor necessary but based on choice. Since choice may harden into custom or dissipate into whim, one asks for justification; it is a way of assuring that teaching will periodically pass muster. In justifying their actions, people give reasons. For teachers, personal reasons can be appropriate when explaining a given action to others, but they carry less weight in considering the wisdom of an action or decision. In other words, some contexts call for explanation and others for justification. When one wants to understand why someone did something, one wants to know what actually motivated him or her. But if one wants to know whether what was done was right, one wants to hear and assess justifications. Here it is important that the reasons are good reasons, and it becomes less important whether they were operating at the time.

The question, then, is what counts as good reasons in teaching. I argue that for many teacher actions, personal reasons are subordinate to external standards and that the scope of these actions is much broader than people often assume. Providing acceptable justifications requires the existence of a community to both set standards for adequacy and to determine a set of rules for guidance. The role obligations of teachers as members of such a community forge bonds that not only ensure compliance but also generate effort and involvement.

Curriculum decisions may be at the top of the list of teacher actions for which one should expect adequate justifications, for, as Scheffler points out, it is not 'a matter of indifference or whim just what the educator chooses to teach. Some selections we judge better than others; some we deem positively intolerable. Nor are we content to discuss issues of selection as if they hinged on personal taste alone. We try to convince others; we present ordered arguments; we appeal to custom and principle; we point to relevant consequences and implicit commitments. In short, we consider decisions on educational content to be responsible or justifiable acts with public significance'. (Scheffler, 1977, p. 497)
But decisions about the social organization of the class, how to deal with parents, and how to treat requests (or directives) from school administrators are also examples of teacher actions that are responsible acts of public significance. It is useful to recall the root meaning of responsibility; being a respondent has to do with one's answering for things and defending a position.

Personal reasons – centering on one's habits, interests, and opinions – are relevant for considering the wisdom of actions where the question is what the individual per se wants to accomplish, but not for professional situations where goals (and perhaps a range of means) are a given. People accepting professional roles are in the latter situation, and one must ask whether their particular actions and general dispositions are enacting and conforming to given standards and goals. Such people have no right to decide whether to act on their clients' behalf and in their interests: it is their obligation to teach school, put a leg in a cast, or appear in a court of law. This is why a professional's most significant choice is whether to take on the role (Fried, 1978).

What is close to people is always important to them; the personal will take care of itself. However, professional aspirations, responsibility, and curricular subjects with their pedagogies must be learned. Tendencies in teacher preparation and staff development to stress individualism, self-realization, and the personal – even idiosyncratic – element in teaching are therefore problematic. This would be true in any case. But such tendencies are extremely questionable in American education, where structural features (e.g., recruitment, induction, rewards) and the ethos of the profession already converge in conservatism, presentism, and individualism. The point is that attention to role is especially important for American teachers because it goes against many potent forces.

An understanding of teacher orientations (role versus personal) and their effects is particularly important now when there is a strong press to set policies that will improve schooling in the United States. It is well recognized that teachers have the final word on exactly what will be done in the classroom and what the actual curriculum will be (for a review, see Brophy, 1982). This implies that making good policy requires knowing how teachers are likely to act in answer to policy initiatives and why (Wise, 1979). It requires, furthermore, thinking about those competencies and dispositions that teachers should have as professionals (Kerr, 1983; Sykes, 1983).

Teaching as a Role

It is crucial to appreciate the fact that 'teacher' is a role word. Roles embody some of our highest aspirations and provide social mechanisms for shaping action in their light. They are parts people play in society and do not describe individuals. Teacher obligations – those behaviors and dispositions that students and the public have a right to expect of teachers – actually have three important aspects that have no personal reference or connection. First, these obligations do not depend on any particular individuals (teachers or students). Second, they apply regardless of personal opinions, likes, or dislikes. Third, they relate to what is taught and learned. In schools, teachers are supposed to help students participate in 'the community of subject matter' whose objective contents of thought and experience – systems, theories, ideas – are impersonal because they are distinct from the people who learn or discuss them. They are, to some extent, independent of time and place (Hawkins, 1974; Polanyi, 1962).

In an immediate sense, teachers have obligations toward their students; these obligations center on helping them learn worthwhile things in the social context of classrooms and schools. The view of students as learners underlies the distinctive obligations of teachers; and role orientation in teaching by definition means taking an interest in student learning. Thus, insofar as teachers are not social workers, career counselors, or simply adults who care for children, their work centers on the curriculum and presupposes knowledge of subject matter. This does not exclude their caring about children or being a person in their role.

Roles also indicate obligations toward more remote communities; in teaching, these communities include the profession, the public, and the disciplines of knowledge. For instance, while it is important to communicate the fact that disciplinary knowledge is not absolute, teachers have to recognize and respect the constraints imposed by the structure of different disciplines on their decisions about how to teach, for: If a structure of teaching and learning is alien to the structure of what we propose to teach, the outcome will inevitably be a corruption of that content. (Schwab, 1978, p. 242)

Since teachers are supposed to look after the educational interests of children, they have to learn to live with the fact that they are not free to choose methods, content, or classroom organization for psychological, social, or personal reasons alone.

The teacher educator slogans of 'finding the technique that works for you,' 'discovering your own beliefs,' 'no one right way to teach,' and 'being creative and unique' (see, e.g., Combs, 1967; Goodman, 1984) are seductive half-truths. They are seductive because anyone likes to be told that being oneself and doing one's own thing is all right, even laudable. Conduct sanctioned in this fashion – while consistent with professional discipline for those who already have the necessary dispositions and competencies – allows for both minimal effort and idiosyncrasy in other cases. These slogans are half-truths because – although identifying teachers' personal and commonsense beliefs are important – once identified, these beliefs must be appraised as bases and guides for professional conduct and, where necessary, changed.

Professional socialization marks a turning point in the perception of relevant others and of oneself, yet a reversal of prior conceptions is less clear cut and typical in teaching than in other professions (Lortie, 1975). Formal socializing mechanisms in teaching are few and short in duration, not very arduous, and have weak effects. The lengthy, personal experience of being in school as a student, however, provides a repertoire of behaviours, beliefs, and conceptions that teachers draw on. Where it is successful, genuine professional socialization trains attention on the specialized claims that others have on one. Thus, the teaching role entails a specific and difficult shift of concern from self to others for which the 'apprenticeship of observation' (Lortie, 1975) provides no training. Highet (1966) describes the nature of this shift:

> 'You must think, not what you know, but what they do not know; not what you find hard, but what they will find hard; then, after putting yourself inside their minds, obstinate or puzzled, groping or mistaken as they are, explain what they need to learn'. (p. 280)

In general, a shift of concern from self to others comes more from acknowledging, 'This is the kind of work I am doing,' than from stating, 'This is how I feel,' or 'This is how I do things.' Subjective reasons refer to personal characteristics and preferences. They are permissive rather than stringent, variable rather than uniform. Appraisal requires distance, but detachment is difficult where things are simply seen as part of oneself. A danger is that personal beliefs and preferences are 'no longer easily accessible to reflection, criticism, modification, or expulsion' (Schwab, 1976b, p. 37). This explains the air of finality that many subjective reasons have. Yet it is not that personal beliefs and preference must necessarily be misleading or selfish, but that – where such criteria rule – other and more legitimate concerns may become secondary (Lortie, 1975). This reverses the relation in which personal and professional reasons should stand in teaching.

Subjectivity and reasonableness

When people say, 'This is the kind of person I am,' they mean to close an issue and put an end to debate, whether the issues have been satisfactorily resolved or not. An emphasis on the self can block the flow of speculation, conversation, and reflection by which people shape habits of action and mind that affect others or the self; it means cutting oneself and the collective off from some of the most valuable human resources. Nor are teachers' expectations to the rule that not everything people want is good. Imperviousness and finality – of feeling, belief, or habit – interfere with learning and with getting better at helping others learn.

Justification is always tied to reason and susceptibility to reason; teaching is special in the sorts of reasons that are acceptable. Professional decisions are tied to the public realm where they are constrained by facts and norms, both of public knowledge. Put differently, justification needs to reach beyond the particulars of teachers' own actions and inclinations to consider larger, organized contexts relevant to their work, such as the disciplines of knowledge, laws, and societal issues (Thelen, 1973). Moreover, teachers need not be creative to be reasonable. Rather, they must be willing to act in accordance with rules, submit to impersonal judgement, and be open to change for good reasons. To call an action or person reasonable still is praise, for reasonable people are neither inconsiderate nor rash, and their actions are unlikely to be futile or foolish (Black, 1972).

Caprice and habit cut teaching off from thought, particularly from its moral roots. In cause and origin, caprice is inherently self-contained; it contrasts with cultivation or improvement by education, training, or attentive labor. Habit is the opposite of impulse, and it confines in a different way. Yet, caprice and habit are alike in that they both allow for action without adequate reason, removing teacher actions and decisions from the realm of criteria for judging appropriateness. Part of reasonableness is the habit and capacity of giving due weight to evidence and the arguments of others who may offer new data or alternative explanations.

Workplace isolation and role orientation

Teaching is lonely work in the United States. Controls are weak and standards low, rewards uncertain, often elusive (Lortie, 1975). While an inner transformation from person to teacher may be wanting, one can still get a job teaching school. There is a sense of 'easy come, easy go' in teaching; such transiency does not support a sense of community. Tenure and salary are based on years of service rather than competence or commitment. An active interest in student learning does not come with teaching experience, as some teacher development theories seem to suggest (see, e.g., Fuller, 1969). To the contrary, teaching seems to have a calcifying effect on teachers (McLaughlin and Marsh, 1978; Waller, 1932/1961). The teaching career is flat, not providing sufficient opportunities for changes in responsibilities and professional renewal. Together with the uncertainties of teaching, all these things can affect even dedicated teachers. Thus, Sizer (1984) describes the feelings of Horace, a veteran teacher of 28 years:

> 'He is so familiar with the mistakes that ninth-graders make that he can sense them coming even before their utterance. Adverbs are always tougher to teach than adjectives. What frustrates him most are the partly correct answers; Horace worries that if he signals that a reply is somewhat accurate, all the students will think it is entirely accurate. At the same time, if he takes some minutes to sort out the truth from the falsity, the entire train of thought will be lost. He can never pursue a one student's errors to completion without losing all the others'. (p. 13)

The organization of public schooling in America isolates teachers from one another, and there is a lack of a common language and shared experiences. Hence, it is difficult to develop a role orientation that one would be able and willing to use in the justification of teaching decisions and actions. In addition, what does the 'inner self' do when left unwatched and deprived of rules of conduct based on external standards and role-specific sanctions? Anything that comes to mind? The degree to which one's behaviour can be observed and one's beliefs examined by relevant others is crucial in role performance and professional discipline. As Merton (1957) argues, 'If all the facts of one's conduct and beliefs were freely available to anyone, social structures could not operate' (p. 115); however, insulation can lead people astray, for 'the teacher or physician who is largely insulated from observability may fail to live up to the requirements of his status' (p. 115). Where workplace structure insulates individuals, they are also less likely to be subject to conflicting pressures – simply because what they do is less well known.

With increasing size and a continuing accumulation of formal policies, schools are becoming public-service bureaucracies. Teachers adapt to conflicting policies and endemic uncertainties as best they can. These adaptations can result in private, intensely held redefinitions of the nature of teaching and of the clientele. In resolving the tension between capabilities (often constrained by workplace demands) and objectives, individuals may lower their goals or withdraw from attempts at reaching them altogether. In responding to a diverse clientele, they may reject the norm of universalism and discount some groups as unteachable. Because such private, personal conceptions can help individual professionals placed in difficult situations, they tend to be held rigidly and are not open to discussion. In addition, though modifying one's conception of students is private, the content of typical coping responses is likely to reflect prevailing biases (Lipsky, 1980). There is thus a troubling relation between the development and persistence of inappropriate coping strategies in teaching – including racial, cultural, and sexual stereotypes – and the relative likelihood of staying on the job.

Role orientation as a disposition can steady teachers in their separate classrooms, helping them call to mind what their work is about and who is to benefit from it. A disposition is a special kind of orientation. While 'to orient oneself' means to bring oneself into defined relations to known facts or principles, a disposition is a bent of mind that, once it is in place, comes naturally. Dispositions are inclinations relating to the social and moral qualities of one's actions; they are not just habits but intelligent capacities (Scheffler, 1965). With role orientation as a disposition, no extraordinary resolve is necessary to occasionally take a hard look at what one does or believes in teaching. Instead of instilling role orientation as a disposition, teacher educators often focus on the personal concerns of novices and experienced teachers.

Personal concerns and teacher learning

In examining the process of learning to teach, teacher development, and the adoption of innovations in schools, researchers and educators have identified a shift from personal to 'impact' concerns (how is my action or innovation affecting my students?) as crucial. For example, Hall and George (1978) found that among the teachers who do not use innovations are those most concerned with the implications of change for themselves personally, and Fuller (1969) identifies the emergence of concern for student learning as a culminating point in teacher development. Yet recently, Fuller's concept of personalized teacher education has been questioned, even as an approach that may lead teachers from self-oriented concerns to other-oriented concerns (Feiman and Floden, 1980). The assumption that earlier concerns must be resolved before later ones can emerge confuses readiness and motivation. Just because some concerns carry more personal and affective charge, it does not follow that other concerns – less immediate, more important – cannot be thought about. These criticisms also apply to the work of Hall and his associates (e.g., Hall,

Loucks, Rutherford and Newlove, 1975; Hall and Loucks, 1978), who base the content of interventions in staff development on teachers' concerns. Actually, teacher preparation and staff development that focus on personal concerns may have the undesirable effect of communicating to teachers that their own comfort is the most important goal of teacher education.

Zeichner and Teitelbaum (1982) draw attention to the political attitudes that a personalized, concerns-based approach to teacher preparation may promote.

> 'By advocating the postponement of complex educational questions to a point beyond preservice training and by focusing attention primarily on meeting the survival-oriented and technical concerns of student teachers, this approach (while it may make students more comfortable) serves to promote uncritical acceptance of existing distributions of power and resources'. (p. 101)

One form of conservatism is to take the given and rest – an attitude that bypasses an important source of learning and change, namely, to take the given and ask. An emphasis on personal concerns is unlikely to change the ethos of individualism, conservatism, and presentism in teaching. There is, moreover, recent empirical evidence that both elementary and secondary teachers base significant curricular decisions on personal preferences. This empirical backing for my claim that role orientation is not getting sufficient emphasis in education is sketched below.

Teacher preferences and the curriculum

At the elementary level, Schmidt and Buchmann show that the allocation of time to subjects in six elementary classrooms was associated with teachers' personal beliefs and feelings concerning reading, language arts, mathematics, science, and social studies. Briefly, average daily time allocations went up and down in accordance with (1) teacher judgements on the degree of emphasis subjects should receive and, (2) indications (self-reports) of the extent to which teachers enjoyed teaching these curricular areas. When projected over the entire school year, differences in time allocations associated with teacher preferences amounted to significant differences in the curriculum, for example 45 hours more or less of mathematics instruction, 70 of social studies, and 100 of science.

Researchers also asked teachers to indicate how difficult they found teaching the five areas of the elementary school curriculum. Findings here were mixed and thought provoking. For instance, in the area of reading, the six teachers studied did not seem to spend less time on reading just because they found it difficult to teach. However, some such tendency could be observed in language arts, social studies, mathematics, and science. However, even here the results were less than clear. The mean differences between the teaches who found it difficult to teach social studies or mathematics and who found either subject easy to teach, for example, were small. It is possible that personal difficulties experienced in teaching a subject will to some extent be neutralized by external policies or a sense of what is an appropriate emphasis on a particular subject. Also, these unclear results may be due to the fact that 'finding something difficult to teach' has two alternative senses, (1) the difficulty for children of the subject, and (2) the difficulty of the subject for the teacher. [18]

---

[18] This idea was suggested to me by Joseph J. Schwab (personal communication, December, 1984).

In a related exploratory interview study (Buchmann, 1983), 11 out of 20 elementary teachers showed some form of role orientation as they explained the ways they typically organized curricular subjects in their classrooms (integrated versus not-integrated). What united the responses of role-oriented teachers was the fact that they placed themselves within a larger picture in which colleagues, the curriculum, and accountability figured in some fashion. They looked outward rather than inward. This is not to say that they had no personal interests or preferences that influenced what they taught and how they taught it. Nevertheless, they felt bound by obligations; the personal element in their responses was framed by a sense of the collective.

Teachers demonstrating a personal orientation in their responses did not go beyond the context of their own activities. Most of them (six out of nine) explained their classroom practices by reference to themselves as persons. Their responses tended toward the proximate: affinity to self, immediate experience, the present characteristics of children. The 'language of caprice' (Lortie, 1975, p. 212) pervaded several of their responses. In cases where they recognized that the needs of some children were unlikely to be met by their approach to teaching, these teachers would still explain what they thought and did by reference to personal inclination or habitual ways of working.

A three-year study of 14 fifth-grade classrooms examining curriculum and learning in science (Smith and Anderson, 1984), concluded that teachers' reliance on personal beliefs and teaching styles hindered student learning. For example, in using a text with an unusual and sophisticated teaching strategy, teachers did not pay attention to critical information provided in the teachers' guide, depending on their previous ideas instead. In general, the researchers distinguished three approaches to teaching science that they identified by observing how teachers used textbooks and materials.

Activity-driven teachers focused on management and student interests rather than student learning; while following the teacher's guide rather closely, they omitted or curtailed class discussions meant to help students think about the science activities they were doing. Didactic teachers stayed even closer to the text that they regarded as a repository of the knowledge to be taught; their presentations, however, made little room for children's expression of their naïve scientific conceptions, which therefore remained largely unchallenged. By contrast, discovery-oriented teachers avoided giving answers and encouraged students to develop their own ideas from the results of experiments; yet, this distorted crucial intents of the text, which required direct instruction at certain points. While the texts were not perfect (failing, for instance, to spell out assumptions about teaching and learning science in the teacher's guide), the fact remains that these teachers relied on their personal approach to science teaching, with the result that the curriculum miscarried.

Cusick (1982) studied two large secondary schools, one predominately white and suburban, the other racially mixed and located in the central part of a smaller industrial region. Though there were exceptions, a self-oriented and laissez-faire approach to curriculum and student learning was typical in both schools. An American history class with a teacher who had served in World War II became a class on that European war; in a class on speech and forensics, the teacher encouraged students (mostly black) to talk about the seamier side of their personal lives – with no one listening, or teaching about speaking. A premium was put on 'getting along with kids,' and this reward structure combined with isolation from colleagues, lack of scrutiny, and an open elective system turned these schools into places where teachers and students did what felt comfortable or what allowed them to get by. Though there was a pattern to these adaptations, they happened privately. These schools were not normative communities.

Cusick concludes that the secondary teachers he studied constructed 'egocentric fields': They treated their job as an extension of self. The presumed needs of students accounted for most justifications of teaching practice ("This is the way to teach these kids"; "This is what they relate to"; or "I'm getting them ready for life"). However, curriculum and student needs were never discussed among teachers in these schools. This raises at least two important problems. First, though the freedom teachers enjoyed may bring high effort in some, other teachers can get by with doing little. Second, while able students with adult guidance may still learn worthwhile things under such conditions, others will pass through high school without learning much of anything.

In Teaching, Self-Realization is Moral

Autonomy and self-realization are indisputably personal goods. Schools, however, are for children, and children's autonomy and self-realization depend in part, on what they learn in schools. Thus, self-realization in teaching is not a good in itself, but only insofar as pursuing self-realization leads to appropriate student learning. The point is that in professional work reasons of personal preference usually will not do; this applies to nursing, soldiering, and managing a stock portfolio as well as to teaching. The idea of a surgeon keen on self-realization at the operating table is macabre. A nurse who brings up personality and preference in explaining why he changed standard procedures in dealing with a seizure would not get very far. There is no reason why such things should be more acceptable in teaching. The fact that we may have come to accept them is certainly no justification.

Everyone likes to be comfortable, free of pain and bother. However, the perspectives of psychology and profession are not the same. Things charged with personal meaning may lead nowhere in teaching. Even the integrity of self depends in part on suspending impulse. Simply declaring 'where one comes from' makes justified action a matter of taste and preference, which expresses and reinforces a massive moral confusion (MacIntyre, 1984; Shklar, 1984). In general, conscience does not reduce to sincerity: While the 'heart may have reasons of its own,' when it simply chooses to assert these without critical inspection, then reason must condemn this as complacency. (Gouldner, 1968, p. 121)

A deeper analysis of self-realization shows plainly that the self people aim to realize is 'not this or that feeling, or any series of particular feelings' (Bradley, 1876/1952, p. 160). Bradley maintains that people realize themselves morally: So that not only what ought to be is in the world, but I am what I ought to be, and so find my contentment and satisfaction. (p. 213)

The self has a peculiar place in teaching as a form of moral action; it is at once subdued and vital as a source of courage, spirit, and kindliness.

Profession requires community

What is characteristically moral presupposes community, both on conceptual and pragmatic grounds. The concept of community is logically prior to the concept of role. The very possibility of the pursuit of an ideal form of life requires membership in a moral community; it is extremely unlikely that minimal social conditions for the pursuit of any ideal people are likely to entertain would in practice be fulfilled except through membership in such communities (Strawson, 1974; Schwab, 1976a; for an excellent review of empirical literature relevant to this topic, see Purkey and Smith, 1983). Membership in moral communities is realized in action, conversation, and reflection. As a moral community, a profession

> 'is composed of people who think they are professionals and who seek, through the practical inquiry of their lives, both alone and together, to clarify and *live up to* what they mean by being professional'. (Thelen, 1973, pp. 200-201; emphasis in original work).

The quality of aspiration – of aiming steadfastly for an ideal – is supported by the normative expectations of others. Individual and collective learning in the teaching profession depend, in particular, on norms of collegiality and experimentation. Norms of collegiality can reduce workplace isolation and help develop an orientation toward the teaching role. Norms of experimentation are based on a conviction that teaching can always be better than it is. If it is expected that teaches test their beliefs and practices, schools can be places where students *and* teachers learn.

Norms of collegiality and experimentation are moral demands with intellectual substance. They are not matters of individual preference but based, instead, on a shared understanding of the kinds of behaviours and dispositions that people have a right to expect of teachers (Little, 1981). These norms require detachment – a willingness to stand back from personal habits, interests, and opinions. What one does or believes in is not talked about as part of one's self but as something *other* – it becomes a potential exemplar of good (or not so good) ways of working, or of more or less justified beliefs. In teaching, what people do is neither private nor scared but open to judgements of worth and relevance in the light of professional obligation.

Community provides not only constraints and guidance but also succor. Collegiality, however, also depends on the degree to which another person is deserving and one's equal in deserts; it is not just loyalty and mutual help, but also the enjoyment of competence in other people. Essential to collegiality in teaching is the degree to which its practitioners are good at talking with one another about their work and can be confident about their own ability, and that of others', as teachers and partners in the improvement of teaching. Without mental, social, and role competence, norms of collegiality and experimentation cannot take hold. Some uncomfortable questions need to be confronted here:

> 'What effect does the relative exclusion of ordinary teachers from the wider governance of education, their restricted access to educational theory and other kinds of school practice, and the consequent overwhelming centrality of classroom practicalities to teachers, have on the kinds of *contributions* they make to staff discussion?' (Hargreaves, 1982, pp. 263-264; emphasis in the original)

Morality and authenticity in teaching

Of course, teachers are persons. However, being one's self in teaching is not enough. Authenticity must be paired with legitimacy as opposed to impulse and inflexible habit, and with productivity or a reasoned sense of purpose and consequences (Thelen, 1973). Thelen places authenticity in the context of action (authentic activities make teachers feel alive and challenged) and gives legitimacy and productivity the accent of thought:

> 'An activity is legitimated by reason, as distinguished from capricious-seeming teacher demand, acting out impulse, mere availability, or impenetrable habit. An activity may be legitimated by group purposes, disciplines of knowledge, career demands, test objectives, requirements, societal issues, laws, or by an other larger, organized context that enables the activity to go beyond its own particulars.
>
> An activity is productive to the extent that it is effective for some purpose. It is awareness of purpose that makes means-ends thinking possible, allows consciousness and self-direction, tests self-concepts against reality, and make practice add up to capability'. (p. 213)

Legitimacy and productivity are entwined, capturing social expectations and aspirations central to teaching and to learning from teaching. People's ordinary conception of morality describes this interplay between ideals and the rule requirements of social organizations (Strawson, 1974).

To the extent that roles have moral content, their impersonality is not inhuman or uninspired. But rules, norms, and external standards alone cannot account for moral action in teaching. First, role orientation must be lodged concretely in someone's head and heart. Where one's solid and full response to obligations is withheld, the claims of others are not acknowledged livingly (James, 1969). As Dewey (1933/1971) stressed, thoughtful action does not only depend on open-mindedness and responsibility, wholeheartedness is also part of it. To the extent, then, that the content of role has been absorbed into the self, role becomes a personal project – shaping the inner self and the self as it appears to others. Thus, moral aspirations cannot be separated from the question of personal identity, but conversely, responsibility for oneself, as a person, does not mean that anything goes (Taylor, 1970).

Second, the moral quality of role relations between professionals and clients draws on loyalty to concrete persons and analogues to friendship in enacting role (Fried, 1978). The warmth and selectivity of feeling implied by this contradicts the impersonality of role. Loyalty as abstract duty is not the same as actually taking faithful care of the particular people put into one's charge. All this is complicated by the fact that, in teaching, professionals face groups of young clients, not in school by choice. The role of the classroom teacher, therefore,

> 'puts the major obligations for effective action on his shoulders; it is the teacher's responsibility to coordinate, stimulate, and shepherd the immature workers in his charge. ... Task and expressive leadership in classrooms must emanate from the teacher, who, it is presumed, corrects for the capriciousness of students with the steadiness, resolve, and sangfroid of one who governs. The austere virtues, moreover, must be complemented by warmer qualities like empathy and patience. It becomes clear, then, that the self of the teacher, his very personality, is deeply engaged in classroom work; the self must be used and disciplined as a tool necessary for achieving results and earning work gratifications'. (Lortie, 1975, pp. 155-156; see also Waller, 1932/1961, pp. 385-386)

In sum, the moral nature of teaching – which also requires being genuinely oneself – does not remove the need for role orientation. Instead, a proper understanding of authenticity in teaching builds in the idea of external standards within which teachers make authentic choices. The need for authenticity hence supplies no argument against role orientation, but suggests that there are some teacher decisions that will be completely determined by role, some that are constrained by role but not determined, and some – not many – for which role does not and should not provide guidance.

Chapter 19

Forming Judgements in the Classroom: How do Teachers Develop Expectations of Their Pupils' Performances?

H. Manfred Hofer

Summary

This experimental study investigates the question as to how teachers relate to two different information factors about pupils and arrive at a prediction. Anderson and Butzin (1974) suggested that it is by multiplication that adults integrate information about the motivation and the aptitude of a specific person to reach a judgement on his/her ability. Here it was assumed that there were major differences in the patterns by which individual teachers integrate information. An experiment was designed to allow a statistical evaluation for every teacher. The dependent variable was a prognosis of pupils' performances in the next examination of the teacher's subject. The findings of this study contradict the results mentioned above. Teachers arrive at their judgement by means of differentiated considerations, best explained by a two-stage process of information integration.

Problem

How teachers process information to reach a prognosis on the future achievement of a pupil is the question under discussion. Because of research, we already have sufficient knowledge of the variables which teachers use to reach a prognosis of the pupil: effort, aptitude, and home environment. It is not known, however, in which way these variables are interrelated and how they are combined to arrive at a prognosis.

Expectancies of future pupil performance play an important role in guiding teacher behaviour. In accordance with notions on the theory of action (Hofer, 1986), these expectancies represent cognitions, which are then compared with the teachers' aspiration level. The result of this comparison provides the starting-point for future activity.

This paper deals with the investigation of the rules teachers adopt when combining information about a given effort of individual students, in order to arrive at expectancies about their future achievement.

One possibility to elaborate such combinations is offered by the Information-Integration Theory by Anderson (1981). This theory was originally developed to represent judgements, which can be measured in objective terms. Wilkening (1979), for example, presented test persons rectangles with various heights and widths and had them estimate the surface area. Children arrived at judgements by summing up the height and width of the individual stimuli. With adults, on the other hand, the judgement was reached by means of multiplying the information. Similar results were produced by the test persons who were given stimuli containing time span and speed of moving objects and who had to make judgements about the distance covered. This theory can also be used for psychological phenomena, which are not measurable in objective terms (Anderson, 1974).

The questions contained in this experiment have already been investigated in two other experiments, but not with teachers. Anderson and Butzin (1974) differentiated between four stages of motivation (low, moderate, moderately high, high) and intelligence (IQ = 100, 115, 125, 135). The tested persons were given descriptions of stimulus persons who were each characterized by a combination of these two variables. They were told that it was an applicant for admission to graduate school. On a twenty-centimeter long graphic rating scale, the test person was to indicate which performance this stimulus person would probably achieve. The statistical analysis showed an extremely significant interaction, 75% of the interaction showing a bilinear component. Kun and others (1974) were able to show that six-year-old children processed aptitude and effort information by addition to reach a prognosis on performance, while on the other hand eight and ten year olds, as well as adults, increasingly processed the information by multiplication.

In the present paper, effort and aptitude were also used as indicators. The patterns according to which teachers combined these information factors were investigated statistically for each individual person (Rathje, 1982).

Method

At first, a preliminary inquiry with seven teachers was instigated, in which the various aspects of the experimental structure was tested. Twenty-five teachers involved with all educational levels participated in the main inquiry. Fourteen test persons were masculine and eleven feminine; the average length of teacher experience was 10.04 years. The teachers taught pupils from ages of 11 to 18 in the following subjects: Mathematics (11), English (6), German (6), Biology (11) and Sports (1). As a rule, the teachers were contacted by a notice on the schools' bulletin board. The investigation was done in individual sessions. The design was a 3 by 4 variance-analytical design, with repeated measurements on both factors: Effort (very low, average, very high) and aptitude (well below average, slightly below average, slightly above average, well above average). The dependent variable was the expected achievement of the pupil in the next examination taken in the subject taught by the teacher.

The instructions said:

> 'With this experiment I would like to obtain information about the influence of certain traits of students on their future performance. Let us presume you have just finished presenting a six-week block

> in your subject, and to finish off the unit you would have your pupils take a written examination of average difficulty. A day before the exam you ask yourself how the pupils will perform.'

For each student a card was prepared with his name and his distinctions (for example: Markus; ability: well above average, effort: willing to exert great effort).

> 'On the basis of your observation and experience, deriving principally from the completed block unit, you believe Markus (the given example) to have well above average ability and to be willing to put in great effort. Each time you have read a card please estimate the expected performance of the pupil in the projected examination by means of indicating on this scale by use of the pointer; on the scale there is a possibility of obtaining a maximum 15 points.'

As a dependent variable, a numerical scale was used. It was mounted on a wooden construction. The pointer was moveable and so constructed that the decision of the teachers could be read by the examiner on the reverse side. In each cell, the test persons were confronted with three fictitious pupils, on whom they were asked to make judgements. A prior run through was regarded as practice and therefore not used in the evaluation. In addition, the aptitude information was given four times without the effort information. Each test person had thus to give 64 single judgements. Sixty-four registration cards were each labeled with a different masculine first name. The sequence of information within each stimulus was systematically varied in all the tests to eliminate a sequential effect. The complete processing time for the 64 stimuli was between 25 and 35 minutes. Together with a final interview, the experiment took about an hour. Before beginning, the test persons were asked by the administrator to register spontaneously the impressions and thoughts that came to their minds while they were working on the cards.

Results

First, the opinions expressed by the teachers were evaluated and interpreted by the researcher.

a. Spontaneous comments:

More than half of the test person reported that they had imagined their own pupils. Some teachers commented on particular students, which they noticed particularly with remarks such as: "That will not do him any good either." (Combination of low aptitude – high effort). The comments show that the teachers translate the shallow information into personal experiences and realistic school situations.

b. Comments based on the stimulus material:

The test persons were most doubtful about the cases in which only the aptitude information was given. Well over half of the test persons had problems with the evaluation of the fictitious student and were unsure in their judgement or tried to evade the problem with the aid of assumed effort.

For some test persons there was not enough information given on the students so that they felt a judgement was very difficult based on these conditions; added information as to traits of character and home environment was desired.

c. Comments suggesting a prediction:

Aptitude and effort were regarded as the main factors for achievement. For various test persons, the aptitude of the student corresponded with an upper achievement limit; the attainment of the highest achievement is then dependent on the effort willing to be exerted by students.

The comments that were made about the combination of information are particularly interesting. In half of the cases, the presumption that one factor had greater importance turned out not to be valid. The prognosis about the compensating ability of the factors was proved predominantly correct. As a rule, it was stated that the factors could compensate for each other.

More than half of the test persons stated that they based their valuation mainly on the aptitude information. Only two reported that the information on effort was more important to them. All that implies that a two-staged strategy has been used: Starting with the factor, which is regarded as principal, the information of the second factor is added to reach a judgement.

Altogether qualitative results of this type show that:

- The persons tested could cope well with the instructions
- No definite judgement could be given on the basis of only one piece of information
- Even with two factors of information the judgement was regarded as indefinite
- Aptitude and effort were regarded as very important achievement factors
- The test persons are only partially conscious of their routine reasoning of combining information.

Statistical analysis

The evaluation of all the test person simultaneously follows the randomized-block-factorial-design (Kirk 1969). Figure 1 shows the mean values. The effort factor is significant and constitutes 25% of the total variance. The aptitude factor constitutes 68% of the total variance. The proportion of significant interaction, on the other hand, is 2%. An analysis of the components of the interaction (following Anderson 1970) shows that the non-linear component alone is significant.

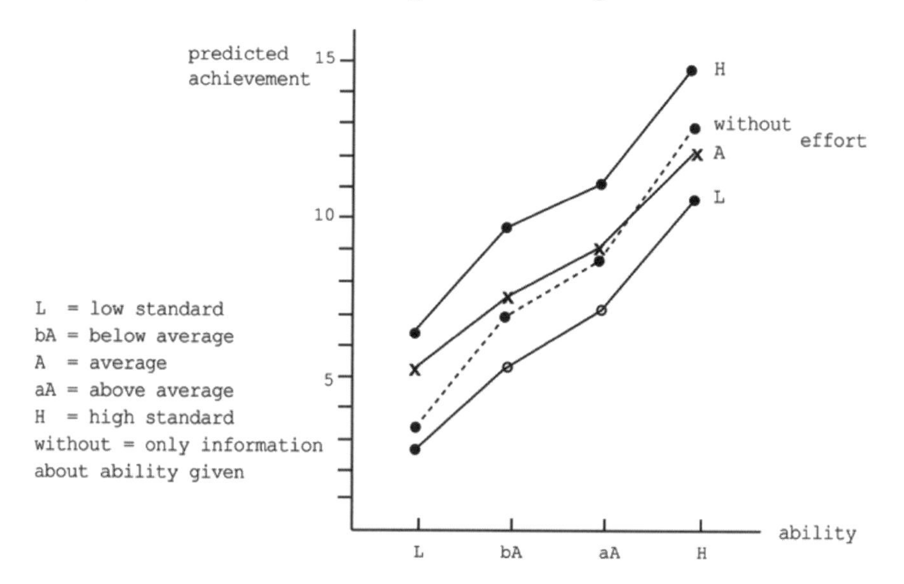

Figure 1. Overall result of information integration.

We are especially interested in the evaluation of each individual test person. The analysis followed a two-factor "Within-Subject-Design" (Anderson 1976, p. 685), in which each person was subject to all treatments. As an error term the average variance of the data within the cell of the design was used: the so-called 'pooled error term.' To test the multiplicative model, interaction variance was divided into a bilinear component and the residual variance. The calculation of the numerical values of the bilinear component followed Anderson (1970, p. 157).

Table 1 shows the results for each person separately. In all cases, both main effects were very significant. This was expected and indicates only that effort (A) and aptitude (B) were used by teachers as indicators for school achievement. Graphic illustrations reveal monotony: a higher degree of effort, or aptitude, on average leads to better prognosis. The relationship, however, is not linear; the stimuli were not observed as equally distant. The type of combination between both variables is shown in the form of interaction.

For seven teachers the interaction effect was not significant. The average model is confirmed in this case. These test persons assumed a stable influence of the factors: the effort has the same effect with low aptitude as it has with high aptitude.

The theory that judgement can be described by means of an additive model (Fishbein and Aijzen, 1972) presumes that the first information results in a positive value, as does the following, and that each single piece of information contributes to the judgement in a quantitative way. If this were the case, then judgement based on only one information factor (here: aptitude) would have to be lower in any case, than when two separate factors of information were given (aptitude and effort).

The figures show for all test persons that the judgement line, which was based only on information of aptitude, does not fall below the other lines. It intersects the other lines in such a manner that the following interpretation can be

	TEST PERS.	df=2 A		df=3 B		df=6 AxB		Df=1 LINxLIN		Df=5 RES		TYPE
CS	1	77,80	**	95,00	**	2,37	*	3,71	ns	2,70	*	DIF
	2	78,25	**	491,11	**	5,36	**	5,36	*	5,35	**	DIF
	4	12,03	**	151,35	**	3,83	**	3,99	ns	3,80	*	DIF
	5	210,70	**	524,90	**	2,70	*	2,39	ns	2,76	*	DIF
	6	14,32	**	136,11	**	2,27	ns	0,17	ns	2,96	*	DIF
	7	1050,99	**	1326,55	**	27,88	**	12,16	**	31,03	**	DIF
	11	27,84	**	128,19	**	1,70	ns	1,55	ns	1,73	ns	ADD
	12	45,00	**	311,35	**	1,62	ns	2,79	ns	1,38	ns	ADD
SM	13	163,00	**	432,08	**	4,91	**	4,80	*	4,93	**	DIF
	14	95,13	**	49,43	**	4,00	**	6,75	*	3,44	*	FIF
	15	63,76	**	43,22	**	3,50	*	8,07	**	2,59	ns	MUL
	16	48,64	**	122,24	**	0,64	ns	0,25	ns	0,71	ns	ADD
	17	117,13	**	138,14	**	3,39	*	4,72	*	3,12	*	DIF
ST	8	38,86	**	128,07	**	0,41	ns	0,58	ns	0,37	ns	ADD
	9	342,16	**	468,61	**	7,27	**	0,071	ns	8,71	**	DIF
	10	384,12	**	269,12	**	9,62	**	33,35	**	4,87	*	DIF
	18	123,35	**	138,13	**	0,48	ns	0,005	ns	0,57	ns	ADD
	19	85,16	**	343,27	**	4,27	**	13,61	**	2,41	ns	MUL
	20	135,50	**	171,31	**	2,09	ns	1,42	ns	2,22	ns	ADD
	21	103,22	**	109,66	**	4,29	**	4,57	*	4,24	**	DIF
	22	228,00	**	542,66	**	4,38	**	16,05	**	2,04	ns	MUL
	23	79,40	**	1282,40	**	5,00	**	11,74	**	3,65	*	DIF
	24	187,35	**	121,21	**	4,78	**	18,23	**	2,09	ns	MUL
GS	3	162,92	**	293,80	**	11,59	**	2,89	ns	13,33	**	DIF
	25	16,91	**	19,06	**	3,28	*	16,73	**	0,59	ns	MUL

TYPE = Type of Information Integration
DIF = Differential Type
ADD = Additive Type
MUL = Multiplicative Type
CS = Comprehensive School
ST = Secondary Technical School

** = p < .01
* = p < .05
ns = not significant

SM = Secondary Modern School
GS = Grammar School

Figure 2. Results of the variance analysis.

drawn: The test persons overall presumed with low aptitude are lower and with high aptitude are higher efforts. (Except persons number 1, 10, 21, 23).

In a multiplicative model, only the linear part of the interaction should be significant and the lines should spread out to the right like a fan: the higher the aptitude the stronger the effect of the effort on the prognosis. With five teachers (persons number 15, 19, 22, 24, 25) only the linear part of the interaction is significant. The lines do not run parallel. Only with four of them do the lines run close to the form of a fan. These teachers believe that with high ability a strong effort has greater effects on the achievement than with less able pupils.

There remains a large number of test person with whom the non-linear part of the interaction is significant. Their judgements are to be described as differentially weighted models, since the weighting of the information combinations does not exclusively run in a linear fashion. There are six cases in which only the non-linear part of the interactive effect is significant, and eight cases in which both components reach the significance level. Various individual patterns occur. Five test persons considered the influence of effort as strongest with average aptitude. One teacher, on the contrary, regarded the influence of effort with average aptitude as less important in comparison to high or low aptitude.

In addition, there is the extreme form of judgement from test person number six who saw at the two lowest levels of capability an average effort as more successful than a higher effort. This teacher seemed to believe that with lower aptitude an above average effort results in lower efficiency.

Discussion

This paper has investigated by means of individual cases – but with the aid of quantitative and statistical methods – the thought processes of teachers integrating two kinds of information in making judgements of future pupil performance. It was shown that the way in which aptitude and effort information are combined in order to reach a judgement, varies substantially between different teachers. The relationship between predictors and criterion was monotone – with one exception. By no means did the integration follow the multiple regression model. Here the contribution of the one variable to the criterion is independent from the distinction of the other. Compensation is possible. With 72% of the test persons, the interaction effect between the indicator variables was significant. Within this group, there were several different types. Sixteen percent of all persons judged (at least in tendency) in the manner of the multiplicative model in which the effect of the one variable increases with the presence of the other.

On the other hand, a further 20% considered the influence of effort to be strongest with average aptitude. One test person went even so far as to give a lower prognosis with lower aptitude and high effort than with low aptitude and average effort. Other judgement patterns were difficult to identify.

On the whole, the findings clearly run contrary to the results of the study of Kun and others (1974) and Anderson and Butzin (1974), which concluded from tests with adults that aptitude and effort information were used according to a multiplicative pattern.

From the verbal comments of the test persons, one can assume that the teachers did not process the information simultaneously but sequentially. First, the most important information is used, and then the second information is taken into account. Aptitude is the most important information, because it contains a statement about a stable disposition. The information about effort then states in which way the persons can effectively control their dispositions. This two-staged process can be interpreted in two different ways:

a. The person processes the aptitude information and bases his/her final judgement upon the effort information.

b. A judgement is rendered based on aptitude information and is corrected in respect of the differences offered by the second information.

One could possibly test this by means of presenting the persons with both factors of information in different sequences, asking them to give judgements after each one. The result, that the teachers' judgements are not, as a rule, based on the compensatory model of regression analysis, does not of course imply that their judgements are more valid than a statistical regression analysis.

Chapter 20

Conflicts in Consciousness: Imperative Cognitions Can Lead to Knots in Thinking

Angelika C. Wagner

Summary

Not all cognitions are alike – some of them have acquired imperative qualities in the individual's mind. This means that in some cases s/he feels that "I must do … ", "It must not happen that," or "They really ought to...." Sometimes these imperative cognitions are violated or in conflict with each other; then the individual feels to be in a subjective dilemma and then her/his thinking processes start going round in circles. This phenomenon we have called a 'knot'. Fifty-six students and seven teachers in 6th grade were asked to recall what had been going 'through their mind' in class using the method of 'Retrospectively Thinking Aloud' which had been developed for this purpose. The resulting data were analyzed sentence-by-sentence with respect to imperative and non-imperative cognitions on actions and the resulting 'knots' in thought processes. Some of the results of the study and implications for further research are presented.

The Emotional Side of Action

Every day teaching is much more complex than action theory (e.g. Miller/ Galanter/Pribam, 1960) leads us to believe. Teachers not only have plans and strategies that they may or may not carry through. They also, at times, feel anxious, angry, confused and under stress as well.

The basic hypothesis of this paper is that such emotions and conflicts are the result of a special *mode* of cognitions - i.e. imperatives. An individual thinking in terms of imperatives gets caught in a certain type of conflict which we have come to call knots and which are subjectively experienced as anger, anxiety or stress. How do such conflicts arise?

Miller/Galanter/Pribram's TOTE model revisited

In their now famous book on 'Plans and the Structure of Behavior' - a book which has been credited with launching the 'cognitive revolution' Miller, Galanter and Pribam begin as follows (1960, 5-6).

> 'As you brush your teeth you decide that you will answer that pile of letters you have been neglecting. .... After lunch, if you remember, you turn to the letters. You take one and read it. You plan your answer. You may need to check on some information, you dictate or type or scribble a reply, you address an envelope, seal the folded letter, find a stamp, drop it in a mailbox.'

This kind of plan seems to be straightforward enough. A person plans something and later on proceeds to do what s/he has planned unless some other plan prevents that. Miller, Galanter, and Pribram's analysis of plans and the structure of behavior is based on the assumption that most human plans are of this straightforward nature.

As we discovered in our research project, however, this does not hold true for all plans. In real life, many of them seem to become entangled, confused and subjectively stressful, sometimes teachers appear to be 'unable' either to follow through on them or to drop them, and so they become angry, confused or anxious.

Some of this confusion can already be seen when taking a closer look at the plan Miller et al. talk about in the beginning of their book.

If we were able to listen in on what Miller's hypothetical subject is actually muttering to himself while brushing his teeth and planning his day, we might for instance discover that what he says to himself, sounds somewhat different from what Miller et al. describe:

> 'I am going to write letters after lunch. Damn why haven't I written to Bill already?! I keep postponing this stupid letter... I really MUST sit down and do it; not that he couldn't wait much longer, may be it would even serve him right to be kept waiting, but I really SHOULD learn to get things off my desk fast. It is awful how I postpone even letter writing ... (sigh) well, as I know myself, I will probably find another excuse for not doing it today ... it is terrible ... I never get myself to do things on time ... Well, (with a lot of emphasis) I must write this letter this afternoon!'

The end result of this thought process is a sentence that only superficially resembles those intentions about which Miller et al. write. The cognition 'I MUST write this letter!' has a different quality to it. With respect to other, more ordinary plans, the subject makes a simple statement (e.g. I will drive to work); with respect to the letter writing business, however, he commands himself to do it.

In our research project, we have called these kinds of cognitions subjective *imperatives*. Such imperatives are basically different from the goals and test criteria described by Miller et al. (1960). Even though they may superficially resemble each other, imperatives have a different impact on the individual, especially when they are violated.

Suppose our hypothetical subject does *not* write the letter that very afternoon (as some of us may have suspected all along). In this case, he will most likely experience some inner turmoil: guilt, anxiety, anger -, and this thinking will go around in circles (e.g. 'I did not write the letter - I should have written it - I did not write it ... etc.').

The violation of a subjective imperative leads to a conflict in consciousness that has a quality of its own. It is like being trapped in a 'knot' - thinking goes around in circles without finding an exit. And looking back to how our hypothetical subject set up his 'good intentions' in the morning, one discovers that the very imperative itself already implied a conflict in consciousness which was only smothered over by saying "I must ...".

The central hypothesis of this paper is that the distinction between imperative and non-imperative cognitions is a necessary step towards understanding and explaining the dynamics of teachers' (and students') thinking in class.

Imperative and Non-Imperative Cognitions

Definition

Subjective imperatives were defined by us as cognitions that for the individual himself/herself have the character of a subjectively compulsory 'MUST' or 'MUST NOT'.

Phenomenological description of imperatives

On a cognitive level, imperatives are like sentences with an exclamation mark, i.e. sentences which, grammatically speaking, are in an imperative mode. In whatever form an individual may privately express these imperatives; in each case, they can be restated without loss of content in the form of 'X MUST happen...', respectively X MUST NOT happen'.

On an emotional level, imperatives are cognitions that carry with them a feeling of urgency, of being compelled, of 'having to do something about it'; this tension may sometimes be experienced as rather intense anger or fear.

On the level of subjective theory, imperatives are often linked with irrational assumptions as Ellis calls them (Ellis, 1977; Beck, 1979; Meichenbaum, 1977). These are assumptions that something 'terrible' would happen if the imperative were to be violated. However, one also finds frequently that an individual 'knows' that, of course, it would not be terrible if X is to happen - and yet the imperative continues to exist in her/his consciousness.

On a connotative level, imperatives function as injunctions. These injunctions are manifestly directed either toward oneself, toward others or toward life in general.

Actually, and this is where the confusion comes about, they are directed against part of the very content of consciousness itself. When an individual tells himself or herself 'X MUST NOT happen', than consciousness attempts to deny X which at the same time is already part of consciousness. So in a sense consciousness attempts to wipe out that which is already part of it. Subjective imperatives are not necessarily identical with social norms. The content of any given imperative may, of course, correspond to prevailing social norms and does so in many cases. However, this *is* not necessarily so. An individual may follow certain social norms without making them into a subjective imperative. Teachers are well capable of following mandatory regulations in school, as our data show, without making them into subjective imperatives. On the other hand, some of the individual imperatives have a content which deviates from social norms or even may run counter to them.

Analysis of imperatives

How do subjective imperatives differ from non-imperative goals and test criteria as Miller et al. (1960) describe them? What is the difference between a goal and an imperative in the consciousness of an individual?

At first sight, they resemble each other closely. Both contain a statement of some desired (resp. some undesired) state-of-affairs for the future. On second sight, however, there is an important difference: a goal is simply a goal in the eyes of the individual that may or may not be reached. Even though she/he probably prefers to reach this goal, stating the wished-for state as a goal implies the acceptance that the goal may not be reached. A subjective imperative, on the other hand, does imply that it is forbidden not to obey the imperative.

Here, an imperative closely resembles those paradoxes that Bateson (1973) and Watzlawick (1969, 1974) have analyzed in great depth. The underlying assumption (level 1, according to Bateson) with an imperative is different from that of a goal: an imperative implies that its violation simply MUST not happen. Violating the imperative is declared off limits, while a goal implies that it may not be reached.

However, as everybody knows, imperatives of course can be violated; otherwise, there would be no need for them. There is an anecdote from the time of pre-World War I that illustrates this dilemma quite aptly.

> 'At that time, political attempts on the lives of politicians were quite frequent. So, one day, a new edict was published:
> Shooting at the minister of inner affairs - 2 years of hard labor!
> Shooting at the minister of external affairs - 5 years of hard labor!
> Shooting at the prime minister is forbidden!'

The dilemma of any imperative is that on the one hand it is based on the assumption that it may be violated – and on the other hand, this violation must be prevented from happening at all cost. An imperative implies that the occurrence of a discrepancy between WHAT IS and WHAT SHOULD BE is forbidden.

Imperatives and the 'Tote' Mode

Going back to the original TOTE-model by Miller et al. (1960), it is possible to demonstrate what the appearance of imperatives means in terms of that model of the structure of behavior. The original TOTE-model as a two-dimensional indicates this, as the following replication is drawn .

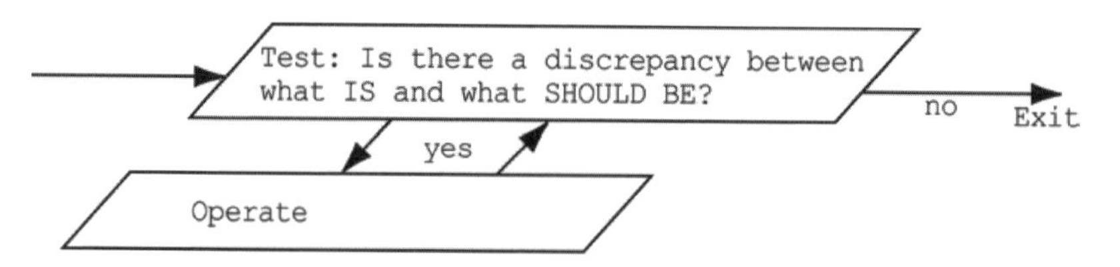

Figure 1. The basic structure of behaviour without imperatives: the original TOTO-model by Miller et al. (1960)

The introduction of a subjective imperative changes the model by introducing a third dimension. An imperative implies the injunction a discrepancy, between WHAT IS and WHAT SHOULD BE, may not occur. According to Russel (1956), this is a second-order statement.

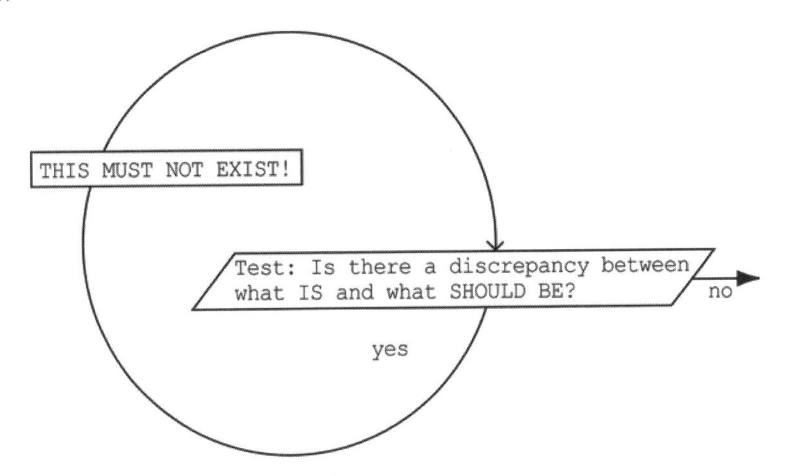

Figure 2. The basic structure of behaviour when a subjective imperative is involved: Model of a conflict in consciousness arising from the vioation of a subjective imperative (VIC)

This model demonstrates what happens in case a subjective imperative is being violated. Since an imperative implies that it should ALWAYS be obeyed, the violation of an imperative continues as long as the discrepancy between WHAT IS and WHAT SHOULD BE is conscious.

The model also shows the reason why – as long as the discrepancy between WHAT IS and WHAT SHOULD BE continues in the consciousness of an individual – there is no exit.

Moreover, indeed, this is why an individual feels trapped when caught a knot.

Knots are Conflicts in Consciousness

Arising from the Violation of a Subjective Imperative

Originally, we have termed the conflicts that arise in consciousness from the violation of imperatives 'knots' (Barz et al., 1981; Wagner Weidle, 1982; Wagner et al., 1981; Wagner et al., 1977). This corresponds to everyday use of language. In Germany, one occasionally talks about someone 'having a knot' in the sense of being caught in an unsolvable psychological problem.

The phenomenology of knots

How does an individual experience being caught in a knot? There are several typical symptoms of being caught in a knot:

1. Thinking goes around in circles without finding an 'exit'' - at least as long as the subjective perception of the situation does not change.
2. There is a feeling of arousal that can be experienced as fear, anxiety, anger, aggression, *or* depression, depending on individual strategies in interpreting it (Ulich et al., 1981).
3. The individual feels under stress.

4. The individual feels, at least briefly, that there is no way out. And this is correct; there is no way out of a knot as long as the violation of imperatives continues to exist.

How strongly these symptoms are experienced depends on various circumstances. In some instances, a given individual may experience these symptoms only fleetingly for a few seconds, especially when s/he is preoccupied with other problems; in other cases, s/he may have a sleepless night, tossing around in bed, or s/he may even feel trapped in a knot for months which can become so stressful that s/he finally goes to see a psychotherapist. Most knots however are luckily of the short-term variety, as our research data show.

The Definition of Knot-Type Conflicts

Definition: Knots are conflicts in consciousness arising from the perceived or anticipated violation of subjective imperatives. This definition implies the following:

1. Knots are located in *consciousness;* this is to say that they *do not* exist independently of an individual's consciousness. The situation itself is not a knot, but rather the knot comes into existence only by the way a given individual deals with reality.
2. Knots originate only through *thinking;* without thought, there would be no knots.
3. Knots originate only through a particular way of thinking, i.e. *imperative cognitions;* where there are no imperative cognitions, there is nothing to be violated.
4. Knots are conflicts existing in consciousness, *the very moment* the violation of a subjective imperative is perceived as having occurred or is anticipated for the future.
5. Knots arise because *consciousness is in conflict with itself.* Knots arise from the unsuccessful attempt of consciousness to push away something (i.e. the discrepancy between WHAT IS and WHAT SHOULD BE) that is already in consciousness. (For further discussion, viz. Wagner et al., 1984).

Six types of knots

In the course of our research project, we have identified six basic types of knots. These types are not mutually exclusive; two or more of them may appear together in one actual conflict in consciousness.

Type 1: Reality knot: An imperative is violated by perceived reality.
The simplest form of knots occurs when the subjective perception or attribution of present reality collides with a subjective imperative (e.g. "I am screaming at my students - I MUST NOT do that!").

Type II: Past-event knot:
An imperative is violated by perceived past events.
A second form of knots arises when the remembrance of past events collides with a subjective imperative (e.g. I did X - I SHOULD NOT have done X). Feelings of guilt are examples for this type of knot. Nietzsche describes the 'solution' of such past-event knots quite aptly

"I have done that", my memory says.
"I cannot have done that", says my pride and remains merciless.
Finally my memory gives in.

Type III: Anticipation knot: An imperative is violated by anticipated future events.
A third form of knots arises when the anticipation of possible future events collides with a subjective imperative (e.g. X may happen - X must not happen) [19].

In any given moment, the number of possible anticipation knots is almost unlimited because one may worry about almost everything in life. "My life consisted of a series of catastrophes" Mark Twain said, "most of which never happened".

Type IV: Imperative dilemma knot:
Two imperatives collide with each other in a given situation.
The fourth type of knots arises from the collision of two imperatives in a situation where the individual assumes that s/he will be unable to follow both of them. Hence, the violation of at least one of them appears to be inevitable - at least in her/his mind. (For example, "I MUST call upon all students who wish to contribute to the discussion" – "I MUST finish this lesson in time". At a certain moment in class, it may be impossible to obey both imperatives at the same time.) Difficulties to make decisions often involve knots of this type.

Type V: Counterimperative knots:
An imperative collides with its own counterimperative.
This fifth type is a special case of the imperative dilemma knot; however, since it is much more 'vicious' than Type IV, it merits a number of its own. This type of knot arises when in consciousness two subjective imperatives collide which demand just the very opposite of each other. An imperative that demands just the opposite from an already existent subjective imperative we have termed counterimperative. An example for such contradictory imperatives would be: I MUST NOT feel anxious about the exam - I MUST BE anxious about the exam (because otherwise I will not study hard enough). Imperative and counterimperative collide with each other *every time* they are activated. Many of the problems that people present in psychotherapy are of this kind. Counterimperative knots feel especially vicious because there actually is no way that an individual may obey one of the imperatives without violating the other. As opposed to imperative dilemma knots, counterimperative knots arise in all situations where the imperatives apply.

---

[19] Strictly speaking imperatives already entails TYPE III-knots, because they imply that they may be violated in the future while they at the same time assume that this MUST NOT happen. However, in everyday life, we usually succeed in making ourselves believe that a violation of imperatives can be avoided and hence, for all practical purposes, the appearance of a subjective imperative in consciousness is much less stressful than being actually trapped in a knot. For practical purposes, then, the difference between an imperative and an anticipation knot lies in the subjective probability a subject assigns the violation of the imperative involved. Someone who assumes that he/she will never have to face an exam any more will not feel in a knot even though the imperative - I MUST NOT fail in exams – is still there. If this person, on the other hand, has to pass a language exam in the near future and s/he believes it possible that s/he may fail, then there will be an anticipation knot.

Type VI: Double bind knot
The sixth kind of knots arises when a subjective imperative in itself is a paradox. A paradoxical imperative is an imperative that is formulated in such a way that it is impossible for the individual to carry it out. (For instance: I MUST be spontaneous. Or: "I MUST NOT blush!")

Bateson (1973) and Watzlawick et al. (1969, 1974) have studied these paradoxes extensively. They pointed out the trap involved in "double bind" situations.

Such "double binds" also occur within one individual. An example for such a paradoxical imperative is "I MUST be spontaneous now!" Whenever such a paradoxical imperative is activated in a given situation, there is a double bind knot.

The basic dilemma of knots
In the final analysis, the basic dilemma of a knot arises because consciousness is reacting on itself. At the very moment a subjective imperative is violated, consciousness attempts to obey the demand that this MUST NOT happen and yet, the perception/cognition of this discrepancy is present in consciousness.

The oscillating nature of knots results from consciousness trying to push away something that is already part of its own content. This is the basic knot. Hence, conflicts arising from the violation of imperatives are unsolvable,
- as long as consciousness continues to register a discrepancy between WHAT IS and WHAT SHOULD BE,
- As long as the individual believes it cannot afford to ignore this discrepancy, and
- As long as the subjective imperative still is an imperative for the individual.

The basic dilemma of a knot is that consciousness attempts to push away something that is already part of it. Knots arise only when a discrepancy between WHAT IS and WHAT MUST BE is detected - and in that very moment, this cognition is already part of consciousness itself.

An Empirical Study of Imperatives and Knot-Type-Conflicts in Teachers, and Students' Thinking During Class

In an extensive research project at the Padagogische Hochschule Reutlingen from 1976 to 1982, we[220] studied what teachers and students think about in class.

Subjects
For the purpose of this study, seven sixth grade classrooms were selected with a predominantly student-centered approach to teaching, in various schools in Baden-Wurttemberg. In each class, the homeroom teacher and eight students participated in the study.

There were a total of 7 teachers (five females, two males) and 56 students; among the students, there was an equal number of males and females and of high and low achievers.

[20] The research project was carried out by Susanne Maier-Stormer, Ingrid Uttendorfer-Marek, Renate Weidle, Monika Barz (1980-1982) and the author. It was financed by a research grant of the Deutsche Forschungsgemeinschaft in Bonn from 1976 till 1982).

Methods
The first task to be solved was the development of an appropriate research method. For the purposes of the study, the method of Retrospectively Thinking Aloud was developed by adapting the method of Thinking Aloud (viz. Claparede, 1971; Luer 1973) to the needs of the classroom. A similar approach has been used since by other researchers in this field (Lowyck, 1979; Bromme, 1980; Heymann, 1980; Treiber, 1980; Wahl, 1979; Weidle and Wagner, 1982).

Using the method of Retrospectively Thinking Aloud (RTA), teachers and students were shown a video recording of their own classroom lesson on the very same day the lesson had taken place.

The video tape was stopped quite frequently, approximately every 40 seconds, and the subject was asked then to tell the experimenter "what had been going through your mind at this point in class". The experimenter herself attempted to conduct the interview in a non-directive approach (Rogers, 1973); however, when necessary she asked the subject to clarify whether a given thought process actually had occurred during class or afterwards. (For a more detailed discussion of this method, see Wagner et al., 1977 and 1981).

One important prerequisite for the method of Retrospectively Thinking Aloud is a fairly open and trusting relationship between experimenter and subject. For this, it turned out to be necessary first to conduct an in-depth extensive interview on attitudes and experiences with school in general and this class in particular.

Procedure
After initial contact had been established, the experimenter first conducted the extensive interview with the teacher that lasted up to three hours. Then the experimenter visited the classroom, taking trial video shots in class and carefully explaining the research methods planned. Then, she interviewed the students in pairs, each interview lasting about 45 minutes. A few days later, she returned and videotaped a lesson on a predetermined topic. This videotape then was shown to the student pairs and to the teacher separately using the Method of Retrospectively Thinking Aloud. The same procedure then was repeated with a second lesson a few days later.

Analysis of data
All interviews and RTAs were taped and transcribed. For each class thus several hundred pages of interview and RTA protocols were available as well as a transcription of the actual lesson itself.

These extensive data were then analyzed (1979-1982) with respect subjective imperatives and knots in students and teachers' thinking.[21]

In order to do this, an elaborate method of analyzing cognitive structures and action plans in students' and teachers' thinking was developed ("System zur Rekonstruktion von Handlungsplanen und ImperativKonflikten" – "HIK"). This HIK-system of analysis entailed the following steps:

---

[21] This was actually the second phase of analyzing the data. In the first phase, three of the classroom lessons were analyzed in a case study method as to interaction of plans and strategies of teachers and students with the actual flow of the lesson. The results of this analysis are published in Wagner et al. 1981.

1. Encoding the interview and RTA protocols as to basic units (Thematische Interaktions-Sequenz - TKS). Each TKS contained approx. 3-13 sentences of the interview resp. RTA protocol dealing consecutively with the same topic.
2. Each basic unit (TKS) was then categorized according to a list of more than 100 categories according to the content of the TKS.
3. Then, within each TKS, every sentence or equivalent thereof was coded as to whether it functioned as imperative, goal, test criterion etc.
4. Based on this coding, for each TKS the structure of action resp. thinking was reconstructed in a diagram, showing the structure of the action/thinking process involved. These structure diagrams were then coded as to appearance and type of knot.
5. These structure diagrams were then coded as to appearance and type of knot.
6. Finally, each knot was analysed as to the type of thinking strategy with which the subject (unsuccessfully) tried to solve it. For this, another category system was developed (this is explained in more detail by Uttendorfer-Marek in Wagner et al., 1984).

Because of restriction of space, only a few of the results of the empirical study can be presented here. (For more details, viz. Wagner et al., 1984).

Frequency of knots

About one-third (32.13%) of all basic units (TKS) contain one or more knots. That is, about every third issue or topic students and teachers mention has to do with a conflict arising from the violation of subjective imperatives.

Students apparently have significantly more knots than teachers (33% vs. 28%, $p < 0.05$). When students and teachers are retrospectively thinking aloud about their own classroom lesson, they get trapped in a knot less frequently than when they are interviewed about general attitudes and experiences with school (28% vs. 35%, p A 0.01).

This supports our theoretical notion that knots are basically unsolvable conflicts (see above). Hence, when subjects are asked to report on their general experiences, it is to be expected that they talk about issues which involve knots for them more often than when they are simply asked to report what had been going through their mind at such and such point in class.

Types of knots

As was expected, both teachers and students most frequently report knot resulting from the violation of an imperative by (past or present) reality (TYPE I and III). 81% in the interview and 69% in the RTA of students' knots and 61% resp 47% of teachers' knots are reality and past-events knots.

The second most frequent type of knots was the violation of a subjective imperative by anticipated events; 8% resp. 27% of the students' knots and 21% resp. 32% of the teachers' knots were such anticipation knots. This again supported our assumptions. Due to their professional role, teachers are expected to think and plan ahead more than students do, and in thinking ahead they very well may get caught in a knot.

In similar vein, teachers also have a higher percentage of knots resulting from two conflicting imperatives. Particularly during class (to be more specific: in RTA protocols) teachers significantly more of than students ($p = 0.0001$) get caught in such dilemma knots. This may be due to those conflicting imperatives they have learned during teacher training.

What issues are most frequently associated with knots?
The analysis of the data showed that a few of the topics are significantly more often associated with knots than others did. While generally speaking only about 30% of all TKS were associated with knots, the following topics apparently bother the teachers much more than average: colleagues (83% knots), anger/irritation (80% knots), and principal (81%). Then followed attention/concentration of students during class (65%) and giving or withholding permission (70%). On the other side, teachers appear to be significantly less bothered by the content of the lesson ($p \leq 0.01$) than by anything else. These findings are surprising in that the German teachers who were studied apparently are most bothered by their colleagues and the principal whereas dealing with students is only moderately associated with conflicts.

However, the teachers selected for this study were above average in using student-centered methods. So, at this point it remains an open question whether this is why this group of teachers has fewer problems with students and more with the administration and colleagues, or whether this represents a more general situation of teachers in Germany. Some of the informal complaints of teachers seem to support the hypothesis that they actually are more bothered by what is going on outside the classroom than by what the students do.

If this can be verified, it would mean that teacher education should better prepare teachers for this aspect of their professional life.

Implications For Teacher Training

The implications of this study for teacher training are however much more far-reaching than simply to introduce a new topic into the curriculum.

Much more prevalent is the question of how teacher training can aid in the dissolution of conflicts arising from the violation of imperatives.

So far, many textbooks proudly present a host of imperatives, of what a good teacher should or should not do. So far little has been done to help the student to integrate these pieces of "good advice" into her/ his repertoire of teaching strategies without becoming caught in knots.

On the contrary, our data show that, during class, teachers are caught in knots more frequently than students, especially in conflicts where two opposing imperatives collide with each other. If this is the result of teacher training, then it seems necessary to do something about it.

Author Reflection Twenty Years On

Looking back at this 1984 paper, I find myself today just as fascinated by the phenomenon of imperatives as I was then. In the meantime, there has been an upsurge of newer studies dealing with the structure and causes of excessive worrying (e.g. Davey and Tallis, 1994; Wenzlaff and Wegner, 2000; Hoyer, 2000).

From today's point of view, the unique contribution of the 1984 paper was to point out the close link between worrying and subjective imperatives. In order to find out why this is so, my coworkers and I have been conducting a number of studies since 1984. We wanted to learn more about three major questions: (1) What is the empirical evidence of the frequency, structure, and content of knots and their occurrence in other contexts? (2) What is the function of imperatives and emotions within mental self-regulation? (3) What are the practical consequences for teacher training and counseling – i.e. how can knots and the concomitant worry be resolved effectively?

In more than a dozen empirical studies, my coworkers and I have analyzed the intricate patterns of imperatives in different situations (e.g. test anxiety, learning and memory, females in science, fear of public speaking, hearing loss, depression, counseling). They are closely interwoven into complex patterns with various circumvention strategies, cognitive confusion and emotions.

At the same time, we developed a new counseling method directed at effectively resolving knots by "undoing" imperatives based upon *Konstatierende aufmerksame Wahrnehmung* (Non-Evaluative Inner Perception; compare Wagner, 1986, 1995, forthcoming). The effectiveness of this approach was tested with depressive outpatients (Iwers-Stelljes, 1996); further studies are in progress: e.g. of resolving fear (Schöning, 2002, Albrecht 2002), reducing inner conflicts in counseling (Iwers-Stelljes, forthcoming) and in sports (Petersen, forthcoming), and improving hearing ability after hearing loss (Saure, forthcoming, Wagner, forthcoming).

The most intriguing challenge during the last five years has been to understand what it means to resolve imperatives within the context of information-processing theory. Imperatives somehow seem to "behave" rather strangely and quite differently from other, i.e. non-imperative cognitions. They are closely linked to high arousal and oftentimes connected with emotions. Sometimes they are rather difficult to deactivate, even if the individual wants to do so. Thinking deeply about these issues has led me to develop a general *theory of mental self-regulation* (Wagner, forthcoming).

In hindsight therefore the discovery of imperatives in the thinking process turned out to be an important challenge for us to learn more about how to unravel "knots" and improve mental self-regulation - in teaching and learning as well as in every-day life.

Section C

Teacher and Teaching: Theories of Practice

This section deals with teachers and teaching and their theories of practice. Chapter 21, Ten Years of Conceptual Development in Research on Teacher Thinking is contributed by Christopher Clark. He describes the conceptions of the teacher, of students, of subject matter, of context and of the research process that characterizes the field of research on teaching in 1975 as contrasted with contemporary views. This chapter provides an interesting contrast when further compared with today's contemporary views. Miriam Ben-Peretz provides Chapter 22, Kelly's Theory of Personal Constructs as a Paradigm for Investigating Teacher Thinking, in which she outlines her research on how teachers construe particular curricula. To elicit these constructs she developed a 'Curriculum Item Repertory'-Instrument (CIR) based on Kelly's personal construct psychology. Using this instrument, she was able to compare the constructs of experienced and inexperienced teachers.

In Chapter 23 Elbaz, Hoz, Tomer, Chayo, Mahler, and Yeheskel present The Use of Concept Mapping in the Study of Teachers' Knowledge Structure. Their work sees the development of teacher thinking in terms of knowledge structure, and declarative and procedural knowledge. A framework of three interacting factors: knowledge, the location in which it develops and the role assumed by the teacher or student identifies transition points between each factor when knowledge structures are expected to change. Concept maps are used to illustrate the process.

A Cognitive Model of Behaviour and Learning as a Heuristic Method of Theory-Based Teacher Training is contained in Chapter 24 where Hartmut Thiele presents a model for the training of teaching skills based upon a cognitive view on teaching and his research project evaluating this model. Alan Brown, in Chapter 25 How to Change What Teachers Think About Teachers: Affirmative Action in Promotion Decisions reports that administrators involved with promotion decisions changed their promotion criteria in use after having been confronted with them. He uncovered their criteria using techniques based on Kelly's theory. Chapter 26, Arguments for Using Biography in Understanding Teacher Thinking by Richard Butt pleads for the use of teachers' biographies in understanding teacher thinking, a well-known method within historical studies which could contribute to understanding teachers' professional development

In Chapter 27, Access to Teacher Cognitions: Problems of Assessment and Analysis, the problems of gaining access to teaches' cognitions is discussed by Gunther Huber and Heinz Mandl. Normally these verbalized cognitions are validated by the teacher's approval of the researcher's interpretation of his/her verbalized cognitions. Huber and Mandl wonder whether this communicative validation is sufficient in respect to the explanation of teacher's actions. If these cognitions are valid, then they must stand the test of a criterion of predictive adequacy: that is, they will then influence future actions. So they proposed action validation to follow communicative validation of teachers' verbalized cognitions.

In Chapter 28 On the Limitations of the Theory Metaphor for the Study of Teachers' Expert Knowledge, Rainer Bromme discusses the limitations of the theory-metaphor for the study of teachers' expert knowledge. This theory-metaphor assumes that teachers acquire and test their practical knowledge as scientists do their theories. Although this framework aids in the description of the way in which subjective theories about reality are tested, it does not facilitate in the analysis of the function of knowledge in respect of teaching activities. Another problem as for Bromme is the conceptualisation of subjective theories as systems of propositions. These are assumed to be accessible to teachers for verbalization yet this is not always so.

Chapter 29 Post-Interactive Reflections of Teachers: A Critical Appraisal by Joost Lowyck, extends the ideas of teacher thinking by acknowledging that much attention is paid to teachers pre-interactive thoughts, less to the interactive and even less to the post-interactive. Teachers think about meaningful content rather than in chronological terms regarding planning, interaction, and post-interactive reflection. Chapter 30 contains Jan Gerris, Vincent Peters and Theo Bergen's presentation of Teachers and Their Educational Situations They are Concerned About: Preliminary Research Findings and reports on teachers' descriptions of the teaching situation that worry them. The descriptions were also categorised and in this way related to teaching experience, which gives some insight into how problems change with increased teaching experience.

Chapter 21

Ten Years of Conceptual Development in Research on Teacher Thinking

Christopher M. Clark

Summary

This paper describes conceptions of the teacher, of students, of subject matter, of context, and of the research process that characterized the field of research on teaching in 1975 and contrasts these with contemporary views. Concerns are raised about the meaning of these conceptual developments with regard to multi-disciplinarity, competition for research funding, absence of attention to populations of teachers and students who are most at risk, focus on thinking processes to the exclusion of content, and insufficient attention to unintended side effects of research on teacher thinking.

Introduction

In the last decade, many researchers have become active in research on teacher thinking. We have done tens of studies, invented new methods and designs, formed special interest groups and organizations (like ISATT), and published books, journal articles, monographs, and reviews. Now the time has come to ask ourselves what we have accomplished in the service of teachers and students – what does this work add up to and in what directions should we proceed?

This is a critical paper. It is critical because I care deeply about education and I care deeply about researchers on teacher thinking and the work that we do. The work, research on teacher thinking is exciting and important. And I believe that today we have reached a crucial decision point ourselves – a choice between continuing to be merely curious about teachers' thought and action, or to also do good for teachers, students, communities and for ourselves. I hope to influence that choice and to suggest ways in which to act on it in the years to come.

I have organized this paper into two major sections. The first and longer part traces the development of conceptions that have influenced research on teacher thinking during the past ten years. In that section, I contrast views of the teacher, of the students, of subject matter, of context, and of research that characterized our field ten years ago with contemporary views. In part two of the paper, I raise some nasty and unsettling questions about what the last ten years of conceptual development in our field imply. Part 1 celebrates how far we have come. Part 2 calls us to reflect and (perhaps) to repent.

Celebration of Conceptual Progress

Conceptions of the teacher

How did the founders of research on teacher thinking think about the teacher? Who was this person in 1975? What roles did teachers play and what tasks were they expected to perform? In addition, what was the status of teachers' knowledge at that time? My reading of the report of Panel 6 of the National Institute of Education Conference on Studies in Teaching (NIE, 1975) offers the following portrait: The teacher is a decision-maker. The teacher is in a clinical relationship with students. A major role of the teacher is to diagnose needs and learning problems of students and to wisely prescribe effective and appropriate instructional treatments. The metaphor of teacher as physician was alive and well in 1975, and many of the methods of inquiry proposed for use in the study of teacher thinking were first developed to study the clinical decisions of medical doctors. In 1975, the teacher was also portrayed as a kind of business executive who operated in a 'boundedly rational' world by defining a problem space in an infinitely complex task environment and by seeking merely satisfactory solutions to problems and dilemmas rather than working toward optimal solutions (Simon, 1957).

This, then, is our composite portrait of the teacher in 1975. Perhaps it tells us as much about the scholars who wrote the Panel 6 Report as it does about teachers ten years ago. The medical metaphor certainly reflects the fact that Shulman and Elstein (1975) spent many more hours working with physicians than listening to teachers in the years immediately preceding the NIE conference. This image of the teacher also reflects our tendency to borrow theory and method from psychology, in this case, cognitive psychology, with a heavy overlay of information processing theory. The heroes of the day were Simon, Brunswick, Hammond, and Kleinmuntz.

In the past ten years, the image of the teacher implicit or explicit in research on teacher thinking has evolved. Decision-making has been replaced with 'sense-making' as the central cognitive activity of teachers. (This is not to say that teachers do not make decisions, but rather that teachers' decision-making is now seen as one among several activities in the service of making meaning for themselves and their students.) The metaphor of teacher as physician is giving way to the image of teacher as reflective professional (Schön, 1983). In a sense, this change represents a move toward a more abstract conception of teacher and teacher's role. Medicine is a profession, and so are architecture, law, business, and the military. We have moved away from the primarily diagnostic-prescriptive way of thinking about the mental life of teachers toward a more general view of teaching as a profession that calls for extensive knowledge of the content to be taught and of the psychologies of learning and of students, all of which must be interpreted, adapted, and artfully applied to particular situations by the reflectively professional teacher. This view of the teacher supplements or subsumes role definitions that emphasize the technical skills of instructing as the defining characteristics of effective teaching.

The teacher of 1985 is a constructivist who continually builds, elaborates, and tests his or her personal theory of the world. The notions of bounded rationality, task environment, and problem space are still invoked, but with some important developments. First, we are coming to recognize that it is unfair and incomplete to impute all of the responsibility for defining a problem space to the teacher. Powerful influences outside the control of individual teachers play parts in defining the problem space. A classroom with fifty hot and hungry children of poverty calls forth a rather different problem space than does a class of twenty compliant and comfortable children of the middle class. A further twist on the notion of bounded rationality in teachers is the realization that teachers, like their students and other human beings, can and often do hold multiple conflicting theories and explanations about the world and its phenomena (Roth, 1984). Teachers (and researchers) tend to switch back and forth among these inconsistent and incompatible ways of thinking and explaining. Moreover, the amazing thing is that this kind of inconsistent, imperfect, and incomplete way of thinking works rather well in the complex and practical world of the classroom. Teachers accept merely satisfactory solutions to problems not because they are lazy or ignorant, but because many of the problems that they face are genuine dilemmas (Berlak and Berlak, 1981; Lampert, 1985; Wagner, 1984) that have no optimal solutions. In short, we have begun to move away from the cybernetically elegant, internally consistent, but mechanical metaphors that guided our earlier work.

Conceptions of Students

Ten years ago, the image of students held by the founders of research on teacher thinking had to be inferred, because little direct attention was paid to writing about the image of the student. My inferences are that students were seen as puzzles to be solved or as patients to be diagnosed. While a great deal of attention was paid to the teacher as an active processor of information, students were still seen as relatively passive and manipulable recipients of information. Early research on teacher thinking was justified by claiming that teacher thinking controlled teacher behavior and that teacher behavior was what produced student learning. Students, by implication, were objects to be acted upon by teachers with more or less thoughtful technical skill. Students were also seen as the source of cues that teachers might notice and decide to use as a basis for adjusting their own teaching behavior (NIE, 1975; Marx, 1978; Cone, 1978; Shavelson, Cadwell, and Isu, 1977; Peterson and Clark, 1978).

In the past ten years, our way of thinking about students has also evolved. There is some recognition that, like teachers, students are thinkers, planners, and decision makers themselves (Wittrock, 1986). Students are now seen as transformers of knowledge – active learners who enter the classroom with robust preconceptions about the world and how it operates (Posner, Strike, Hewson and Gertzog, 1982; Roth, 1984). Students are now called ‘novices’ who have responsibilities for cooperating in their own learning. The metaphor of the individual medical patient to be diagnosed is giving way to a more dynamic and social view of students who learn from one another, who build on and draw from what they already know and believe, and who necessarily make a somewhat different sense of what teachers teach them than the sense that the teacher intended to communicate. In short, students are constructivists also, and our research has just begun to explore what happens when two sets of constructivists, differing in knowledge, experience, motivation, and authority, put their heads together. Shulman and Carey (1984) have recently suggested that the educational research community has evolved in its thinking toward a view of humans as collectively rational. This view acknowledges the importance of attending to how teachers and students exercise their abilities "... to participate intelligently and reasonably in groups, to pursue multiple mutually shared goals through the exercise of reason jointly produced and collaboratively exercised" (Shulman and Carey, 1984, p. 157).

Conceptions of Curriculum

At the beginning of this era of research on teacher thinking, mention of curriculum was noticeably absent from our literature and our deliberations. The psychologists were in charge and our focus was on how teachers dealt with the 'givens' of schooling. Subject matter was seen as one of these givens – a condition or set of variables to be described or controlled. Research by DeGroot on chess masters was cited in this early literature (Shulman and Elstein, 1975), but the way in which that work was cited did not include much attention to the content and organization of subject matter knowledge in the minds of either chess masters or, by extension, school teachers.

Two lines of research in the past decade have done much to enrich our sense of the role that curriculum and subject matter knowledge play in understanding teaching generally and teacher thinking in particular. The first of these lines of research is that on distinctions between the knowledge and performance of novices and experts. Research in this area by Greeno (Greeno, Glaser and Newell, 1983), Leinhardt (Leinhardt, 1983; Leinhardt, Weidman and Hammond, 1984), Larkin (Larkin, McDermott, Simon and Simon, 1980) and others demonstrates that the amount of knowledge in a richly interconnected conceptual network is rather different from the amount and organization of knowledge in the minds of novices. These differences have profound implications for the ease with which experts approach and solve both familiar and novel problems, including problems of pedagogy. We have come to believe that there are qualitative differences in the ways in which experts and novices know and think about what they know.

A second line of research that bears on teachers' subject matter knowledge is that on teacher planning (see Clark and Peterson, 1986, for a detailed review). It was a surprise to me in 1978 that my own research on teacher planning, which I thought of as primarily psychological, aroused so much interest among curriculum theorists. As I began to correspond with curricularists and to publish in their journals, I came to see that teacher planning is a means for organizing and transforming subject matter knowledge and curriculum into pedagogically useful forms and routines. I came to understand teacher planning as a link between thought and action that, in many respects, defines and sometimes distorts the content to be taught. I came to see a connection between what process-product researchers call 'opportunity to learn' and the dynamics of teacher planning that shape opportunity to learn.

Both the research on novice-expert contrasts and that on teacher planning support the idea that teachers' knowledge of content to be taught and the ways in which that knowledge is organized are crucial influences on teacher thinking and action. More recently, ethnographers and philosophers have shed even more light on the importance of what teachers know and how they hold and use their knowledge. Here, I think of the important work on personal practical knowledge done at the Ontario Institute for Studies in Education (Elbaz, 1983; Connelly and Clandinin, 1984; Clandinin, 1985; Kroma, 1983), of Lampert's analysis of the nature of teachers' knowledge and needs (Lampert, 1985), and philosophical analyses of conceptions of knowledge and education by Buchmann and her colleagues at Michigan State University (Buchmann, 1984; Buchmann and Schwille, 1983). Along with the phrase 'subject matter knowledge' the word 'epistemology' has become much more common in the literature of research on teacher thinking. That is, we have come to appreciate that analysis of teachers' mastery of the facts and procedures of the disciplines they teach is insufficient. We must also come to understand teachers' *ways of knowing* and their beliefs about the nature of knowledge itself before we can begin to understand the role of knowledge and curriculum in teacher thinking and in education more generally (Floden, 1983).

One intriguing hypothesis about the way in which teachers' knowledge is organized is that it takes the form of cases (Shulman, 1986). This line of thinking helps us to distinguish between knowledge of a discipline in the forms commonly represented in textbooks (i.e., abstract principles, general laws, fundamental ideas, and puzzle forms) and replaces it with images of cases, vivid experiences, and good examples. This line of thought also suggests that there may be important differences between the ways in which, say, scientists or mathematicians hold and express their knowledge and the ways in which teachers of science or mathematics hold and express their knowledge of the same phenomena. Perhaps the distinction between disciplinary knowledge and pedagogical knowledge is overdrawn, but I think that it holds some promise. In part of my own work of the next several years, I intend to pursue a program of research on the psychology, epistemology, and pedagogy of good examples; how teachers learn to create, recognize, and use good examples, when good examples can get us into trouble (conceptually speaking), and how students' understandings of and memory for what they are taught are affected by the quality of the examples used in teaching.

Conceptions of Context

Perhaps the most dramatic set of conceptual developments in research on teacher thinking relate to the changes in how we have come to think of the context of teaching. My impression of our 1975 image of context is that the school classroom was seen as the unit of analysis: a clearly bounded yet complex task environment, the major purpose of which was to produce or foster student learning. This is essentially the same image of the context of schooling found in the process-product and teacher effectiveness literature. Phillip Jackson (1968) alerted us to some of the invisible complexities of the classroom almost 20 years ago. But, as researchers on teaching, we tended to believe strongly in the power of the classroom walls as boundaries for the physical, social, and psychological context of education.

The active presence of anthropologists of education in our field during the past 8 years has done a great deal to broaden and enrich our conceptions of context. We have moved away from a rather impoverished and fragmented notion of context as an aggregation of 'background variables' to a richer, more dynamic, collectively defined and negotiated understanding of context (Shultz, Florio and Erickson, 1982; Erickson, 1986). However, most of the anthropological and multidisciplinary studies of teaching in our literature illuminate the *internal* dynamics of classroom context. We have learned how classroom rules and routines are negotiated, discovered, and enforced. Field workers and socio-linguists have shown us how cultural differences and similarities between teacher and students can impede or enable teaching and learning (Phillips, 1972). But with all this progress in our thinking about context, I believe we have a need for more conceptual development in this area. My developing view is that the school and classroom are rather more permeable settings for teaching and learning than our literature and our thinking suggest. Schools and classrooms are the locus of social, psychological, physical, political, and metaphysical action, embedded in the world and affected by it. The purposes of schooling and of classrooms are many and varied, and only occasionally is student learning the first and overriding priority. Context is not a 'variable' or a collection of more or less mechanical components, any more than a river or the temperate zone of the northern hemisphere are aggregations of variables. I believe that we must think more synthetically and holistically about context if we are to continue to make progress in understanding teacher and student thinking and their ramifications.

Conceptions of Research

My conception of research on teacher thinking has certainly developed in surprising ways during the past ten years. I would like to summarize these conceptual developments under four headings: a) Methods of Inquiry; b) Concepts and Models; c) Aims and Products of Research; and d) The Relationship of Research to the Practice of Teaching.

In the past ten years, I believe we have moved from a dependence on methods of the psychology laboratory to a dependence on methods of fieldwork. The psychology laboratory contributed techniques such as stimulated recall interviewing, clinical interviews of various sorts, experimental designs for policy capturing studies, and think-aloud and protocol analysis methods. Use of these methods has sparked interesting and important debates about their validity and reliability, and raised fundamental questions about the limits of what can be learned from introspection, recollection, and self-reports of cognitive processes (Nisbett and Wilson, 1977; Eriksson and Simon, 1980; Yinger and Clark, 1982; Huber and Mandl, 1984). While this debate is certainly not yet settled, our approach to the study of teacher thinking has tended to move from relatively well-controlled and researcher-defined conditions and settings to the more representatively complex teacher- and school-defined world of real classrooms. Anthropological description and fieldwork analysis methods have been added to evolved versions of laboratory methods. The emphasis has moved from hypothesis testing about cognitive processes to what Erickson (1986) calls interpretive analyses, in which we become more explicit about the role of the investigator in making sense of his or her experience. Thick description, triangulation, and collaborative interpretation of descriptive research have become more common (Erickson, 1986). We have begun to adopt the canon of 'disciplined subjectivity' in place of the myth of 'scientific objectivity.'

Ten years ago, the majority of our concepts used to frame questions and describe teachers' thinking came from psychology, and our models for describing processes had a decidedly cybernetic quality (e.g., Peterson and Clark, 1978; Shavelson and Stern, 1981). In the ensuing years, the limitations of these primarily psychological concepts and cybernetic models have become more apparent. Perc Marland has written a nice analysis of the relative strengths and limitations of models of teachers' interactive thinking (Marland, 1983), and Clark and Peterson (1986) discuss the shortcomings of both their own and Shavelson and Stern's (1981) models of teacher thinking during instruction. Similarly, the once-unquestioned rational planning model has been challenged by empiricists and theorists such as

Robert Yinger (1977). I agree thoroughly with Yinger's more recent exhortations that we should put energy into developing and discovering what he calls 'the language of practice' (Yinger, 1985), by which he means a language using concepts and terms that mirror and express life in classrooms as teachers and students see and experience it, rather than as visiting social scientists see it. As a field, we have become somewhat more sensitive to the dangers of reification of our own invented concepts and models, but there is still abundant need for caution. We must remind ourselves that it is possible for many false and even dangerous models to 'fit' the intrinsically partial data that we have about teachers' thought processes.

Ten years ago, researchers on teacher thinking were primarily concerned with increasing the depth of our understanding of the mental lives of teachers in order to be able to a) explain why teaching operates as it does, and b) to improve the practice of teaching both by direct training interventions and more indirectly by organizing schools, curriculum, policies, etc. to fit more smoothly with the operations and limitations of the mental lives of teachers. In short, the early promise of this work was to produce fundamental psychological knowledge that had immediate implications for practice. Recently, I closed an extensive review of research on teacher thinking with the assessment that "A decade of research on teachers' though processes has taught us as much about how to think about teaching as it has about teachers' thinking" (Clark and Peterson, 1986). I still believe that this is so; that is, that the psychological processes that we have learned somewhat more about are not unique to teachers but are generally human qualities, strengths, and limitations. However, we have gained a new and more detailed appreciation for teaching in all of its complexity; of this, there can be no doubt. In addition, beyond this appreciation, I believe the field has begun to adopt the aim of empowerment of individual teachers. That is, the aim of providing the reflectively professional teacher with tools and encouragement to frame and solve his or her own unique professional challenges in much the way other professionals do. We are now on a mission of advocacy and service to teachers rather than on a quest of discovery in a strange and unfamiliar land.

Finally, we take up a continuing and troublesome issue: that of the relationship between research and the practice of teaching. The early literature of our field is rather silent on this topic, except for the ever-present analogy to the study of medical decision-making. However, it has rarely been the case that researchers on teacher thinking actually see themselves in the same kind of power and respect relationship to schoolteachers as they do to physicians. The implied relationship to teaching practice was essentially explanatory and prescriptive. The implication was that effective planning and decision-making would be discovered, described, modeled, and taught to less experienced and less effective teachers. Hand-in-glove with this implication was the idea that successful research will make the effective practice of teaching easier.

Today, I think we are in a different, though still incomplete relationship to the practice of teaching. I have advocated thinking about research on teaching as providing *service* to the practice of teaching (Clark, 1984). My work with Susan Florio-Ruane in the MSU Written Literacy Forum (Clark and Florio, 1983) is one example of an ambitious, expensive, and rewarding way to work out and discover appropriately helpful relationships between research and practice; between researchers and practitioners. Others, particularly in science education (e.g., Posner, Strike, Hewson and Gertzog, 1982; Roth, 1984), have attempted to improve the practice of teaching by designing curriculum materials and teacher's guides that are intended to make use of what we have learned about students' preconceptions and about teachers' planning and subject matter knowledge. Teacher education programs at Michigan State University have been reorganized to reflect new discoveries about the formerly invisible domain of teacher thinking and to incorporate and draw implications from contemporary views of knowledge and knowledge change (Confrey, 1982). In addition, perhaps most importantly, I see changes in the relationships between teachers and researchers. Teachers' knowledge is now more respected by researchers than was the case in 1975 (Clark and Lampert, 1985). Teachers have begun to become more active as full partners in the research process (Burton, 1985; Florio and Walsh, 1981), and a few courageous researchers have spent a year or more shouldering the responsibilities, demands, and rewards of classroom teaching themselves. All this has led to a constructive turning away from the goal of 'making good teaching easier' to that of portraying and understanding good teaching in all of its irreducible complexity and difficulty. Quality portraiture may be of more practical and inspirational value than reductionistic analysis and technical prescriptiveness.

Reflections and Implications

What do these conceptual developments of a decade of research on teacher thinking mean, and where do they lead? First, I believe that they reflect progress, in the best sense of that word: Gains in what we know about teachers and the tasks they face, improvements in our methods of studying and describing thought and action, more complete preparation of prospective teachers. Ten years of research on teacher thinking have made possible a constructive combination of research on instruction with research on curriculum. Practicing teachers have been brought more fully into the important enterprise of understanding, describing, and improving teaching and learning.

But there are also less positive and more worrisome ways to interpret these conceptual developments of the last ten years, and I raise these now – because it is time to ask big questions; to take a larger perspective on our work; to ask what we have really accomplished, and at what costs. We have a rare opportunity to confront these larger questions, to think and reflect together, and possibly, to chart a new course for research on teacher thinking. To do so, I believe that we should examine the darker side of progress in research on teacher thinking. Please consider these hypotheses about what my list of conceptual developments in research on teacher thinking could imply:

First, these conceptual and methodological developments could reflect merely application of the theories and methods of more and more academic disciplines. Multi-disciplinary research was called for early in the life of our field. This was because the early writers realized that the complexities of schooling would not yield to the point of view of a single discipline. Psychology was followed by anthropology, sociology, economics, and philosophy. To these were added the practical wisdom of experienced teachers. However, at some point, multi-disciplinarity may have become an end in itself. Borrowing increasingly heavily from other disciplines will certainly add to the variety of concepts and tools for inquiry at our disposal. Nevertheless, we should ask, does this variety represent progress, or merely proliferation?

A second possible explanation for our conceptual progress is that of competition between paradigms, people, and institutions. Competition for research funding (at least in North America) is vigorous. In addition, one of the principal grounds on which research proposals are judged is originality. Thus, there seems to be more of a premium placed on making our research distinct from that of our predecessors than there is on synthesis, replication, and conceptual coherence. This pressure for originality and distinctiveness may have been self-imposed. It also may have produced a richer variety of approaches and tools for inquiry than would have otherwise developed in ten years. But now may be the time to reject originality and distinctiveness, return to our fundamental research questions, and proceed to answer those questions by using the small number of research designs and methods of inquiry that seem to have served us well.

A third implication of our short history is most unsettling. As I have reviewed the literature of research on teaching, I have developed the conviction that this work represents a failure of moral courage on our part. In my judgement, none of the research on teaching thinking has directly addressed serious and difficult problems and crises in education. For example, the work on teacher planning and decision making has been done almost exclusively in nice, well-organized, upper middle class suburban elementary school classrooms. We have omitted any attention to all to more volatile and challenging problem areas in our education systems including poverty, nationalism, cultural conflict, racism, sexism, discrimination, and massive failure to learn in certain quarters of our educational systems. Have we doomed ourselves to triviality by our lack of moral courage? Are we satisfied with serving our own intellectual curiosities while teachers of the poor and the handicapped struggle without help? On the other hand, have we just reached the point at which our own understandings of teacher thinking permit us to sail in more troubled waters?

A fourth possible reading of our recent history is that we continue to prize process over the content and substance of instruction. We have described and come to better understandings of various planning, decision-making, and reflective processes used by teachers. However, there is considerably less attention in our work to the quality and organization of what is being taught. Even the curricularists among us seem more caught up in a focus on process than on the content of teaching. To the extent that this is so, we are no different and no better than the process-product researchers who can tell us how to increase time on task, but who have nothing to say about the quality and worth of the tasks themselves.

Fifth, I am concerned that our research reflects a narrow parochialism, in which we stare so intently at the teacher as the source, linchpin, and dynamo of education that we have become insensitive to the other very powerful forces and constraints that shape and influence schooling. Like solar astronomers, we have become partially blind by staring too long at the sun. Where are the students, the curriculum, the community, and a dynamic theory of their interactions? The teacher is certainly an important and central agent in education. However, let us not lose all sense of proportion as we frame and interpret our research.

A sixth interpretation of our work is that, unconsciously, we have promoted an insidious form of elitism. We have elevated and lionized those few schoolteachers who are most like ourselves (reflective, analytic, verbally articulate, sophisticated in their knowledge, liberal and worldly in their values). These are the teachers whose planning, thinking, and decision-making we study and, unreflectively, portray as ideals for all other teachers, experienced and novice alike. While our rhetoric sounds a call for 'power to all teachers', our research is cast in such a way that only those few teachers who are already most like us can identify with it. In what ways does this work serve those who need our support most? (A related point made recently by Roger Simon of OISE is that our advocacy of reflection, self-examination, and attention to private and individual cognitive activity of teachers may have an unintended and paradoxically conservative effect. For by urging teachers to focus mainly on their inner lives, we draw their attention away from the larger, collective, external forces and entities that may be manipulating and controlling them and the entire system of education.)

Seventh, and finally, our research on teacher thinking may entail the assumption that the end justifies the means. That is, that we need not examine the moral, social, and psychological costs of a method of instruction (e.g., direct instruction for children of the poor; conceptual change teaching of science) if that method leads to higher achievement test scores or 'more correct' conceptual organization of ideas. If we are on a mission of discovery, we have responsibility to discover and count the costs of unintended and unexamined side effects of our suggestions and prescriptions, as well as documenting the ways in which our discoveries have solved old problems.

Conclusion

I could have pointed to many promising, exciting, and praiseworthy implications of our ten years of research on teacher thinking. I have done so in other places, at other times (e.g., Clark and Peterson, 1986). However, my purpose here is to set a self-critical tone for researchers on teacher thinking. As you reflect on your own research, read the reports of colleagues, and plan future studies I urge you to ask yourselves and your colleagues: What are the social and personal dangers in pursuing this line of inquiry? How should your ways of working be changed to better serve the interests of all educators, school children, and the world community? How does knowledge about teacher thinking fit into larger, more dynamic, and more complete conceptions of education?

In my judgment, we need to put at least as much as energy and creativity into answering the question 'How shall we do good?' as we have into inventing methods for probing and describing the hidden world of teacher thinking.

Author Reflection Twenty Years On

Reading my chapter today, in the year 2002, is a bittersweet experience. When the chapter was first published, 16 years ago, researchers on teacher thinking were full of spirit, optimism, and energy. As a group we had accomplished enough to be proud and were mature enough to encourage constructive self-criticism. The year 1986 may well have been the developmental peak of research on teacher thinking. Those were the good old days.

I am pleased, on reflection, that some of my exhortations bore fruit. Researchers did indeed turn toward studying the subject matter knowledge of teachers and how they bring that knowledge to life in classrooms. Social constructivism has been embraced by many of our colleagues as a theory that could move us away from myopic focus of the teacher alone, bringing students and the learning community into the picture.

But, disappointingly, my most critical assessment of research on teacher thinking in 1986 rings as true today as it did then: the vast majority of contemporary studies of teaching an learning take place far from the most needy schools and communities. We are learning more and more about teachers and children blessed with social and economic advantages. The children of the poor and their teachers are distressingly absent from our literature. The achievement gap widens, and high-stakes accountability schemes narrow the curriculum to mimic the all-important standardized tests. The mental lives of teachers, so celebrated in 1986 are now as much characterized by stress and anxiety as they are by planning and decision making.

Even so, I am still filled with awe and hope about the future of schooling, inspired by heroic teachers working in under-resources schools and communities. Let us collaborate with them to come to an understanding and appreciation of teaching at its most challenging.

Chapter 22

Kelly's Theory of Personal Constructs as a Paradigm for Investigating Teacher Thinking

Miriam Ben-Peretz

Summary

The appropriateness of Kelly's theory of personal constructs for investigating teacher thinking is analyzed and some of the insights to be gained are noted. Focusing on teachers' thinking about curriculum materials, research methodology and instruments are described and results of some research projects are presented. Implications for further research, as well as for professional development programs, are pointed out.

Personal Constructs of Teachers

According to Kelly (1955) human beings grasp their environment by means of interpretations through the use of a system of personal constructs. Personal construct theory examines the thoughts behind the actions of individuals, and can be used for the analysis of action in terms of subjective categories. Kelly's theory as a general theory of thinking and action provides us with a framework for viewing professional thinking and actions as one instant of a general paradigm. The theory emphasizes interaction between man and the environment as an experiential cycle in which people develop their personal construct system.

A construct is a way in which some things are seen as being alike and, simultaneously, different from others. Each construct consists of a single bipolar distinction, e.g. moral-immoral. One pole of the construct represents the basis of the perceived similarity; the other pole represents the basis of contrast.

A construct is unlike a logical concept in that its boundaries are personally defined because of individual and personal experience. One may well perceive teacher thinking as a development of personal construct systems through the experiences teachers have in interaction with their environment, inside and outside the educational establishment. According to Bannister (1970), no two systems are exactly alike, although similarities may exist because people inhabit similar internal and external worlds. Teachers are, on the one hand, inhabitants of similar worlds – there exists an astonishing resemblance between classrooms viewed as ecological units in which teaching learning situations occur. Teachers also share some fundamental similarities in the process of teacher training. On the other hand, every educational situation has its unique characteristics, beyond the features common to all, and the divergence may be quite pronounced. Therefore, one may expect to find commonalities as well as differences, between construct systems of teachers. Personal construct theory attaches a significant role to individual past histories, which may have a crucial function in the process of teacher development. Instead of searching for general stages of professional development (Fuller, 1969), Kelly's theory provides a tool for investigating the personal and unique process of development of individual teachers.

Kelly's experience corollary, that a person's construct system varies as he successfully construes the replication of events, provides us with a frame of reference for viewing teacher development as a personal process of learning. Constructs are used by a person to describe present experience, to forecast events - thus building theories, as well as to assess the accuracy of previous forecasts after the events have occurred, thereby testing their predictive efficiency. Teachers are, by the very nature of their profession, in a situation of constantly forecasting events and testing these forecasts. This process of testing forecasts, attempting to validate idiosyncratic construct frameworks, constitutes teachers' learning in the context of their profession.

Kelly claims that individuals should be encouraged to see their ideas as hypotheses, or representational models, open to refutation. Representational models are viewed as composed of a series of interrelated personal constructs or tentative hypotheses about the world. Thus, constructs are not to be conceived as isolated components in the thought processes of teachers, but their interrelationship becomes a crucial factor in teacher thinking.

Hunt (1982) used Kelly's approach to assist teachers in identifying their own personal implicit theories of teaching and learning, which are shaped by their personal experience. This experience comprises the acquisitions of theories about man, society, learning and teaching. Using Kelly's theory for investigating teachers' thinking may serve to reduce the gap between theory and practice in education, as personal constructs may have their roots in formal theories as well as in classroom experiences or personal histories.

The theory of personal constructs provides us with a large array of research questions related to the nature of teacher thinking. Appropriate research instruments and data processing procedures have been developed. The theory permits in depth investigation of construct systems of individuals as well as comparison between individuals and between groups.

Insights into similarities between construct systems of different individuals, or groups of teachers, may be a basis for the understanding of professionally unique ways of interpretation of educational environments. These professional unique ways of interpretation of reality are viewed as regulating teachers' classroom behaviors. Analysis of the personal construct-systems of teachers may thus become a source for defining their professional identity.

Interesting questions may be pursued through the application of Kelly's theory to teacher thinking. Thus, one may ask questions in relation to Kelly's range corollary. This corollary stipulates that each construct is convenient for the anticipation of a finite range of events only. A distinction is made between the "focus of convenience" of a construct, which is defined as the sector of convenience wherein it is most useful, and its "context", all events to which it is ordinarily applied. Teachers may use constructs in different ranges of convenience. For example, the focus of convenience of a teacher's use of the construct successful-unsuccessful may be limited to scholarly achievements of pupils, or may be extended to their social behavior. Investigating ranges of convenience of teachers' use of constructs may yield insights into their ways of thinking in professional situations.

Other questions may be derived from the application of the organization corollary, which implies that a person's constructs are interrelated. Positive relationships between constructs applied to the same set of events may illuminate the thoughts behind teachers' actions. For example, if we were to observe a teacher who designates as 'industrious' most of the pupils he categorizes as 'bright', we would gain insight into his approach to learners.

Kelly's fragmentation corollary permits the differentiation of an individual's construct system into independently organized subsystems, thus allowing a person to handle several events simultaneously. This seems to be an important quality in the 'on the job' thinking of teachers and calls for investigation. However, the extent of differentiation is limited by the degree of permeability of superordinate constructs. According to Kelly a construct is permeable if new experiences and new events can be admitted to its context of application. A stimulating line of research is the investigation of the *permeability* and change of teachers' constructs.

Personal construct theory originated in Kelly's experiences with clients in psychotherapy. Kelly emphasized interpersonal relationships and the interaction between different construct systems. The sociality corollary stipulates that to the extent that individuals construe the construction process of other persons, they may play a role in social processes involving these others. It follows that it may be important to inquire into teachers' abilities to make inferences about the construct systems of pupils, parents, colleagues and principals.

The following are some of the theoretical insights to be gained by using Kelly's theory of personal constructs for investigating teacher thinking;

(1) Knowledge about a variety of cognitive maps of personal construct systems of teachers with different background characteristics;

(2) Understanding of the development of personal construct systems of teachers through the experiential cycle of teaching;

(3) Definitions of characteristics of the teaching profession based on a distinction between professional versus non-professional components of the personal construct systems of teachers.

Investigating teacher thinking in the framework of personal construct theory may have practical implications. Making people aware of their own construing patterns and processes plays an important part in allowing them to change, i.e., to learn. Thus, participation in the research may become an educative process for teachers. Moreover, workshops and exercises for identifying personal constructs may be planned as part of teacher's professional training and development.

Focusing on Curriculum Materials

Research on teaching thinking sometimes aims at recovering processes of teachers' lesson planning (Clark and Yinger, 1977). Curriculum materials are an integral part of the professional environment of teachers and may serve as the basis of their lesson planning. Much of teacher thinking goes on in relation to curriculum materials, their interpretation, and transformation into lesson plans and class realities. Teachers may be perceived as using their personal constructs for examining, making sense of, and using available curriculum materials. Constructs are considered as forerunners of action. Thus, the teacher who construes an activity included in curriculum materials, such as "block building", as an exercise for muscle development will make different predictions about this activity, and will undoubtedly act in different ways, from one who construes this activity as play, or from another who construes it as a child's concrete representation of thought (Bussis et al., 1976). Although teachers' personal constructs may coincide with those of their colleagues, the construct system of any individual teacher determines his, or her, professional activities in the planning and implementation of their lessons.

Several studies were carried out focusing on teacher thinking about curriculum and curriculum materials, exploring the personal construct systems of teachers (Ben-Peretz and Katz, 1980; Ben-Peretz et al., 1982; Ben-Peretz and Katz, 1983).
The following research questions were posed:
(1) What constructs do teachers use in relation to curriculum materials?
(2) What impact do personal and contextual factors have on construct systems of teachers?
(3) Are "teacher" constructs different from those of academics, who are not teachers, relating to the same curriculum materials?
(4) Does teacher education change the construct systems of student teachers in relation to curriculum materials?

Methodology and Research Instruments
The Curriculum Item Repertory (CIR) instrument developed for the studies reported here is an extension of the Role Construct Repertory Test Form and is based on repertory grid methodology. According to Adams-Webber (1979) repertory grid technique is basically a method of quantifying and analyzing relationships between the categories used by a respondent in performing a sorting task. The data are yielded in matrix form. The methodology is idiographic as it leads to the discovery of the unique pattern of relationships between the personal constructs elicited from an individual. In the original Role Construct Repertory Test each respondent is shown a standard list of 'role titles' such as 'father', 'boss', 'a person with whom you feel least comfortable' and asked to nominate for each role the individual who seems to fit best. The name of each nominated person is recorded on a card, together with the role title. A triad of these figures is then presented to the respondent who is asked to think of some important way in which two of the persons are similar to one another and different from the third. The basis of the perceived similarity and contrast is recorded as a single bipolar construct, for example, honest-dishonest. A series of triads is used to elicit twenty to thirty constructs from each respondent. This procedure is known as 'triadic sorting'.

The Curriculum Item Repertory (CIR) instrument presents curriculum items instead of role titles, as elements in order to elicit constructs used by teachers in their thinking about curriculum. The instruments consist of a set of twenty curriculum items, selected from a set of curriculum materials. These items include paragraphs from student textbooks, illustrations, worksheets, etc. Participants are asked to sort random triads from the set of items and decide for each triad which two of the three items differed from the third according to a self-generated construct. A grid form was prepared with numbers along the top representing the numbers of the curriculum items. The two poles of each construct, the positive and negative poles, are written down alongside the raw. Participants are then asked to use the construct chosen by them for classifying all items. The letter X indicates the positive pole and the letter Y the negative pole. The grid form can be scanned vertically to examine relationships between constructs and horizontally to examine relationships between curriculum items. The large number of personal constructs generated by participants necessitates the setting up of a communicable categorization system without losing the genuineness and validity of the primary constructs. The categorization is referred back to participants for validation. Classification of constructs into categories makes the comparison of sub-population possible. This seems important in order to gain understanding about the specific teaching environments. These categories are perceived as superordinate constructs subsuming other constructs. In order to understand long-term commitments of teachers in the ways they make sense of, and interpret curriculum materials, the superordinate constructs in their construct systems have to be investigated. Such constructs are likely to be more stable than specific constructs and therefore may have a more lasting impact on teaching. In order to identify differences between professional and non-professional personal constructs, participants are asked to generate constructs elicited by a set of cartoons. The differences of construct frequencies in each of the categories, as elicited by various sub-populations, are analyzed using the chi-square test. Construct frequencies are calculated in percentages.

Examples of Some Research Projects

(1) The impact of personal and contextual factors on personal constructs systems of teachers relating to curriculum materials

Five samples of secondary school teachers (N=34) participated in the study. The first comprised experienced English teachers (N=8); the second non-experienced English teachers (N=8); the third Arabic Language teachers (N=8); the other two comprised English teachers who teach advantaged students (N=5) and disadvantaged students (N=5).

Examination of the distribution of constructs in the various categories showed significant differences between subgroups. Experienced teachers produced more constructs related to teaching strategies and learning tasks demanded of pupils, than their inexperienced colleagues. Interesting differences were revealed between personal construct systems employed by English Language teachers and Arab Language teachers, the latter focusing more on formal constructs. This emphasis may be interpreted as related to the nature of the subject matter. Visual elements of the Arab language pose learning difficulties in the early stages of study. Thus, in Arabic, letters may have different shapes according to their position in the word. This difficulty may account for teachers' preoccupation with the visual format of curriculum materials.

(2) The impact of teacher education programs on the personal construct systems of student-teacher in relation to curriculum materials.

A set of 20 curriculum items was selected from a set of curriculum materials in Hebrew (Literature and Language). 10 groups of student teachers, altogether 145 individuals participated in the study. Five groups were first year students and five groups were 3rd year students. All had finished high school and army service; all except one were females. Participants of two groups, one first and one third year students, were Kibbutz members. Four groups of students, two first and two 3rd year students, specialized in special education, and four groups, two first and two 3rd year students, prepared themselves to teach in regular classes. The student teachers participating in the study were asked to sort random triads of items from the pool of 20 items, deciding each time which of the three was different from the others, according to a construct suggested by them. They were asked to repeat that procedure ten times. The constructs elicited from the subjects were then classified by investigators using seven categories. Inter-rater reliability for this classification was 0.95. The categories were as follows:

1. Format of materials (i.e., "lack of illustrations")
2. Method of teaching advocated by materials (i.e., "Inquiry")
3. Pupils' task specified in materials (i.e., "self directed reading")
4. Level of cognitive domain (i.e., "understanding", "evaluation")
5. Difficulty of items (i.e., "difficulty in relation to age")
6. Administrative aspects (i.e., "time needed for activity")
7. Field of knowledge incorporated in materials (i.e., "science")

The same method of triadic sorting was used with a set of 20 cartoons, chosen randomly, dealing with various fields of life such as – weather, army, medicine, sport, that served as a control instrument for eliciting personal constructs.

No significant differences in construct systems were found among five groups of first year students, which indicate a high level of similarity. Their emphasis was on "Format of materials". No significant differences were found among construct system of five groups of 3rd year students. Their emphasis was on "Methods of Teaching".

Non-significant, subtle differences were revealed between third year special education student teachers and those intending to teach regular pupils. The special education student teachers related more to pupils' tasks and pupils' difficulties, while the others related more to methods of teaching. Significant differences were found between first and 3rd year students relating to curriculum materials. The differences in frequencies of constructs in categories between first and 3rd year students were highly significant, at the .01 level of confidence.

First year student teachers were mostly concerned with "Format" while third year students were mostly concerned with "Methods of Teaching," "Pupils' Tasks" and "Level of Cognitive Domain". These results are explained in terms of a developmental process from the relative simplicity of the first year students' construct system towards the higher level of complexity of the 3rd year students' construct system.

Complexity of personal construct subsystems of student teachers relating to curriculum materials is determined in the first place, by an increase of construct frequency in those categories that represent professional aspects such as "Methods of Teaching," "Pupils' Tasks", and "Level of Cognitive Domain".

There is a distinct shift of emphasis of the personal construct system from mainly one category in the first year to three categories, almost equally emphasized, in the third year. This indicates not only an increase in construct number, which is one expression of complexity, but is also an expression of power of imagination and divergent thinking.

These findings strengthen the conclusion that three years of experience in teacher education programs cause a change in construct systems in relation to curriculum materials. This change cannot be attributed to a process of maturing because the construct system of participants relating to cartoons has not been affected by their three years of study.

Some Implications for Further Research And Professional Development Programs

Niels Bohr is quoted as having said: "The opposite of a correct statement is a false statement but the opposite of a profound truth may well be another profound truth". (Heisenberg, 1971, p. 102).

Kelly's theory of personal constructs offers one valid and interesting approach to the study of teacher thinking which may yield some true statements in this important research area.

Several questions lend themselves to further investigations such as:

- What relationships between cultural and ethnic contexts and teacher constructs can be identified?
- Do teachers use "professional" constructs in construing their experiences outside the educational environment?
- How do personal constructs of teachers relating to subject matter and curriculum materials change over time?
- What is the relationship between observed teacher actions, implementing curriculum materials and their construct systems?
- What are the construct systems of pupils in relation to curriculum materials? Are these different or similar to teachers' construct systems in relating to the same materials?
- What is the relationship between the nature and complexity of pupils' construct systems, relating to curriculum materials, and their achievements?

Teacher education and professional development programs could benefit from the recovery of personal constructs in relation to various components of the educational environment, such as pupils, peers, parents, physical settings, and curricula. Teachers' awareness of their own implicit theories would enable them to be more reflective in their professional decision-making. Teachers would be able to analyze their own thinking about curriculum materials, old and new, and their modes of curriculum interpretation would be enriched and more complex, yielding more of the educational potential of these materials. In-service teachers education programs, focusing on personal constructs, would seem appropriate in the following cases:

- Implementation of innovative curricula
- Matching curriculum materials to the needs of diverse pupil populations
- Identification of curricular trends as expressed in curriculum materials.

It may well be that a process of comparison and validation of constructs between teachers could contribute to the introduction of change into schools.

Chapter 23

The Use of Concept Mapping in the Study of Teachers' Knowledge Structures

F. Elbaz, R. Hoz, Y. Tomer, R. Chayot, S. Mahler and N. Yeheskel

Summary

In this paper the development of teacher thinking is conceived in terms of knowledge structures, and declarative and procedural knowledge. A framework is set that comprises three interacting factors: the kind of knowledge, the location in which it develops, and the role assumed by the teacher (or student). This framework was used to identify transition points within each factor, at which changes in knowledge structures are expected to occur. Representations of knowledge structures are obtained by using a general analysis scheme for cognitive maps, which are constructed by individual students from a set of domain-related concepts. This procedure is exemplified by the application of the above scheme to the maps of one biology student teacher, which resulted in the specification of both his knowledge structures and their development over a short period of time.

Introduction

A substantial number of studies on teacher thinking to date involve either experienced or beginning teachers. This paper is concerned with the development of teacher knowledge from the onset of formal training i.e., during preservice education. We propose to study this development using a conceptual framework drawn largely from cognitive psychology. Within this framework, we will speak of the knowledge structures (or cognitive structures) of human beings as the organized body of knowledge, which is stored in long-term memory in the form of concept hierarchies, schemas, prepositional networks, and production systems (Ausubel, Novak, and Hanesian, 1978; Schallert, 1982; Anderson, 1985). In particular, we will use the distinction between declarative and procedural knowledge. *Declarative knowledge* refers to the internal representation in the form of propositional (semantic) networks containing concepts and their relations. It is an essential component of expertise in a variety of domains, teaching included (e.g., Chi, Glaser and Rees, 1982; Lesgold, 1983; Leinhardt and Smith, 1984), and may also take the form of scripts (Schank and Abelson, 1977), plans (e.g., Hayes-Roth and Hayes-Roth, 1978), and images (Elbaz, 1983; Clandinin and Connelly, 1984) which guide both covert (conceptual) and overt behaviours. *Procedural knowledge* refers to the operations that can be performed on the declarative knowledge by the application of cognitive skills. Anderson's (1982) theory suggests that declarative knowledge is the origin and source from which procedural knowledge is developed by practice; on the other hand, work on teachers' personal practical knowledge suggests that the opposite relationship also holds following teachers' reflection on their classroom practices.

The study of the development of teachers' knowledge structures in the course of preservice programs and afterwards must take into account these important issues:

1) The interaction of three major features: (a) the kind of knowledge acquired by prospective teachers, (b) the locations where training and instruction take place (university and school), and (c) the various roles of persons involved in training (university teachers, school teachers, student teacher, beginning teacher, school pupils).
2) The difficult transitions between the categories within each factor, which must be made (a) from theory (declarative knowledge) to practical classroom skills (procedural knowledge), (b) from the university to the school, and (c) from student to teacher.

We will first discuss the factors involved in the development of teachers' knowledge structures, and this will be followed by description of the methods by which cognitive maps are obtained and analyzed to represent knowledge structures, and detect changes in them.

Factors Involved in the Development of Teachers' Knowledge Structures

The development of teachers' knowledge structures is a complex process in which several factors interact, in particular 1) the *kinds* of knowledge acquired, 2) the *location* where training and instruction take place, and 3) the roles assumed by *participants* in the training program. (The factor 'role' is nested within the factor 'location', and these factors are crossed with the factor 'kinds of knowledge'.)

A. In studying the knowledge structures of prospective teachers, we will look at two main aspects: content and use. By 'content', we mean that knowledge which is represented in memory mainly as *declarative*, but in the disciplinary domain, it also includes *procedural* knowledge, e.g., inquiry skills and techniques. By 'knowledge-in-use', we mean the operations, both conceptual and physical, that teachers carry out in the course of their work, e.g., planning instructional sequences, and managing class activities. The nature of and relationship between the substantive knowledge, or content, and the knowledge-in-use of teachers is difficult to specify at this stage in research on teachers' thinking.

1. The content knowledge to be acquired by the prospective teacher will be classified, for convenience, into three domains (some overlap between these domains may exist); Disciplinary

knowledge – subject matter knowledge which is to be taught in the classroom; pedagogical knowledge – knowledge of principles involved in teaching the subject matter; and contextual knowledge – knowledge both explicit and tacit which orients the prospective teacher in the school setting.

Within each of the domains, obstacles exist in the provision and acquisition of pertinent and usable content knowledge. For disciplinary knowledge, the student will be called upon to make a transition from using the criteria of valid knowledge, which he was taught as an undergraduate, to using criteria appropriate to teaching the subject matter. Furthermore, while disciplinary knowledge is usually taught and learned by direct instruction, and is mostly declarative it may include procedural components that are taught and learned by indirect (as well as direct) means, e.g., knowledge of inquiry or problem-solving skills in the discipline. The prospective teacher must not only acquire these components but also translate them into a form appropriate for students.

Pedagogical knowledge is a major component of any teacher training program but few training programs have solved the problem of how to provide theoretical pedagogical knowledge which will both engage the student intellectually and be amenable to eventual transformation into knowledge-in-use, which is strictly procedural. Prospective teachers are still lectured on the discovery method, and conversely we find an emphasis on competencies, without provision of tools to reflect on one's teaching and adapt it to varied situations. Contextual knowledge is subject to similar problems of transition. Prospective teachers may acquire formal knowledge of the school system, for example, but the encounter with reality is usually a shock nevertheless, and it is rare for prospective teachers to be given help in articulating their own understanding of the school system based on their experience.

2. With respect to 'knowledge-in-use', both declarative and procedural knowledge are involved, with complex causative relationships. Student teachers come into training with a stock of knowledge-in-use developed from their previous experience as students and from the variety of informal teaching situations, which abound in everyday life. This knowledge-in-use sometimes interferes with the different knowledge-in-use to be acquired in the program. This phenomenon is in line with other findings that knowledge structures are stable (e.g., Preece, 1976, 1978; Champagne, Hoz and Klopfer, 1984), and student misconceptions are highly resistant to modification (e.g., McKlosky, Caramzza and Green, 1980; Novak, 1982).

B. The location is a central factor in the development of teachers' knowledge. In one location, viz. teacher education institution, certain types of knowledge are acquired by prospective teachers for use in the other one, viz. school, where professional operation takes place. We expect changes to occur in the nature of this knowledge with changes in location. For instance, pedagogic knowledge, which was acquired as declarative knowledge, should be transformed into procedural knowledge in order to be used by the same person who is now a schoolteacher in the other location. Further, the location would seem to affect the manner in which knowledge can be conveyed and acquired in the first place: universities impose certain styles of teaching and learning, whereas on-site training in the school invites other kinds of instruction. The fact that universities and school are cut off from one another within the larger society further complicates the transition.

C. The change in role from student to teacher is a complex one, involving many factors: there is a cognitive change from the acquisition of pre-structured content to the shaping of content for students, a change in level of activity from (relatively) passive to active; a change in degree and scope of responsibility, a change in social status, and a personal change in how one views oneself as a result. The change in role from student to teacher is likely to be significant as far as the nature of knowledge is concerned. For example, the same subject matter may be viewed differently by the prospective teacher in the program and after graduation, during school teaching. Likewise, the pedagogical and contextual knowledge, which is available to the teacher, is different from that to which he had access as a student. The prospective teacher does not really learn how to talk to parents until he or she sits behind the desk as the one who will decide on the fate of their child.

D. We have sketched the major factors involved in the development of teachers' knowledge and have indicated some of the transition points at which changes should take place in the knowledge of the prospective teacher if he or she is to function effectively in the classroom. In the rest of this paper, we focus on a particular, limited aspect: teachers' knowledge structures in one discipline.

Concept Mapping as a Probe of Knowledge

Knowledge structures can be abstracted from cognitive maps, which are obtained from our revised ConSAT (Concept Structuring Analysis Task, Champagne and Klopfer, 1981; Hoz et al., 1984) interview. ConSAT was developed from Novak's (1980) concept mapping technique, originally designed to help depict the hierarchical organization of concepts in a text, in order to better understand the text and to learn it meaningfully (Novak, 1980). Several probes of knowledge structure have been developed (see Champagne et al., 1984, for comprehensive discussion and comparison). These probes are characterized by (i) producing representations in the form of a clustering (categorical) or spatial arrangement of a given set of concepts; (ii) the application of mathematical procedures to proximity matrices (produced by either individuals or a group), which were obtained either directly or indirectly from the subjects' responses, to produce these clusterings; and (iii) lack of subject labels or explanations. Such probes thus constitute high inference measures that need further interpretation by the researcher. It is preferable to have low inference probes yielding knowledge structure representations, which require lesser interpretation, such as ConSAT. Using this probe each student constructs a cognitive map that shows the spatial arrangement of a given set of concepts, which includes the student's groupings and explanations, thus avoiding or minimizing possible inference leaps.

Initial findings indicate that the revised ConSAT interview has satisfactory task test-retest reliability, and the classification scheme for the bi-concept link has a high between-judges agreement (95%) (Hoz et al., 1984). The reliability of the analysis scheme developed for the biology domain was also assessed and a close agreement was obtained for it (see next section). The ConSAT interview is administered individually, requiring about an hour to complete the first time. Additional administrations once the interviewee is acquainted with the task take about 30 minutes.

In the next section, we describe the procedures by which knowledge structures are abstracted from cognitive maps, and illustrate them by analyzing the disciplinary cognitive maps of one preservice biology student (participating in our larger study of teacher education). The concepts for the ConSAT were chosen by the methods course teacher who considered them central to both high school curriculum and the methods course. These concepts were given expert confirmations by several research biologists at the university. The students in this study were interviewed on two occasions and produced cognitive maps in both biology and pedagogy.

Abstracting Knowledge Structures of Cognitive Maps

Concept maps are analyzed at several levels. The structural organization is characterized along several domain-specific dimensions that were identified by the discipline experts. Links between the concepts will be classified by a semantic scheme, developed from that of Dansereau et al., (1979), who devised it for the semantic classification of links in textual concept maps.

Cognitive maps are assessed on three dimensions pertaining to the elements of the map (concepts and groups), their composition, organization, and relationships, as follows:

A. The *quality of concept groupings* is determined by the total number of groups, their nature and similarity to the experts' groupings, and the presence of focal concepts and kernel groups, which are congruent with the expert's.

B. The *quality of links among concepts* is determined by their disciplinary characterization and relevance, and by their congruence with the experts' total set of links for these concepts.

C. The *semantic nature of links among concepts* is identified by a system developed for this purpose (Hoz et al., 1984).

We illustrate this scheme by applying it to the cognitive maps of one student on two occasions, 3 months apart. During this period, he had completed the second two semesters of a biology methods course.

B. The quality of groupings in the cognitive map is determined by three dimensions (1) the total number of groups, (2) the nature of groups and their similarity to the experts', and (3) the existence and nature of kernel groups.

(1) Regarding the existence and number of concept groups which are formed on a characteristic common to its component concepts, and labeled by this characteristic, in our example, on the first occasion the student formed three groups which he labeled, and three concepts remained ungrouped with no reasons given; on the second occasion he formed a large single group which he labeled.

(2) Three biology experts who served as consultants agreed on two 'legitimate' partitions of the concept set, each having different structure and number of groups. *Partition A*. Group 1: hormone,

enzyme, metabolism, photosynthesis, energy, diffusion, and feedback. Group 2: biotope, evolution, classification. Group 3: cell, DNA. *Partition B*. Group 1: Cell, enzyme, metabolism, photosynthesis, energy, DNA, diffusion. Group 2: biotope, evolution, classification. Group 3: hormone, feedback. These experts classified the students' cognitive maps according to their similarity to the experts' groupings on a three-level scale. Level 1: strong similarity, indicated by nearly complete group overlaps and labels. Level 2: moderate similarity, indicated by some group overlap and labels. Level 3: weak similarity, indicated by a multi-group partition, with numerous small groups that do not overlap any of the experts' partitions.

(3) To identify focal concepts and kernel groups the three experts defined the 'strength' of links between all pairs of concepts (with 13 concepts there are 78 pairwise links altogether). The strength of links between concepts was conceived as being directly proportional to the biological functionality of the two concepts. Link strength was measured on the following three-level scale, on which the between-expert agreement was 99%. Level 1: necessary links between concepts that are tightly connected and interact strongly. Twenty-five of the 78 possible pairwise links are at this level. Level 2: possible links between concepts that are moderately connected and interact somewhat. Twelve of the 78 possible pairwise links are at this level. Level 3: links between concepts for which a connection is biologically meaningless. Forty-one of the 78 possible pairwise links are at this level.

This classification of links were inscribed in a 13 X 13 array, termed the *experts' matrix*, on which the analysis of bi-concept links was based. It enabled the identification of errors or misconceptions about concepts and their links. The following features of knowledge structure were obtained using the experts' matrix:

1. The number and quality of links. A good map is characterized by a high percentage of first level links.
2. Focal concepts and kernel groups. We defined a focal concept as one, which is characterized by a high percentage of both inter- and intra-group links of the highest strength level. We defined a kernel group as one in which all, or most intra-group links are of the highest strength level.

In our case, in each cognitive map (i) four concepts that were focal for the student were also focal for the experts, and (ii) two kernel groups were highly similar to the experts' kernel groups.

Using focal concepts and kernel groups enabled the identification of disciplinary misconceptions, by (i) detecting differences in focality between students and experts, and (ii) locating differences between strength levels for students and experts. In our case, 'metabolism' and 'diffusion' were misconceived: for the experts but not for the student they are focal concepts, and the link-strength of the concept 'metabolism' is at the third level whereas it is at the first level for this student.

C. The quality of bi-concept links was assessed by the disciplinary characterization and relevance, measured on a three-level scale. Level 1: correct, precise and clear (9 points). Level 2: Correct but partial (7 points). Level 3: indirect and general, or imprecise and lacking in certain aspects (5 points). In our case, the median scores for the links in each map were 7 and 9 on the two occasions, indicating an improvement in the quality of biological relevance.

D. The experts' conception of the domain was obtained by using the experts' matrix to derive two indices, one external and one internal, to further assess the quality of cognitive maps. The external index is convergence with the experts' links, which is the percentage of links in a particular map, which are at a given strength level out of the total number of experts' links at this level. The internal index is salience, i.e., the percentage of links of levels 1 and 2 out of the total number of links in the map. Its complement is the percentage of links of level 3 (i.e., misconceived links), which indicates overall misunderstanding of the relationships between concepts.

In our case, salience values were 61% and 81% on the two occasions respectively, with the respective misunderstanding values of 39% and 19%; and convergence values are 44% and 56% for level 1 links, and 42% and 25% for level 2 links. The changes in convergence and salience enable us to identify rather substantial improvements in the biological understanding of the student.

E. The semantic nature of the links between concepts was identified by the following system comprising 18 kinds of links: Part of, type/example of, leads to, analogy, characteristic, evidence, description, influence, precedence, relation, contingency, operation, aim, sameness, formation, neutral, nonexistent, and unclassifiable. (For more detailed description, see Hoz et al., 1984.)

In our case, the most frequent kinds were as follows: on the first occasion, 4 'part of' links and 4 'operation' links, and on the second occasion, 8 'operation' links and 4 'formation' links. This result reflects a better biological understanding of the relatedness among these concepts.

Conclusions

In this paper, we have illustrated the use of concept mapping as a means of identifying and representing changes in the knowledge structures of student teachers. Here we have focused on the knowledge structures in the disciplinary domain, but the analysis schemes have potential for the other domains as well. In our larger study, the students produced cognitive maps in pedagogy as well as biology, and in addition to the ConSAT the students took part in semi-structured clinical interviews designed to elicit their explanations regarding certain aspects of the training program and their impact on observed changes in the cognitive maps. We are currently developing classification schemes to analyze the clinical data to yield knowledge structure representations in every domain (along the lines of work by Leinhardt and Smith, 1984). We anticipate some difficulties in abstracting knowledge structures in certain areas (e.g., pedagogy) but the proposed methodology shows promise of further progress to be made in the understanding of teacher thinking.

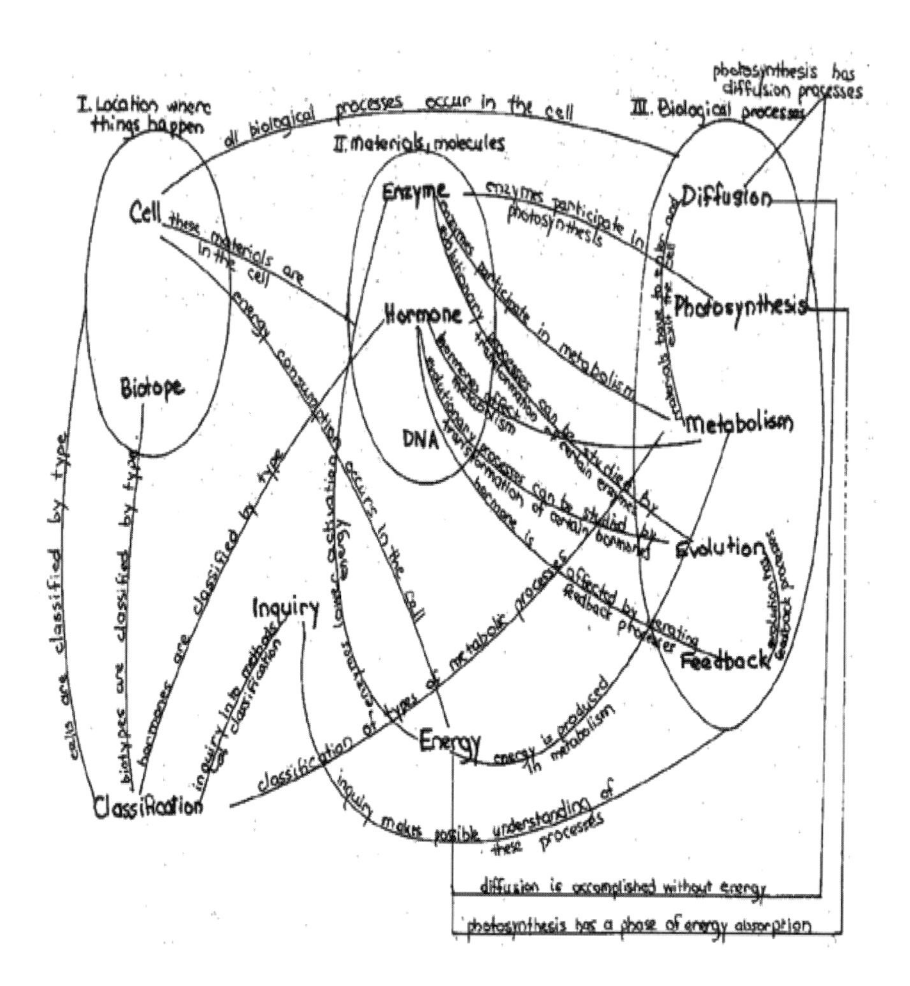

Figure 1. First cognitive map

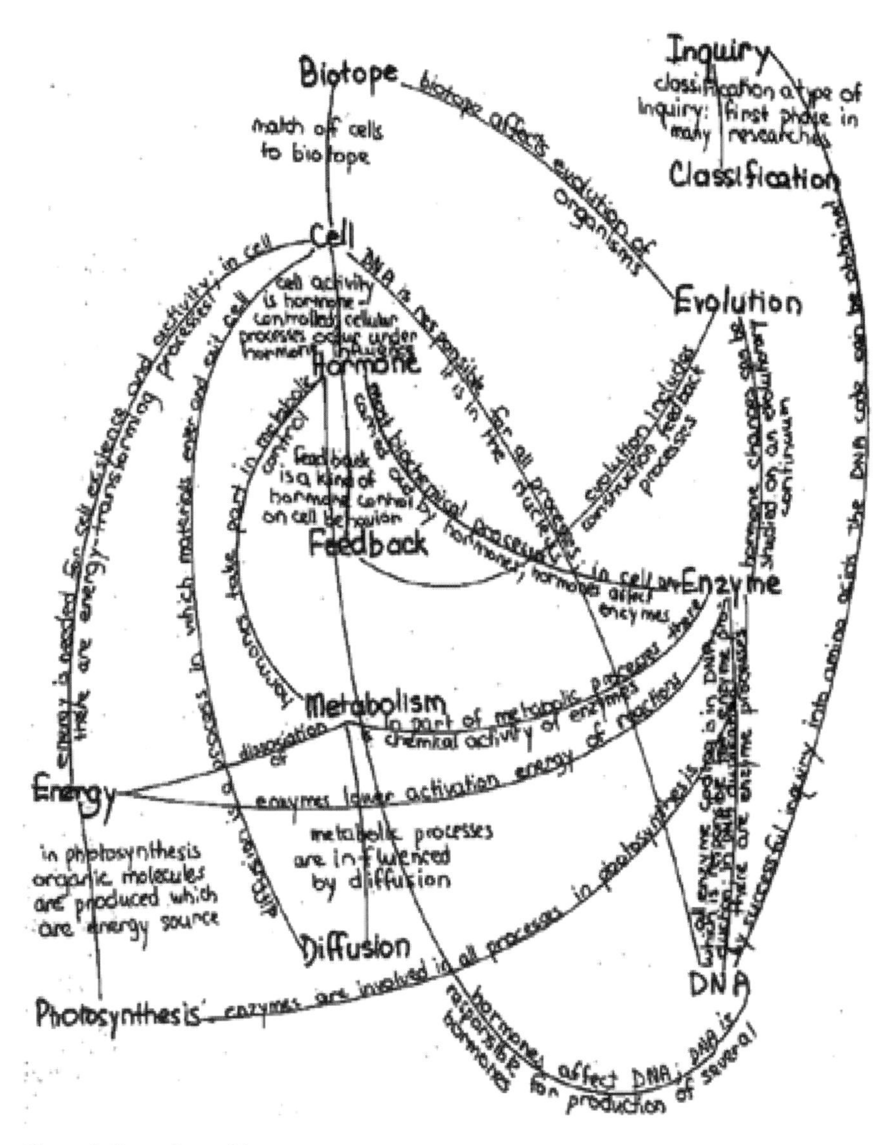

Figure 2. Second cognitive map

Chapter 24

A Cognitive Model of Behaviour and Learning as a Heuristic Method of Theory-Based Teacher Training

Hartmut Thiele

Summary

The subject of the study is the development and empirical examination (field experiment) of a theory-based model of practical teacher training for the acquisition of skills in conducting a discussion in class. The concept of the training model is based on a cognitive model of behaviour and learning (Woodruff, 1967) which specifies four phases:

1. Intake of information (selection and categorization);
2. Information processing (concept and rule acquisition);
3. Plan of action and decision-making (cognitive construction processes, decision to act);
4. Act of teaching (feedback to plan of action).

For each of the four phases assumptions were made concerning the type of learning and the corresponding training methods. These were:

1. the imparting of theoretical knowledge that is relevant to action;
2. cognitive training (training in discriminating and decision-making);
3. training for action (modified microteaching).

The most important result of the survey was that microteaching proved more effective than cognitive training and theoretical instruction. Contrary to the results of other investigations, cognitive training turned out to be less effective.

A theory-based procedure of practical pre- and in-service teacher training should – according to the results of this investigation – include a treatment that fulfils the conditions of the development of a cognitive and behavioural construction competence (Mischel) as shown in this study.

Further studies are planned to find out how the efficiency of cognitive training, especially the training in decision-making, can be optimized by means of an improved method of presentation.

An Outline of the Problem and the Questions it Raises

A common characteristic of the training procedures, which have been developed to mould the behaviour of teachers and educators, is their mostly pragmatic orientation and the lack of any foundation in an integrative theory of behavioural changes that has been checked experimentally. The training concepts are based rather on various principles taken from the field of learning psychology and tend to emphasize particular theoretical aspects.

The training model outlined in the following paper has been conceived as an attempt to go beyond the point already reached and should, in several respects, be viewed as a further development.

The idea of the training model is based on a cognitive model for learning and behaviour (Woodruff, 1967), which provides the theoretical framework.

- It explains the formation and alteration of teacher behaviour as cognitive learning
- It describes the didactic performance of the teacher as a series of interactions based on decisions he has made
- It gives reasons for certain training methods and their combination as well as justifying the examination of them experimentally.

The inquiry aims to compare the effectiveness of three training methods. These are:

(1) The transfer of theoretical knowledge which is relevant to the teachers' performance (the substitution of a scientific theory of instruction for a subjective one)
(2) Cognitive training: practice in discrimination and decision making
(3) Performance training or microteaching.

The overall aim of the inquiry and my present research into training methods could be called: "The development of the ability to perform in class according to a particular theory" and, consequently: "The development of a flexible, didactic attitude to decision-making in the class (interactive decisions)".

A Cognitive Model for Learning and Behaviour

The theoretical framework of the training model which I have constructed and tested in experiment derives from the cognitive model of learning and behaviour published by Woodruff in 1967, who assumes the following phases in the behaviour-learning cycle:

(2) Intake of information: selection and categorization
(3) Information-processing: storage and internal manipulation, acquisition of concepts and rules
(4) Plan of action and decision making: selection from available plans, or rules of action, action-generalization or cognitive constructive processes, the decision to act
(5) Action: feedback to the rules/plans of action.

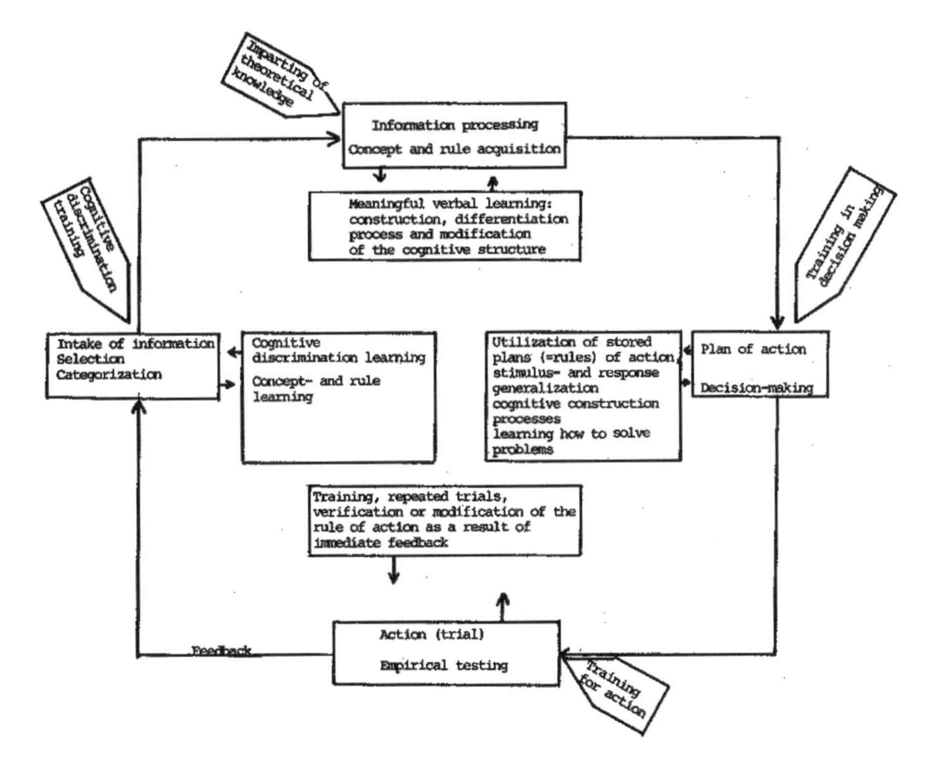

Figure 1. Cognitive cycle in behaviour and learning.

Because of its heuristic value, the cognitive model of learning and behaviour has proved to be fruitful for the conception and cognitive justification of the training and molding of teaching-behaviour (c.f. Hofer, 1977);

- It provides the theoretical framework for a description and explanation of the formation and adjustment of teaching behaviour as cognitive learning.
- It provides indications of what form the treatments must take, that is the mode of learning and training method, which can influence certain phases of the model.
- It shows the interconnections of the treatments and the combination or sequence in which the treatments can be ordered and integrated in order to achieve an optimal training effect.
- It is flexible enough to integrate elements of other cognitive models.
- The model is based on concrete and verifiable phenomena and thus embedded in reality. With this model, one can make assumptions about how to devise the best training methods and these assumptions can be empirically tested.

The following sections will show what this model can achieve.
We will try to formulate some statements that can be used for a theoretically based method of teacher training (c f. Hofer, 1977).

We can see that:

1. Each phase of the model contains the names of the skills to be accentuated and the relevant mode of learning, which can be converted into operationalized training goals (column 2 and 3).
2. The appropriate training method is determined for each mode of learning (column 4).
3. The training method can be arranged in a sequence based on the cyclic progression of the phases. Thus, we have produced a training model that is based on the model of action and learning and that takes the relevant learning processes into consideration (figure 2).

The Phases of the Model

Intake of information

A teacher who is interested in structuring his observations of the teaching process i.e. in identifying, classifying and categorizing observable events must have at his disposal a differentiated conceptual frame of reference with clear conceptual notions. He must be able to classify the situation, the aims and the action or performance. This is the prerequisite for a clear, more deliberate observation of the microstructures in the teaching process and for the ability to better understand and check one's own behaviour.

Training can influence the participant to become more sensitive to the perception and development of clear conceptual notions concerning teacher or pupil behaviour as well as other variables within the didactic and educational situation. During this procedure, known as *discrimination training*, the teacher is offered particular behavioural units, which are operationally defined and related to the rule of action. This occurs in model films, tape recordings, or written protocols of classroom-lessons. The teacher should be able to recognize the behavioural units observed and classify them accordingly.

Similar results to those in discrimination training can be achieved through practice in observation with systems of characteristic items and categories developed for the observation and analysis of classroom behaviour. The effectiveness of both methods with regard to behaviour adjustment has been shown in a number of empirical studies. For example, Wagner (1973) and Borg and Stone (1974) report results where cognitive discrimination learning produces the same behaviour adjustment as microteaching.

A critical reading of the results concerning discrimination training, however, reveals some irregularities – one can only speak of a general tendency in favour of discrimination training. Further research is necessary to explain the effects and results arising from discrimination training.

In my inquiry, I was unable to repeat the results of Wagner (1973) and Borg and Stone (1974). Microteaching rather than cognitive training proved to be the more effective training method (Thiele, 1978).

It is assumed that the explanation for the fact that this inquiry failed to find any observable results of cognitive training lies in the complexity of the criteria, the relatively rigid design of the experiment and the consequent restriction of the didactic presentation.

Phases of the model	Training goal/accentuated skill	Accentuated mode of learning	Method of training
Intake of information	- Differentiation of the categories for the perception of situations - Formation of distinct concepts - Discrimination and categorization of the actions (teaching skills, pupil behaviour)	- Discrimination learning - Concept- and rule learning	- Cognitive training: Discrimination training with protocol materials and/or video recordings - Classroom observation, interaction analysis
Information processing	- Acquisition of extensive body of theoretical knowledge relevant to teaching - Substitution of implicit theories by scientific theories - Acquisition of rules of action (teaching skills) - Command of information relevant to action, that can be used to interprete situations and for cognitive construction processes	Cognitive learning - Formation/construction of concepts, rules, strategies for problem-solving - Construction, restructuring, widening, differentiation of the cognitive structure relevant to teaching	Imparting theoretical knowledge - Lectures and seminars - The study of texts dealing with the theory of whatever training content is involved
Plan of action and decision-making	- Construction of plans of action by using the theoretical knowledge - Acquisition of a cognitive and behavioral construction competence - Generalization of rules of action - Command of knowledge about the appropriateness of actions to teaching objectives and instructional situations - Decision-making by weighing up the likely consequences	- Utilization of stored rules of action - Generalization of situations and responses - Cognitive construction processes - Teaching as learning how to solve problems (Experimenting with one's own actions)	- Cognitive training training in decision-making - Training for action (modified microteaching)
Action	- Trial (testing the hypothesis of action - Confident active execution of the action	- Training, repeated trials of action with video-feedback - Confirmation or modification of the rule of action as a result of immediate feedback	- Training for action: Microteaching; training that ist integrated into a teaching-method (Lehrverfahrensorientiertes Training, THIELE 1981)

Figure 2. Assumptions about skills, modes of learning and methods of training that are derived from the model.

Further investigations will show whether cognitive training with a different didactic structure can be more effective.

Further improvements could be made by:

- Linking the cognitive training directly with microteaching
- Achieving a favourable "training dosage" and alternating the cognitive training with microteaching over a longer period of time and
- Developing other forms of practice in which video recordings are used more often.

Information processing

By information processing, we mean, in general, processes where the teacher stores, organizes, and integrates the information-input in the cognitive structure. The new input is combined, differentiated, and developed along with previously stored or self-generated information in order to acquire new plans of action and problem-solving strategies.

My outline of the cognitive procedures during this phase of information processing is restricted to a description of the most important processes that are significant in the elaboration of the relevant cognitive structure in this context.

Woodruff (1967, p. 77) gives particular significance to the rules, as far as they represent the basis for the generation and generalization of action, for the decision-making processes, and the modification of action. In this function, complex concepts and rules are described as "predictive variables" with more help events and actions can be anticipated and predicted.

A conceptual, theoretical frame of reference is therefore an essential prerequisite for the transfer of theoretical knowledge into practical, classroom behaviour.

During the training period of teaching-behaviour, this phase of information processing is influenced by the transfer of theoretical knowledge, whose main aim lies in the acquisition of rules of action, in the formation and in the alternation or elaboration of the theoretical frame of reference. These should form the foundation of action that accords with a theory. The influence consists of knowledge gained from instruction theory, didactic recommendations, and the experiences the teacher has made.

Plan of action and decision-making

An analysis of this phase of the model affords us an insight into the formation of a plan of action, which is initiated after the decision to act, has been taken.

During the period of planning and decision-making *cognitive construction, processes* must be assumed. Here, the teacher's achievement consists in constructing new rules of action by, first, reflecting on the situation i.e. by thinking productively and, second, by constructing new rules of action through a synthesis of the knowledge of action and performance that he is able to recall in the situation. In order to explain these cognitive constructive processes Mischel (1973, p. 265) introduces the concept of "cognitive and behavioural constructive competence".

According to this, the teacher has the ability to construct new rules of action by combining pieces of information within what we call his "if-then"-knowledge. Furthermore, he has the ability to produce or construct through deduction plans of action from rules of action already available.

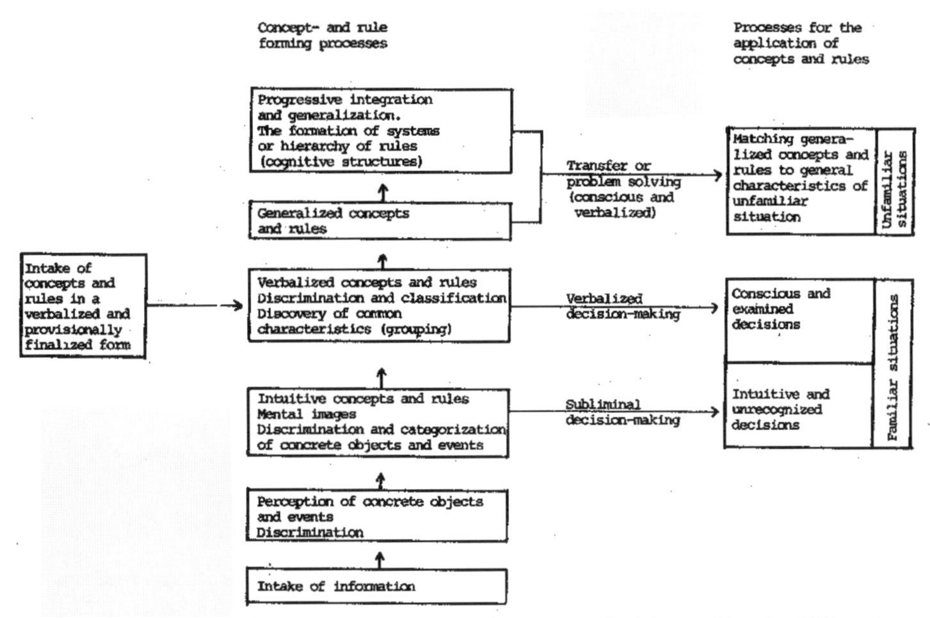

Figure 3. Levels of concept and rule formation, influences on decision making for differently formulated concepts and rules (Woodruff, 1967, modified by the author).

It is the description of teaching as a process in which teachers continually make decisions that has drawn attention to the area of decision-making in teacher-training, an aspect hitherto neglected. Decision-making is a central teaching activity, a basic teaching skill, that overshadows all other (Shavelson, 1973). Flexibility in decision-making is regarded as the decisive prerequisite for appropriate performance in the classroom.

In training that is to affect the flexibility of decision-making should consider the following skills:

1. An interpretation of the situation: the analysis or diagnosis of the teaching situation and corresponding categorization (diagnostic skill).
2. Finding, or deciding on, an appropriate plan of action.
3. The development of a plan of action or deciding on a teaching activity appropriate to the plan.

I am at present trying out an approach that is designed to influence these skills through cognitive training (Thiele, 1983). The aim is to develop the teacher's "cognitive and behavioural constructive competence" and his ability to make rational decisions during a classroom discussion. This is done by introducing the participants to cognitive simulation training with the aid of written material (protocol materials) concerning interactive situations.

The first example of this course of decision training describes a classroom situation where the discussion comes to a halt after a pupil's remark. The participant in the course has information about the teacher's intentions and must decide which is, in his opinion, the most suitable teaching activity to enable the discussion to continue.

In another example, the student must produce a plan of action, having first analysed a particular classroom situation and he must define his aims himself. He decides on what he thinks is an appropriate teaching activity. The student has extensive information about the aims and structure of the lesson in order to facilitate his identification with the teacher. In both cases, the student can compare his solutions with model answers.

Action or performance

In the action phase the teacher tests the assumptions and decisions he has made in the planning phase. We do this with a modified version of microteaching.

Thus, in the context of the model I have presented here, microteaching appears not as an independent and isolated training method, but on the contrary, as one element in a greatly expanded training model. Apart from the action phase, this model contains the other essential elements of the information-intake phase, the information-processing phase, and the plan-of-action or decision-making phase.

An Empirical Test of the Training Model and the Results

In an empirical inquiry, I compared the three training methods based on the cognitive model of behaviour and learning.

The lecture-based transfer of theoretical knowledge in traditional teacher training (treatment I) was compared with:

- Cognitive training as discrimination and decision training (treatment II) and
- Microteaching (treatment III).

The three treatments were combined in different ways and their effects were tested experimentally in four samples in the following sequence:

Sample A (11 subjects)	Receives a combination of treatments I, II and III
Sample B (9 subjects)	Receives a combination of treatments I and III
Sample C (11 subjects)	Receives a combination of treatments I and II
Sample D (12 subjects)	Receives treatment I

The general aim of this investigation was that the participants should develop the ability to produce, to select, to generalize, and to execute the teaching skills necessary for conducting classroom discussions. They should do this based on their knowledge and acquaintance with rules of action.

It was hoped that the exercise would result in qualitative changes in the participants' teaching skills and their more deliberate application. A more deliberate application of teaching skills means:

- That teaching activities are examined with reference to the effects they have
- That the effect intended is anticipated and
- That clear-cut and appropriate verbal formulations are thought out.

The training course was concerned with the skills necessary to conduct a classroom discussion (Thiele, 1981). The results were measured in the form of increased performance, the difference between pre-test and post-test, with the aid of nine variables that were viewed as the most interesting criteria. The data for the pre- and post-test were collected in classes of the fourth grade. Discussions about fables or fairy stories were held in the lessons, other variables being kept constant. The pre- and post-test lessons in the normal classroom-situation were recorded, transcribed, and codified according to the variables.

The hypotheses made about the inquiry were checked experimentally and resulted in significant learning progress in the samples in the order A> B> C> D. Thus, microteaching proved to be the more effective method in comparison with cognitive training and theoretical instruction. The results of the inquiry confirm the assumptions about cognitive processes in teacher training which were based on the model of behaviour and learning.

In conclusion it can be said that the demands made on the model of behaviour and learning were met in the following points:

1. In each of the four phases of the cognitive model, the various modes of learning were identified. Appropriate training methods were matched to these and arranged in such a way that the best results could be achieved.
2. Conditions were crystallized for the formation of a cognitive and behavioural constructive competence:
   - The transfer of technological knowledge
   - Cognitive training during which the possibilities for producing rules and plans of action are given
   - Training where plans of action can be put into practice under simplified conditions in connection with decision training
   - The chance to experiment with one's own performance and to check it in action.
3. Use was made of training methods whose effects could easily be transformed into classroom situations.

The results of this inquiry show that in theory-based teacher training there should be a component in the teaching-practice phase which takes into account the conditions mentioned in point 2, conditions that are necessary for the formation of a cognitive and behavioural constructive competence.

Chapter 25

How to Change What Teachers Think About Teachers: Affirmative Action in Promotion Decisions

Alan F. Brown

Summary

The change, growth or development of how a teacher thinks about another teacher is analysed as a form of adult learning. The analysis draws on an integration and extension of personal construct psychology and interpersonal theory. It is facilitated by facsimilating professional experience with, in these instances, the thinking of school administrators when making staff decisions. The process is developed through several briefly reported studies ultimately leading to a model for creating change in personal thinking. The model joins cognition and reflecting and human interaction in a many-stage process used in a currently ongoing project on affirmative action in promotion decisions. Some assumptions behind these studies are explicated: organizations are persons, administration is persons' purposes, culture is what characterizes a group or organization and can be created by people, art and magic have as much to do with understanding teacher thinking in complex organization as does scientific analysis, theory grows out of interpersonal experience, experience can be accelerated by facsimilation, literacy is necessary for growth in thinking and requires ability to interarticulate the explicit and the implicit, unnerving though it may be.

Introduction

As a foreword to this paper it should be noted there is such a thing as the cognitive determinant of personal action. Whether called teacher thinking, personal constructs or personal practical knowledge, this cognitive determinant is unfortunately idiosyncratic and, worse still for the researcher, it is frequently inarticulated. Our interest is because the behavior of a person responsible for the actions of others affects the actions of others, their values, and their lives. In that sense, such a person creates a culture, or builds a school. You build a better school, I seem to be saying, when these influential persons use their own resources to think, to reflect, to compare notes with others, to attend more closely to what they are doing and how they are doing it, to others and for others.

The ideas discussed here lie midway between earlier studies defining "teacher effectiveness" entirely within the personal thinking of those doing the ratings (Brown 1974) and some upcoming studies examining the personal thinking of classroom teachers who have responded to the actions of administrators: Mullaley on how they construe externally-initiated curricular change and Kompf on resistance as a response to authority.

From Dialogue Into Theory

You and I both tend to develop ways of making sense out of our interpersonal world. It is a world of teachers, students, heads of one thing, coordinators of another and we interact with them in schools, classrooms, seminar halls, offices and council chambers. We form our own ways of distinguishing between persons and of organizing our ideas about them. Your way of thinking about others may differ from mine which is what by definition makes us two people not one. We identify each other and ourselves by our systems of personal thinking or our constructs. Your thoughts about other people characterize you, perhaps even more than they characterize the people you are thinking about. You project self. So do I.

The patterns of personal thinking that we employ in differentiating among people also differentiate among ourselves. In attending to this talk, you compare me with previous speakers using your own criteria, criteria that distinguish you from your neighbor both in the substance you use and in relative priorities you give to thoughts you may perhaps have in common. Furthermore, in listening to my argument, your patterns of personal thinking quickly sort out those features you notice but do not value (tone of voice? nervousness?) from those less obvious but of much greater value to you (cognitive support for your own paper?). Those personal constructs that we post, as upon a notice board – say as in the credentialing we all did earlier in this seminar – may or may not be the constructs that operate critically in our actual thinking behavior, particularly upon those thoughts people make about other people. We become facile in concealing our operating criteria especially when they differ from those we post.

Therefore, it is with personal thinking of those teachers that are called administrators, whose personal thoughts are about those other persons who affect or are affected by their work. Their thoughts, just as those of classroom teachers, are about other persons and at certain points some action is indicated. Let us consider a personnel decision as an example of that action. A personnel decision, or the administrative treatment of one person upon another whether arrived at singly or in concert with other committee members, is only the tip of the iceberg. It is visible and readily identified as selection, transfer, promotion, demotion, declaring redundant, and dismissal. But it also includes less obvious treatments such as assignment to specific responsibilities, enlargement of responsibility, room location, little favors or disfavors, or general assessment of effectiveness on the job. All these notions, feelings, hunches and perceptions, the entire iceberg of personal thinking, form an administrator's set of operative criteria for each specific kind of staff decision. So we enter the underwater ice.

Determinants of decisions about people

Operative criteria, or the criteria actually in use, as distinct from those posted or mouthed, are not the only factors entering staff decisions, however. Our work so far suggests at least four determinants:

1. *Lego-institutional requisites*, as within the educational bureaucracies there exists a bewildering array of permits, conditions, letters, licenses, and certificates that affect or recognize personnel decisions,
2. *Demographic discrimants*, or things which although prejudicial, have become culturally tolerated sometimes to the point of studiously ignoring their presence: age, gender, ethnicity, marital status, family background, SES, regional residence,
3. *Self-selection*, or those things a person does to try to get selected such as gaining *visibility* and showing a commitment to the job, a will or desire for the appointment; the prerequisite conditions enabling a decision to be made, and
4. *Operative criteria*, or one's pattern of personal thinking that covers those talents, characteristics, behaviors, attitudes, virtues, knowledges and skills that make a powerful or weak contribution to our administrator's personnel decision.

Now whereas data on the first three determinants are readily available, the discovery of the operative criteria is a tricky task. It is that idiosyncratic, often inarticulated cognitive determinant. Your posted criteria are more out-front than upfront and you may or may not be aware of your operative criteria on any one matter. Some people talk a lot about certain values and virtues but actually can be seen using different values and vices on the job. Whether they do this knowingly or not is a different question. You could make the comparison of the honest or 'smart' hypocrite and the ignorant hypocrite or 'stupe'. Both, by lore and by research data, are legion among administrators. The smarts know that they are lying and when they are lying; the stupes do not and are therefore dangerous phonies. The best are neither but experience shows them to be as rare as are creative geniuses in any other walk of life.

The attainment of professional literacy, or when one is able to make explicit that which is implicitly within one, is more than being able to read and write. It requires one to articulate in some meaningful set of symbols - perhaps language - the nature of your professional experience and what you do with it. Principals and teachers alike make countless daily decisions about and affecting careers of others. These are usually based on a hunch. The articulation of this hunch or gut feeling is the first step in the attainment of professional literacy.

Myth, metaphor, and management

Administration is people. People interacting with people. Interacting, communicating, in verbal, oral, electronic and other means of their choice, with other people individually, in groups and in collectivities.

Administration is people interacting with people for a purpose. Somebody's purpose. Society's purpose. Sarah's purpose. Administration is the fine art of integrating purposes. Administration is magic: that nice slight-of-hand that causes us to respond to the illusion of integration. Administration is the rigourous science of running a gee-whiz line-of-least-squares straight through these polydimensional purposes. By art, magic and science, administration is the creation of a culture.

How to change what teachers think about teachers means then how any individual affects the acceptable interaction patterns that characterize an organization - starting with their own. To the extent that every person affects the norms of interaction, each affects culture, does administration. But the person most influential in this action and with time to do it, gets called the administrator – head - teacher, principal, superintendent, director - so our work begins with that one.

Professional administrators are artists. They get called bores in the same way all new wave music sounds the same to parents fixated at swing. Most attempted art is trash. Most essayed poetry is doggerel. Most administrivia is not aesthetic. Professional administrators take an organization and create abstractions called rules, systems, policies or schools; these abstractions reduce realities to essentials and convert a mess of particulars to a working structure of manageable proportions and dimensions. It is a work of art of varying quality. The level of quality is revealed by the level of sophistication and inter-articulation of your construct systems, for example of your posted and your operative criteria, or perhaps of your public constructs and your private constructs.

This creativity is not without cost. As a creative act, it bears within it the burden of Pygmalion, the mythical King and sculptor who reified his creation, Cyprus, ultimately falling in love with it and successfully persuading Aphrodite to turn it into a living person. I worry that many of our most creative administrators, as our Pygmalion, will reify, may deify, their creations, ultimately paying more attention to them, which means to themselves, than to those persons the policy was created to affect, the members of the culture. This effect is actually even stronger with structures that are inherited (legitimizing one's appointment). The Pygmalion effect is administrators' tendency to behave as though they believe organizations (created or inherited) to have aims, and bureaucracies to have goals that are independent of the persons comprising them. Any meter-long shelf of current management texts, organization theory references and administration studies leaves incumbents and aspirants bedazzled with tempting anthropomorphisms, all carrying the implication that learning these enticing structures and probability paths is more promising than learning about each other. But metaphors don't manage, or get managed. People do.

Some Theory Roots of Personal Thinking

A theory of personal thinking of school administrators will be useful only if it helps them both to discover and to develop their own professional thinking. The discovery is aided by the theory of personal constructs; the development is aided by interpersonal theory and theory of reflecting.

It is possible to infer from personal construct theory that a productive way of analyzing personal experience – in our instance: administrators' professional behavior – is by facsimilating personal experience. A supposedly omnicapable theory of behavior that increasing numbers of scholars would individualize, its fundamental postulate is that "a person's processes are psychologically channelized by the ways in which they anticipate events" (Kelly 1955); your patterns of construing the world are your personal constructs. The theory readily lends itself to several ways of what I call "facsimilating personal experience," usually through some variation of the repertory test or "grid," so that this experience can be analysed and better understood, at least by you.

This is not to say that grids are necessary for either the discovery or development of your bases for making professional decisions. The studies in which our versions of grid or "rep test" have been used (see Rix 1985, Brown and Ritchie 1982) have resulted in increasing their cognitive complexity, their resourcefulness in facing new problems, and have brought several to be more goal-directed (does this mean we only succeeded in turning ignorant hypocrites into smart ones?). But any effects we have achieved have not been because of grids or other convenient forms or instruments. Rather the other way: to understand the process of change in personal thinking our studies attempt to create facsimiles or true copies of professional experience and grids help.

Persons who seek to improve their professional practice, be they teachers or principals or counselors, are in fact seeking to change their patterns of personal thinking. Rix identifies three Kelly corollaries on the change concept: the Experience Corollary states that a person's construction system varies as he successively construes the replication of events; the Modulation Corollary states that the variation in a person's construction system is limited by the permeability of the constructs within whose range of convenience the variants lie; and the Choice Corollary states that a person chooses for himself that alternative in a dichotomized construct through which he anticipates the greater possibility for extension and definition of his system. Rix sees persons moving from Type A to Type B experience by drawing differentially upon the choices available to them as stated in these corollaries as she sets them out below as extreme types (Rix, 1982, p.28).

Table 1. Personal Construction Systems.

Type A Experience	**Type B Experience**
Experience	Experience
1. no reconstruction/no reflection	1. reconstruction/reflection
2. no alternatives	2. learning/development/growth
3. impermeable constructs	3. alternatives
4. no change	4. modulation
	5. chooses to sift and shift
Same Experience	Different Experience
↑	↓
Same Experience Again	Another Different Experience
↓ ↑	↓ ↑
No change in constructs/criteria	Change in constructs/criteria

When a person wishes to move from Experience A to Experience B, or seeks to attain professional literacy, the essential element is the effect of human interaction. The significance of this effect is most noticeable in its absence. In a study of 69 principals (Brown, Rix, Cholvat 1983) we found that one of their promotion criteria – rapport with the community – jumped dramatically in its decision power from pre to posttest but unfortunately it was still used by a very limited number of participants. The error was that no opportunity for peer and colleague interaction had been provided: just lots of cognitive data taken in large doses and contemplated alone. There is more than cognition in construct acquisition just as there is more than mechanism in management.

That a teacher's thoughts about other teachers become changed through peer interaction would be a basic expectation of interpersonal theory, and administration is nothing if not interpersonal. Just as Harry Stack Sullivan held that a person grows through reflected appraisals from significant others, Kelly's Sociality Corollary states, somewhat less smoothly, to the extent that one person construes the construction process of another he may play a role in a social process involving the other person.

Procedures for Facsimilating Professional Experience

One of our versions of the rep test is innocuously called *Personnel Decision Analysis*. It calls upon you to make a real decision about half a dozen persons you now know. It then allows you to make successive inter-person comparisons in triadic combinations until your repertoire of constructs pertaining to that decision is finally exhausted. These constructs, understandable at least to yourself, become your Likert-type questionnaire to yield a structural hierarchy of the actual criteria you used in making that decision, that is it gives you the relative decision power of each of the criteria you actually use. In a sense, the experience is like going back inside the gut feeling, or analysing the hunch that guided your action. One of our group puts it this way:

> The Rep Test and personal construct theory assume that the constructs employed by an individual when perceiving an object, event, or people influence that person's behaviour. The instrument gathers qualitative information, which is analysed quantitatively to reveal the subject's operative constructs. These can then be interpreted. (Rix, 1980, p.40)

The format that helps people become more articulate with their promotion decisions, "Form PROM", is widely used. This is not because of any fascination with getting ahead. Rather it is because coming to an understanding of the anatomy of a recommendation for admittance into the administrators' club – which is what promotion to vice-principal amounts to – gives the best example of the underwater ice: that massive array of implicit thoughts that contribute to any number of formal actions, or tips of the iceberg. Administrators actually like the task and report it is what regularly confronts them at work, a real experience and not a simulation, and that it certainly makes them think.

A theoretical problem is raised at this point: do we assume personal thinking is solely about other persons? This is rather more a hope than an assumption. The fact is that school administrators as any others in responsibility frequently find it necessary or preferable to consider abstracts, especially those of their own creation, than concrete problems. Hopefully they integrate the two.

Thus, our studies of the personal thinking of administrators provide an exercise known as the "shouldbesheet". They are asked to rank-order their own constructs, the ones they had written on the rep test at an earlier date, according to which "should be" more important to that personnel decision.

What makes our model of changing teacher thinking unique is this element of confrontation, cognitive self-confrontation, which is introduced at this point. Administrators are shown (a) a list of the criteria they actually used in making their decision about teachers, and the relative weight of each upon that decision, i.e. their rank-ordered operative criteria and (b) their rank-ordered posted criteria from the shouldbesheet. The two lists seldom coincide; it is quite common for experienced principals to post three or four qualities as most important for promotions and to discover they had never even used these qualities when making real promotion decisions.

For some this cognitive self-confrontation is unnerving. Earlier projects found principals creating imaginative rejections of reality in favor of a return to the ideals. The way it should be is always a more pleasant way of thinking. It is easy not to accept ownership of thoughts about real people. This problem was easily solved by simplifying the grids and letting subjects do all their own protocol processing. Off the end of your own stubby lead pencil, your thoughts are more convincingly owned. When a principal can say "Like it or not, it's mine" the next question will be "What can I do about it?"

Socializing, Reflecting and Personal Thinking

That teacher thinking affects school curriculum is clearer than how administrator thinking affects what they do to teachers. When given time for reflecting on what it is they are thinking about, principals begin to improve their practice. One of our group did a study of the process of reflecting and arrived at a usage of the term that sees reflecting "as a logical, non-logical, analytic and creative process in which the learner views and reviews an interpersonal event in relation to self" (Boyd 1981, p. 104). She discovered there must be two conditions, time for reflecting and positive attitude toward reflecting, or nothing happens. In the following studies, activities were designed so as to enhance the opportunity for reflecting and to vary the format of human interaction.

1. *The interactive structured workshop*, where large groups compare their own patterns of thinking with others of the whole group, under guidance of carefully structured study-guide questions, resulted in one study (Brown 1978) in which forty school principals, over a six-month period became more goal-oriented in their teacher-selection behaviour. The discrepancies between what they were thinking about and what they should be thinking about - when considering teachers for staff selection - were significantly diminished.
2. *The intimate group*, or small group of three members, has been frequently used because of the opportunity afforded for protocol study, one's own and others', for reflecting, and for - by a process of "musical triads" - rotating personal thinking data from group to group, from known to unknown, to allow time and experience to be compressed in a purposive setting. Usually both the protocol of rank-ordered personal thoughts and the accompanying shouldbesheets are rotated. Some curious findings from this process reveal stages of development in personal thinking. People are first impressed by how different they are but soon figure out a level at which they are similar. When doing the musical triads they usually start by building their feelings of inclusion in the small group but, as successive sets of protocols from other groups move before them the feelings move to curiosity and often to outright laughter at the perceived ridiculousness of the criteria being used by the members of other groups. More slowly comes the 'yes but what is this all about' realization: careers are being decided, statuses are being established or disestablished, and the purpose of the organization emerges. It is what can be termed the '*interest-confusion-insight-judgment*' sequence.
3. *The intensive live-in experience* happens at short courses, particularly in staff colleges or government sponsored orientations to the principalship. At one of these (Rix and Parry 1981), sixty would-be principals virtually lived the content and process of their administrative thinking about others. One result was that during four weeks they developed a more critical and more discriminating mentality, one that was goal-directed but not blinkered. Most striking, in terms of patterns of teacher thinking, was how their pretest-posttest comparisons vividly demonstrated a

progression from concrete to abstract. Things like clothing and age, frequent at pretesting, were thrown out and chief vehicles of their thoughts were now professionalism and a means of defining their own professional identity, maturity and responsibility.

4. *Log-writing*, or the keeping of a journal, characterized a recent study (Brown 1982) of three groups of ten administrators or hopefuls in each. The purpose was to emphasize the individual process of reflecting without taking away from the other experiences such as pre and posttesting, shouldbesheets and self-confrontation, rotating personal thinking protocols among intimate groups, the posting of peer group norms and reference group norms, and the like. At the end of the three-month study, each member produced an individually written analysis based on both the quantitative data from their grids and their workgroup interaction experience. Results were that the patterns of personal thinking demonstrated by members of one group became more reflective than before, more concerned with personal and political dynamics of their organizations; their resourcefulness increased the most or by an average of 2.4 more criteria in use than before. A second group became more affective-oriented than before and showed more interest in cozy interpersonalizing; their average number of criteria in use rose by 1.9. The third group developed thoughts of other teachers that are more characteristic of hard-line administrators; they gained an average of 1 construct.

The chief interest in this last study is not why their thinking about teachers changed in different directions - they were different groups that had come together for those purposes - but that they did change and therefore so can the personal thinking of other teachers, thinking about other teachers. The implication is that this can be the intended behavior of an organization, behavior that is now clearly vulnerable to policy change and individual purpose. Nor may we diminish the power of this effect by doubting its permanence: *eighteen months* after the culmination of an eight-month project we carried out to develop and officially formalize administrative promotion policy (Brown, Rix, Cholvat 1983), one of our members went back and found that fully seventy percent of the principals who had recommended someone for promotion in those intervening months were still using those criteria which they developed during the project (Rix 1980). Considering all the other interventions that occur during a year and a half of any school system, this shows personal construct change to be fairly durable.

Running through all these studies however is evidence of the influence of the social context upon patterns of personal thinking. This is shown in two distinct ways, both of which carry implications for the practice of administration as much as for the art of teaching. The first is that the interpersonal development of constructs develops interpersonal constructs. Put more slowly, the substance of what they thought about when they thought about teachers now became more related to the way people related to people. I am supposing that is because our professional development projects themselves are intensely interpersonal in format. So format determines substance. If the essence of administration is people interacting with people, this insight explains why administrators who lead a lonely life behind closed doors or who act out preset roles in staff meetings are such poor administrators.

The second social context effect is more idiosyncratic. It is that the intensity of interaction between group members (weak or strong) seems to confer a joint proprietorship (similarly weak or strong) upon the ideas produced that way. If three persons working on an administrative problem did not get on well together, the solution was soon discarded; those group members who interacted positively and strongly kept their findings in high profile. There may be a lesson here for principals; certainly, the value of the intensity of cathectic investment is known to every good teacher. It is a way of creating a culture or building a better school.

Affirmative Action and the Case of Carol

As if to test the model, an opportunity for carrying it out over a very brief time was provided to us by an invitation from the Affirmative Action Office of the Windsor, Ontario, Board of Education. There was no deliberate intent that our work would result in more women getting promoted. However because prejudice operates best in ignorance, one way of overcoming prejudicial preferences would be to build a pool of experienced or properly trained persons to serve on selection committees (policy requires an equal number of males and females on selection committees). Could this experience-compressing model do the job in only three workshops? Their final question was "when they're finished, will they be different?" So was ours, for a different reason.

Changing operative criteria is changing constructs and that is close to changing ideologies, not collecting jargon and adding buzzwords. Anyone can alter their espoused and posted virtues, they shift with fashions and management training thrives on fashion. But the person whom you eye for career advancement belies your basics.

To compensate for the skimpy time available, batches of 'homework' were prepared, the PROM Form mailed out to and returned by 36 voluntary participants four week in advance, shouldbesheets mailed out and completed two weeks later, interaction at workshops was intensified, no speeches were made and delicious refreshments were available during sessions, i.e. no breaks.

One of the participants was Carol Fathers (real name), an elementary school vice principal. From her pretest protocol, Carol's terms that differentiated most strongly between persons whom she would recommend for a vice-principalship and those she would not, were "unstable personality, strong negative emotions, responsive to situations, always follows through, not predictable in all situations". Differentiating a little less strongly were her "strong physical presence, sober and predominantly quiet, happy smiling enthusiastic, not consistently dependable, fascinating people because of their interests". Of medium strength in sorting promotables from non-promotables, Carol's terms were "more interested in outside interests, powerful personality, direct". Others were too weak to differentiate.

By contrast, Carol Fathers' shouldbesheet, rank-ordering these same criteria but now in the abstract, shows her favoring a calm, bright outgoing personality with powerful leadership, decision and organizing skills who is committed and articulate. Analysis revealed that of nine criteria she had posted here as ones that should bear a strong effect on this staff decision, she had actually used only two of them that way on the protocol describing real persons to promote or not. Another posted strong criterion had medium strength on the protocol, two others were weak and the other four posted strong criteria had not been used at all, possibly did not apply to persons she would actually recommend.

Carol, confronted with this discrepancy, decided to start thinking again. She talked it up with persons in her small group, studied those of the other thirty-five participants, made between-session telephone calls to her small group colleagues, asked questions about jobs, about people, about policies, went through a workshop two simulation, i.e. did everything that others did. They also studied and talked about the implications of the 'back page' on prejudice and visibility. There the twenty women and sixteen men of this workshop alike chose younger people as promotable with women favoring women more than men favored men. They listened to a nonstatistical representation of their pretest criteria. Because this representation, prepared by E.A. Rix, shows the state of their thinking as we found them, it is reproduced here:

'you all talk the scene language, that is, you use the same words. Women tend to use more words, more synonyms. They talk most fluently about the concept of professional knowledge and growth and are quite articulate concerning how people relate with other people. These areas are in the very high priority for women; not so for men - they were not used in one single case with high strength. Women very strongly emphasize the importance of perceptiveness of being aware of seeing a wide scope of issues. Men by far emphasize the importance of one's relationship to the children. These were the areas that had high priority for the men and not particularly for the women - that is, they were used but seldom. Both men and women rate an enthusiastic personality high on their list of priorities; maybe that's why it's not surprising that motivation and commitment rank high too. As a group is it, possible you possess a philosophy that believes that enthusiasm motivates the students? It is clear that the total group (men and women alike) value leadership qualities and in particular these skills that reveal taking charge/responsibility being organized and flexible'

For a posttest during the final workshop the participants used a short form devised more for professional development than research. A five-by-ten grid, it was more limited in constructs elicited but was short enough for the participants to work out during the workshop. They could complete it, do the post-codings and weightings of the criteria they were now applying to their newly selected candidates, and still have ample time to talk openly about what was happening. Overheard were "for me that's mindstretching," "very revealing" and "now I have a better understanding of my own values in the selection process".

But what of Carol? The short form with its limitations does not do justice to the breadth of understanding or depth of insight Carol brought with her to the final session. The thoughts she now wrote down, and which made the biggest difference between those for whom she would now make a positive recommendation and those for whom she would not, were "approachable, influential positively, logical and cold" and negatively "combative, slow and deliberate speech patterns". She also used, but without significant decision power, "directing, aloof, apparently warm and charming, commanding respect, smooth diplomatic".

Carol's case characterizes the process others were experiencing. It is not that they learned new words; rather they were selecting different persons, different both in their identity and in their verbal representations. Some of Carol's newly developed criteria were more in line with those she earlier had posted as those that should be more important. Some were perhaps the reverse as one begins to reconsider ideals. Others changed in relative priority, possibly from the influence of peers, colleagues, the Windsor system's problems, and from her own inner convictions.

On her posttest, Carol was now using five of those criteria formerly posted strong or medium but previously unused in her selection of people. She upgraded another criterion and had decided to drop some criteria previously used, that is, she now selected persons according to a different standard. She also continued to favor persons with strong personalities despite her posted wish to soften it a bit. But no statistics can be as vivid as the expression on the face of Carol, this school administrator who with an active background in women's teacher federation and some hard knocks on the ladder had entered the project with skepticism but who now stared at her processed posttest protocol, then threw it up and enthused, beaming to all who could hear:

"Look my constructs have changed!"

As an afterword to the paper it could be noted that just as the project continues, theory continues to change. By January all Windsor school principals were brought into the selection skills training. By now we recognize the power of personal conviction upon thinking leading to decisions people make about people.

The basic theory called on cognition, reflection and person interaction to build a seven stage model of adult learning to show how permanent is change when you start from the inside. Professional growth, or being able to inter-articulate thoughts and actions, takes all three modes of reflecting:

(1) *Experiential* reflecting
- (a) the cognitive analysis of one's own actions,
- (b) cognitive self-confrontation and
- (c) cognitive query,

(2) Consensual reflecting or comparing your experiences with those of
- (a) significant others,
- (b) peers, colleagues and
- (c) norms of reference groups and

(3) Categorical or reflecting on
- (a) feelings, convictions, imperatives,
- (b) the unarguable needs or dictates of the organization, political reality, and
- (c) the forced-up alternative actions, e.g. the constructive escape.

Chapter 26

Arguments for Using Biography in Understanding Teacher Thinking

Richard Butt

Summary

The teacher is the major arbitrator within the influences on the classroom's curriculum. Teacher thinking, then, is of vital importance in the endeavor to understand how classrooms come to be the way they are and how they might become otherwise. In pursuing the understanding of teacher thinking and its relationship to classroom activity the question arises as to how to proceed: what methodology might be fruitful. This paper attempts to expose the epistemological potential of biography for understanding how teachers feel, think, and act.

A Brief Characterization of Biography

"A biography is the formative history of an individual's life experience" (Berk, 1980, p. 90). It not only focuses on what has happened in someone's life but also addresses how that individual responded to or initiated various events. It not only addresses attitudes, feelings, thoughts and actions but also examines the relation between earlier and later events. It attempts to infer how persons came to be the way they are. A biography is a story of a life or part thereof. It is a construct, an artifact assembled from the record of someone's activity ... "a selection of incidents from a life arranged and linked with respect to an outcome so as to render an intelligible account of how that outcome came to pass" (pp.94, 95) ... "Biographic study is a disciplined way of interpreting a person's thought and action in light of his or her past (p. 94). Berk also makes the important point that a biography is not, as some might think, the chronological record of tapes, field notes and the like. Rarely does the record contain the events that constitute the biographer's story. The events are inferred from the transcribed record. The record being testimony to the inferred events and interpretations construed from the events. It is a deliberate critical procedure that aims to make educational sense of thought, actions, feelings, attitudes and experiences (pp. 97, 98). Until the last few years, biographic studies were conducted in disciplines other than education. They have been narrative, analytic, psychological, psychoanalytic, and existential in character. As well as focusing on one individual, they also have taken the form of collective biography.

The narrative form is on an integrated life history, whereby all experiences in every aspect of the subject's life are candidates for inclusion in the narrative story. If they are considered important to the understanding of that life, they are included.

Analytic, psychoanalytic, or psychological biographies emphasize understanding individual case histories in psychology, through the lenses peculiar to that discipline. They are quite different in appearance and style, therefore, from a narrative life history. The existential form can be biographical or autobiographical. They go beyond a description of events in one's life in the narrative sense to a deeper examination of patterns underlying one's history of living, to reflection, to psychoanalysis. Psychological principles of explanation have also enjoyed a long tradition in biographies within the discipline of history (p. 94).

Again, in history there is another long tradition of research called collective biography, whereby the focus is a group. Various particular principles of explanation might be utilized whether economic, sociological, and demographic. The main aim is that these principles tap what commonalities do exist across many life histories.

The Teacher's Perspective

If good teachers' actions and thoughts are as much (if not more) guided by their own knowledge, intuition, experience, craft and theories-in-action, as by any prescriptive framework or models, (Eisner, 1983) what is the nature of that knowledge? (Elbaz, 1981; Connelly and Elbaz, 1980). How are teachers' theories-in-action and teaching styles formed? How do these guide teachers' thoughts and actions and in what actions are unexamined habits shaped more by circumstances than teacher thinking? In approaching such questions as these how do we best proceed?

Until recently, educational inquiry or prescriptions were seldom based on classroom reality especially as experienced by its participants. There has been minimal dialogue between teachers (as they perceive their professional lives) and scholars of education - not only because of the nature of the relationship of outsiders to insiders, but also because of the lack of an approach to inquiry that effectively grasped and represented what one might call *the teacher's voice.* I maintain, then, that the lack of this essential perspective has seriously hurt the quality of much educational scholarship, resulting in knowledge inferior in its practical and intellectual usefulness.

In order to remedy this situation, curriculum scholars need to engage with teachers, as co-researchers/co-developers of classrooms as an important part of their research endeavours. In the short term, there is a dire need to develop a body of knowledge that represents the teachers' perspective, to redress the current dearth of such discourse. One way in which outsiders can both learn about classroom reality and create knowledge that carries the teacher's voice is through the use of biographical and autobiographical accounts as to what they think, experience and do.

The nature of teaching: qualitative studies as basic research

> 'Any scientific understanding of human action, at whatever level of ordering or generality, must begin with, and be built upon an understanding of the everyday life of the members performing those actions'. (Douglas, 1970 quoted in Janesick, 1981, p. 15).

In the particular case at hand our interest is in teacher thinking and its interrelationship with action. It is logical, then, that we focus directly on the qualitative nature of teacher's thoughts and actions. How researchers see and interpret these thoughts and actions may be assisted in some ways by their detachment and non-involvement in teaching. This also carries with it, however, the disadvantage of a lack of experience of the phenomena concerned and a lack of personal knowledge of the participants' perceptions, motives, and intentions. Human beings, through their consciousness, can be understood in ways that non-human subjects and objects of research cannot (Janesick, 1981, p.20). What teachers do and think within their professional lives depends, as the phenomenologist would say, upon the meanings those individuals hold and interpret within their personal, social, and professional realities and everyday-life situations.

Just observing an event or a phenomenon, even through the eyes of a participant is not sufficient. One needs to go further to understand the relationship among antecedent, subsequent, and consequent events through engaging in dialogue with the teacher. One can pursue meaning, motive, beliefs, and intentions - all those thoughts judged pertinent to the events by *both* researcher and teacher. This process not only permits access on the part of the researcher to a teacher's thoughts and actions, but also permits access to unexamined or habitual aspects of a teacher's life.

As opposed to being a simple set of phenomena which are devoid of motive and interest that are easily manipulated and controlled, teaching situations, as quantitative studies have discovered, are very complex multivariate phenomena. They are exceedingly difficult to generalize about, and even more difficult to control. Each teaching situation is different in significant ways from others, whether due to pupil, teacher, curriculum, and environmental or other characteristics. The teaching act, itself, is not subject totally to a rational treatment but is a combination of science, craft, and art and "on the spot" human ingenuity and intuition (Eisner, 1983). The 200 or more (Jackson, 1968) decisions a teacher makes per hour are just as likely to be determined by dynamic situational factors as any predetermined stance on the part of teachers, indeed less so by experts prescriptive models.

Teaching is riddled with so many competing influences, dilemmas, paradoxes, and contradictions that prediction of specific thoughts and actions of teachers by researchers is at this point anyway, so difficult as to be futile. Each teaching action and the thinking associated with it is nested within uniquely personal, situational and contextual determinants and influences. Which of those influences will hold sway in a particular situation to result in one of many possible and paradoxical outcomes is difficult to predict. The dynamics of the classroom leaves room for myriads of alternative outcomes for similar mixes of variables. In most empirical studies teachers still remain the largest source of variance! At this stage of our understanding of teaching, especially teacher thinking and action, to utilize only hypothesis driven quantitative studies would prove fruitless. "Discrete variables and their relationships do not seem to be sufficient to deal with complex interactions and patterns of human behavior" (Guba and Lincoln, 1981, p. 81).

This is not to say, however, that an empirical-analytic treatment will never be fruitful. There are glimmers of hope within the careful research being conducted that has established links between certain teaching actions and pupil achievement using a time on task approach. I am not claiming here, then, that generalization across teachers is impossible nor that any generalization that might exist is or will necessarily be educationally insignificant. What is apparent is that what are both generalizable and educationally significant is going to be complex and perhaps paradoxical, leaving room for some uniquely individual manifestations within different personal teaching contexts.

From this perspective, therefore, I see qualitative inquiries into the nature of teacher thinking and action as a type of basic research that is necessary to provide the depth of understanding of educational phenomena required for empirical studies based on complex hypotheses. Initially these studies might be applied to individuals, latterly to groups of teachers.

Inquiry into the complex and personal nature of teachers' thoughts and actions is not a task that can be undertaken by the researcher who functions mostly outside the classroom. This type of inquiry requires the active collaboration of the teacher as co-researcher who can engage in a dialogue with the outsider. In the end, who possesses teacher thinking, not researchers but teachers?

On the Need for Biographical Studies

The foregoing arguments besides being appropriately applied to biography can also support the use of qualitative studies of teacher thinking. I wish to focus, in the remainder of this paper, on making a more specific case for the use of biography in understanding teacher thinking for several reasons. Firstly, the cases for the use of phenomenology and ethnography have been well made, whereas that for biography has not received sufficient attention. Secondly, though phenomenology and biography have very different histories, their interests intertwine easily, sometimes usefully, and legitimately, but now at the expense of certain strengths of biography which lie buried or neglected under the trend to phenomenology. Thirdly biography, despite its similarities to certain aspects of phenomenology, does have unique differences which make it better suited than phenomenology to address certain requirements argued previously in this paper for the study of teacher thinking.

One can understand the potential miscibility of phenomenology and biography in that phenomenology "strives to re-introduce the personal elements in knowing in order to understand how human knowledge is influenced by human imagination, desire, and will" (Collins, 1979, p. 11). Phenomenology examines behavior, motive, purpose, or intent and aims to understand humans' lives from their perspectives (Van Manen, 1975). Biography has very similar aims, but possesses some different methodological approaches than phenomenology that can add other insights of equal and sometimes better value.

Many aspects of current quantitative and qualitative methodologies are biographical in nature. The problem here is, without being acknowledged *as biography* and guided by its principles and processes, these efforts might not take full advantage of the biographical approach.

Jackson's (1968) study of life in classrooms and Freema Elbaz' (1981) recent work, a case study of one teacher, are clearly biographical in nature. Similarly, Valerie J. Janesick's (1981) case study of a professors' of curriculum and pedagogy, while drawing from ethnography, phenomenology, and symbolic interaction is "the story of one person" (p. 15), as is the current work of Connelly and Clandinin (1983) that has been presented at this conference. That biography forms an important aspect of these and many other studies is testimony to its usefulness, as instinctively recognized by researchers. It is important, however, that this aspect of the work be explicitly acknowledged so we can begin to make judgments as to how biography might be most fruitfully used in education.

Access to past influences on present actions can only be gained through individual teachers themselves. Biography emphasizes revealing the potential and actual influence, both conscious and unconscious, of the past on the present. This represents one strength of biography in comparison with the preoccupation of phenomenology at present. The fact that this potential is illuminated *directly through the person himself or herself,* as opposed to phenomenology's preoccupation with classroom events, represents another; especially since the art and craft of teaching is a personal statement. A longitudinal approach to biography, whereby past, present, and potentially future are combined, has the significant potential of recording the *development* of teacher thinking, as it happens. I would also claim that the intimate way in which the biographer and the teacher work together, assembling the record of events significant to the development of one teacher's personal practical and professional knowledge, makes the relationship as collaborative as it necessarily should be for fully understanding a teacher's thinking. This provides for a constant and continuous personal verification on the part of the teacher of the biographers, data, inferences, and form of biography. This is in contrast to many phenomenological studies that appear to gather feedback when their interpretations are more fully formed. This constant correction of existing bias may be somewhat safer, perhaps, than the process engaged in by phenomenologists of education, who attempt to "bracket out" their preconceptions without the constant monitoring of the teacher.

The teacher's voice

The very level of personal collaborative work between the biographer and the teacher, while being one of its strengths, provides the context for its major strength. It provides a vehicle for recording and interpreting the teacher's voice. I use the notion of the *teachers voice* in several literal and metaphorical senses. In a physical and metaphorical sense, the tone, the quality, the feelings that are conveyed by the way a teacher speaks are important to consider in investigating the nature of teaching. In a political sense, the teacher's voice attests to the *right* of speaking and being represented. It can represent the views of both unique individuals, and of a number of people - a collective voice. "Voice" also connotes that what is said is *characteristic* of teachers, as distinct from other potential voices.

An argument can be made for biography (in contrast to uncollaborative autobiography) in that teachers have expressed the need for someone outside themselves to assist them in making sense of what they do (Flanders, 1983, p. 147). Conversely, researchers need biography among other methodologies in order to learn more about teaching reality. They also need approaches that structure and facilitate collaborative work with teachers, within collegial rather than vertical relationships (Butt et al 1983). These types of relationships are necessary to gain access to teachers' true feelings and attitudes that shape their thinking.

If, then, one's purpose is to understand teacher thinking and its relationship to action, then the proper subject of inquiry is teaching *as experienced* by teachers. Knowing the quality of experienced thought and action through biography makes it possible to disclose what significantly influences what a teacher does or does not do. How those thoughts came to be, is understood, through making educative experience intelligible. Through examining the transformational quality of significant experiences in personal and professional lives we can apprehend a teachers' formation or development in an educative as well as a training sense. We can focus on how teachers might transcend the stress of daily teaching life - how they grow personally and professionally (Flanders, 1983, p. 144).

Biographies in Education: How They Have Been Done

Biography does not have a particular unwavering "method". The details of each approach are worked out by the biographer - to best investigate and portray the significant aspects of the life history concerned.

The best means, therefore, of understanding how one might conduct a biographical study beyond knowing the type or characteristics appropriate for a particular individual or situation, is through biographical examples from education that have appeared recently. These, it should be noted, represent an increasing trend to the use of scholarly biography in our field.

Obviously conversations, interviews, observations, video and audiotapes, field notes, stimulated recall, "stream of consciousness" journals, and logs, are all legitimate means by which data might be gathered for the purposes of biography. In what order and fashion these are used would depend upon the particular purpose. The manner in which one might interpret this record; how one might make and check inferences; how one might collaborate with the subject are all matters that would depend upon individual needs and circumstances.

The means by which Elbaz (1981), Janesick (1981), and Connelly and Clandinin (1983) pursued the elaboration of the stories of their teachers could be considered as legitimate for biography provided those means are used to investigate what from the subject's past as well as present life significantly influences current thought and action. "Biographic studies of education ... deal with the formation of an individual consciousness through his or her experience" Berk (1980, p. 140).

Whereas the gathering of data in biography, through whatever means and design, can be difficult and tedious, the most difficult and tedious work is the interpretation of the record.

Berk (1980) searches for educational episodes, which are moments of insight, moments that change us, gestalts that provide a leap forward, a way of resolving a conflict, or of surviving. How did what baffled a teacher - a paradox or dilemma - become clear? These "are the marks of an educative experience ... And an educational episode is the story of how one of these insights happened".

So one looks for moments of insight, searches for the problematic situation that required the insight, and the activity it provoked. Lastly "because the experience is supposed to enable later experience, one looks for evidence that the experience has had fruitful consequences" (p.98). Pinar (1980, 1981) and Grumet (1980) provide examples of autobiographies that are existential, literary, and perhaps psychoanalytic in character. Pinar's approach involves providing descriptive autobiographical narrative from one's childhood (regressive), and from one's development (progressive), scrutinizing that data for important educational episodes (analysis) and bringing this together as a statement of oneself (synthesis). I give this all too simplistic description only to provide the reader the choice of whether to investigate this further or not. Pinar's in-depth treatment of autobiography must speak for itself, as must other examples I have only characterized in a superficial way here.

Grumet (1980) explains how she uses reflexive research and analysis to cause students to recapture their own educational experiences, to reconceptualize curriculum in personal and concrete terms. The narrative is then analysed to reveal previously unknown interests and biases, again raising consciousness and thus allowing for conscious reconceptualization of their lives - and their curricula.

Biography and autobiography are also beginning to feature as an approach to portraying the thinking of curriculum theorists. Schubert and Schubert (1982) present reconstructed dialogue between curriculum theorists, with themselves as both questioner and the one who responds. It represents a clarification of their own thinking about various aspects of curriculum. Similarly, a most recent issue of Curriculum Inquiry included an autobiographical piece by Ted Aoki (1983), which describes the interplay between his life's history as a Japanese Canadian and becoming a teacher and teacher educator shortly after being interned as an alien during the Second World War.

Author Reflection Twenty Years On

When this paper was first published there was some resistance to recognizing the biographical character of research into teacher's personal professional knowledge, and more resistance to making full use of biographic methods. The thought of using *autobiography* as method and data and of having teachers fully involved as co-researchers in interpreting this data drew more extreme responses – especially if one advocated inquiring into "ordinary" rather than "extra-ordinary" teacher's lives, careers and professional practices. (We need to understand the "typical" and "representative" rather than just the atypical.) Nevertheless, the powerful potential of autobiography/biography rapidly broke through any residual resistance so that by 1989 my co-researcher, Danielle Raymond, and I could identify at least six clusters of life story inquiry into teaching which evolved during the 1980's (Butt and Raymond, 1989). This whole journal issue and, next year, a whole issue of the *Cambridge Journal of Education* – both were devoted to biography in education. In 1995 Pinar (et al) devoted several major sections of his book to life story work in regard to teachers. From the early eighties, our own work has fully engaged the particular potential of *autobiography* through a method we called 'Collaborative Autobiography.' (See Pinar, 1995 for a review of this work and Butt and Raymond, 1989).

The arguments made in this article, as simple and crude as they are in retrospect, still are pertinent and worth pursuing further. Our original propositions were that how teachers think and behave as professionals is profoundly influenced by their life histories, and that they think and behave as if *their own* perceptions of that life, their life stories, are "true."

Our work and that of many others portrays the ways in which our life histories (particularly early childhood, epiphanies gleaned from professional experiences and our own experiences as learners in school), shape our thoughts and actions as teachers. "Teacher's stories" have now become a popular bandwagon which has raised all sorts of issues. Is this therapy? Who, therefore, is "qualified" to engage in teacher's life story work? Under what conditions should it be undertaken? What are the ethics which should guide this work? What should the relationship be between autobiographer as researcher and facilitator as researcher? How should they co-interpret stories and for what/whose purposes? In order to address some of these issues I have taken graduate courses in counseling as a student. I have also studied teacher stress and burn out in order to extend and apply Collaborative Autobiography to our understanding and avoidance of this burgeoning problem in teacher development, lives and careers.

Chapter 27

Access to Teacher Cognitions: Problems of Assessment and Analysis

Günter L. Huber and Heinz Mandl

Summary

The relevance of teacher cognitions for explanation and modification of teaching is discussed from an action-theoretical point of view. Studies in this area and modification approaches depend on the possibility to get access to teachers' action-directing cognitions. The article describes methodological criteria and ways for validation of teacher cognitions together with examples of empirical studies.

Goals of Research on Teacher Cognitions

Research on teacher cognitions is rooted implicitly or explicitly within action-theoretical reference systems. These reference paradigms all agree on one point: cognitive processes and products such as assessment of situation of ones own action possibilities and results, or expectations based on prior experiences, evaluations, etc., determine the course and the results of action.

These models of human action, which form the basis for the studies concerning teacher cognitions, differ of course considerably in their dimensions of rationality, which are ascribed to their principles.

On one hand studies on teacher cognitions were built more or less upon models of information input and processing occurring regularly during pedagogical action (teaching); on the other hand one can proceed from implicit subjective models or common sense theories of action, as they are used by teachers every day when dealing with routine assignments or confrontation with new problems.

Furthermore the differing theoretical approaches all have one aspect in common, namely their intended goal (purpose): not only to understand teacher cognitions theoretically but to search for possibilities of change in order that teaching techniques may be improved. This approach seems to be suitable in surmounting deficits in teacher education when based only on knowledge dissemination and behavioural training. The conveyance of knowledge and skills will only influence the behaviour in the classroom if one manages to modify the stable common sense theories. More over one has to find ways of integrating scientific theories with naïve theories. Besides, subjective criteria of teachers have to be made accessible when using action suggestions and have to be considered with teacher training approaches. Finally, it is necessary to instruct teachers on situation reflection, possible action alternatives and their consequences, since all these conditions necessary for a successful modification of pedagogical techniques have to possess, as a prerequisite, teacher cognitions.

The effectiveness of a strategic cognition oriented pre/in service training for teachers is questionable. With this theory substantial basic questions have arisen:

Are pedagogical techniques always concomitant with cognition or expressed differently: what kind of action relevance can teacher cognitions have for everyday teaching? When making a rough classification of pedagogical action, reflected planned action, and unreflected routines one has to calculate that teachers in everyday teaching (depending on the complexity and dynamics of the situation) as a rule have to rely upon their routinized actions. If these practices no longer prove to be successful then their advantages of quick availability and subjective security about actions tend to be the greatest obstacle for change. In this situation, strategies dealing with action analysis of the pedagogical situation and their possible action alternative would be especially applicable, i.e. metacognitive processes in problematic situations.

Further substantial objections concern the possibility of modification of teacher cognitions. Meanwhile approaches have been tried which are dealing with changing teacher behaviour through the modification of thought processes.

Methodological arguments are fundamentally directed at the apperception of cognition regarding action relevance and specifically the possibility of determining action relevant teacher cognitions. The latter argument is especially important in connection with the high degree of automation of everyday processes to which teachers are exposed. We shall mainly deal with this methodological problem and its possible solutions.

Cognitions in the Context of Action

Complex models of information processing

Figure 1 presents a schematic view of the process model of social behaviour in the classroom, developed by Hofer and Dobrick (1978, 1981). The state of events encountered by a teacher is the first link in the chain of actions which must be explained. This initial situation is characterized above all by the student involved in it or in its causation. What is relevant here in terms of action, however, is not the "objective situation," but rather the situation as perceived by the teacher. This perceived situation could, under certain circumstances (e.g., selection, accentuation), be substantially distorted. In particular, the teacher's naïve theories (e.g., about the student's personality) are influential here. The teacher assigns importance to the perceived situation. This occurs in correspondence with her/his goals (goal dimensions). A situation can gain importance, for instance, when it represents one condition for the possibility of goal attainment or when experience shows that it is related to goal attainment.

The teacher's causal explanation for the evolution of a situation is called attribution. This attribution is the basis of expectations about the further development of the situation.
The perceived situation is the actual state for which a corresponding anticipated state is projected. The anticipated state is the situation that – according to the teacher's experience and the results of attribution – will follow upon the present situation with highest probability, unless the teacher intervenes.

If this anticipated state does not correspond to the goal state, desired by the teacher, then he or she must think of adequate strategies for preventing the anticipated state or for transforming it into the goal state. The goal state, for which the teacher is aiming, is in part a product of her/his goals. These are combined in the concept of goal dimensions.

The goal dimensions are the basis of decisions. They also provide the dimensions for a comparison between the anticipated state and the goal state. The success levels, on the other hand, are the fixed points on these dimensions, which correspond to the goal state. The success level defines, for instance, when a teacher considers a student's performance to be successful or unsuccessful.

The fundamental prerequisite of Hofer and Dobrick's (1978, 1981) model is the assumption that teachers behave rationally; that the necessary and sufficient conditions for the explanation and prediction of student related teacher behaviour could be explicated by a system of theoretically postulated mediating cognitive processes.

Cognitive theories were developed and became popular because, as Miller, Galanter & Pribram (1973, p. 12) note, it appeared reasonable, "to slip a little wisdom in between the stimulus and the reaction and one needn't even make excuses for this, because it was already there before psychology ever existed."

Weinert (1978, p. 68) asks the question, "whether the cognivistic action model of Hofer & Dobrick does not require and postulate a little too much wisdom between stimulus and response: Do teachers really behave as rationally in everyday situations as the model dictates, so that even their mistakes can be systematically (cognitively) explained?"

Weinert (1978) presents two reasons for his critic:

(1) Largely automatized routine behaviour, in particular, cannot be adequately theoretically conceived of as the result of decision processes.
(2) Causal attribution of actions is doubtlessly important in the action context; however, in everyday action settings the teacher only has access to a limited catalogue of causes.

The problems inherent in this approach are also made clear by the empirical findings: large varieties of cognitive processes were included in the improved replication study (Hofer et al., 1982). Hofer reports an explained variance of the observed teacher behaviour of only 15% for the entirety of the cognitive model variables. This result is not very impressive, especially considering the methodological thoroughness and the range of variables with which the author is working.

Subjective theories

Another scientific approach to teacher cognitions involves the investigation of teachers' subjective theories (subjective, implicit, naïve, private, everyday theories). This approach is not concerned with the investigation of specific and isolated perceptions, interpretation, judgements and so on, but rather with aggregations of cognitions, (1) which comprehend a teacher's subjective access to experiences and actions in relation to classroom events, and (2) which can become actualized, at least in principle, in coping with classroom events. Together these cognitions form systems, which often are compared structurally with scientific theories (c.f. Groeben and Scheele, 1977). For illustration, several examples of subjective-theoretical statements of teachers are presented (c.f. Wahl et al., 1983, p. 26).

The teacher observes a student playing with small plastic objects. The teacher suspects that the student is bored. The subjective explanation "The student is bored" can be transformed into the subjective hypothesis "If a student is bored then s/he busies her/himself with things that have nothing to do with the lesson."

A subjective hypothesis is a general sentence, which can, in principle, be transformed into an if-then statement. Subjective constructs are used in such hypotheses, for instance: The teacher has the hypothesis "If the material is boring and contains nothing new, then the students' attention will wane considerably." Here s/he uses the subjective construct "attention".

Subjective psychological theories can be put to use in different ways: On the one hand as explanatory knowledge (knowledge about stable relationships in the world, if-then knowledge, causal knowledge); on the other hand as productive knowledge (in-order-to knowledge, functional knowledge).

A survey of research on teacher cognitions shows that these subjective theoretical systems are conceived of as serving the same purposes for teachers and their everyday actions as scientific theories do for researchers. At least it has been tried to get access to teachers subjective theories about teaching, preparation for lessons, learning, student behaviour, learning disabilities and behaviour problems, student personality, etc. (c.f. Shavelson and Stern, 1981; Bromme andHömberg, 1980) in order to (1) reconstruct actions-relevant cognitions of teachers, (2) explain teacher behaviour, and (3) modify teacher behaviour.

Subjective theories (c.f. review article by Mandl and Huber, 1983) have been investigated in the areas of student evaluation (e.g. Hofer, 1975; Hanke, Lohmöller, Mandl, 1980; Huber and Mandl, 1979), lesson plans and their execution (Bromme, 1979; Triber, 1981; Wagner, Maier, Uttendorfer-Marek and Weidle, 1981; Rheinberg and Elke, 1979) and teachers' handling of school-related problems, including the teachers' own personal crisis situations (Wahl, 1981; Dann, Humpert, Krause, Olbrich and Tennstüdt, 1982; Becker, Huber, Mandl, Wahl and Weinert, 1981; Ulrich, Hausser, Mayring, Alt, Strehmel and Grundwald, 1981).

The argumentative relations allow deductions and reflection. However, in everyday action events must not necessarily be explained or predicted in every instance. As a rule, action is based on habitual if-then relationships. It must also be critically noted that action depends on other conditions as well. Therefore, it cannot be expected that inferred if-then relationships will stand in an unequivocal relationship to actions.

Subjective theories, as opposed to scientific theories, are not logically coherent systems of statements but rather loosely connected and partially contradictory if-then statements for individual classes of problems. Occasionally, opposing elements of a teacher's subjective theories can be actualized simultaneously in a classroom situation (so called "knots"; Wagner et al., 1981).

On the Significance of Teacher Cognitions

Teacher cognitions have been studied in the most varied areas of the field of educational action: with respect to teachers' judgement the components of background knowledge for educational decisions have been the subject of research; planning decisions themselves were studied as they occurred; interactive decision processes and their elements have also been in the focus of interest for research on teacher cognitions.

Routine actions are in foreground of teachers' daily experiences, not the reflected processes. Does this make studies of teacher cognitions relatively meaningless? Heckhausen (1980) notes that routines often emerge from reflected processes of behavioural control. In addition, routines are usually accessible to reflection. This access is of particular importance for the goal of modification.

The specific way experimental and behavioural data are structured by "common sense" determines or, at least, influences interactions with other people. If one wishes to analyze interpersonal relations, and even more so, if one wishes to modify them (as, for instance, in the numerous teacher training procedures), it is necessary to reach or assess the level of cognitive common sense structures, the level of the subjective theories. It should be possible by means of carefully questioning to ascertain the principles teachers use in constructing their view of their social environment, particularly their own action potential therein, and which help them to orient themselves. This is necessary to understand, predict and, particularly, to alter existing relations (Wahl, et al., 1981; Huber and Mandl, 1977; Wagner et al., 1980; Schlee, 1982; Tennstädt, 1982).

In connection with the significance of teacher cognitions or subjective theories, the veracity of assessed cognitive systems has often been debated. The thrust of these questions, however, seems to miss the core of the meaning of subjective theories of teachers. Discussion about the relevance of teacher cognitions should concentrate less on its accuracy and more on its functional value.

Dann (1982) has named situation definition, explanation of events, prediction of future events, and generation of action suggestions as the most frequent functions of subjective theories.

Whether the common sense psychologist has understood the "real" causes and consequences of human experience and action is, in fact, irrelevant at least as regards that portion of events to which his/her interest is directed. What is important are the ways and the extent to which his/her actions are guided or influenced by the cognitive structures in interpersonal relations and the rule-like processes occurring in these structures available to him/her. Being a teacher, his/her orientation will, in turn, influence others.

If one takes into consideration the fact that teachers, too, derive hypotheses from their interactions with their students and that they try to verify them, albeit in a less methodologically controlled way than the researcher, then it should be possible to identify focal determinants of educational actions. Thus, not the situation as such, in which the action occurs "objectively," is of interest, but rather the situation as perceived by the teacher, the situation in which she/he subjectively considers her/himself to be and in which she/he is acting.

Access to Action-Directing Cognitions

For research on teacher cognitions, the problem arises of how the unnoticed determination of the research object by methodological consequences of an unsuitable object model can be prevented (c.f. Huber and Mandl, 1979). Verbalization methods seem to offer an appropriate answer. While they promise to eliminate the determination problem, they also raise a number of new questions. These can be divided into three areas:

(1) Intrapersonal domain: The actor is encouraged as a rule to verbalize action-relevant cognitions. With this method, the following points must be clarified:
   - To what degree are these cognitions so accessible to the actor that s/he can report them verbally (Nisbett and Wilson, 1977; Wilson and Nisbett, 1978)?
   - What is the relationship between verbalized cognitions and action (Lazarus, Averill and Opton, 1974; Wahl, 1979)?

(2) Interpersonal domain: Ideally, the actor verbalizes his/her action-directing cognitions. In reality, his/her verbalizations contain interpretative cognitions that explain or justify the actions, as well as anticipatory cognitions (expectations) for future action situations. The researcher interprets the verbalizations with the intention of reconstructing the action-directing cognitions of the reporter and fitting them into a scientific system. In this case, the following points must be clarified:
   - How are the verbalizations influenced by the communicative relation between reporter and researcher?
   - How can the necessary process of interpretative reconstruction of action-directing cognitions be theoretically consolidated?
   - How can the resulting interpretative reconstructions be validated?

(3) Societal domain: The interpretative reconstruction of acting persons' subjective realities, if it is to occur in a valid way, can only register those societal determinants of individual actions which are accessible to the actors.

Verbalizations focused on intra-individual cognition-action relations can be differentiated according to the following characteristics:

- Criterion of the perspective from which action-leading cognitions are considered. We have differentiated a pre-actional, a peri-actional, and a post-actional perspective (c.f. Meichenbaum and Butler, 1979; Cacioppo and Petty, 1980).
- Criterion of structuredness. The different methods by which remarks on one's own cognitions can be encouraged or exacted determine the quantity and quality of the communications to different degrees.
- Criterion of temporal space between cognition or action and the verbalization for action-relevant cognitions. Is there a short or long interval between the report and the cognitive processes, which are to be made interpersonally accessible? This criterion has only minor importance for the verbalization of action-accompanying cognitions. The widely regarded arguments of Nisbett & Wilson (1977) against verbalization methods for the study of action-relevant cognitions refer exclusively to retrospective

verbalizations in extremely specific cognition-action contexts within a relatively short time of the action (compare Wahl, 1979; Meichenbaum and Butler, 1979; Smith and Miller, 1978; Cacioppo and Petty, 1980; Ericsson, 1978; Ericsson and Simon, 1978; Bowers, 1979; Huber and Mandl, 1979b).

- Criterion of specificity. On which level of the cognition-action relationship should action-relevant cognitions be verbalized? Problems arise in the verbalization of general cognitions on the action orientation level as well as in the verbalization of very specific cognition of operation control.

Problems in Testing Subjective Theories

In most of the studies to date, either we have classroom observations, or the teachers are interviewed after finishing the preparation or the performance of a lesson or some critical instructional sequence. They are requested to report what has gone through their heads during the process. Although the verbalizations of cognitions relevant to action do not necessarily disclose the actual, objective reasons of the action, they nonetheless reveal the subjective meaning of the context of the action. Hence, we may deduce the orientation of the teacher in comparable situations. Or, in other words, perhaps teachers – like any other persons – are not capable of recalling action-regulating cognitions, but rather verbalize their ideas on goals, strategies and evaluations instead, which they reconstruct on the basis of their common sense knowledge concerning specific situations and the strategies that are socially desirable and tolerated. But even in this case we get to know something about what a person thinks is expected from him/her, and what s/he is capable of doing if landed in this particular situation. There is at least a possibility that this kind of reconstructed explanation might effect future actions.

At this point, it seems appropriate to deal more specifically with the problem of validity, which has received great attention in Germany in connection with the research on teacher cognitions. Differing findings and interpretations are considered valid if the teacher agrees that the cognitions the researcher has reconstructed from her/his statement are correct ("Rekonstruktionsadäquanz"; "adequacy of reconstruction" Treiber and Groeben, 1981). If the requirements of this criterion are met, one assumes that all the thoughts the teacher has reported and which have been inferred from his/her report through analysis and interpretation actually did go through his/her head during the performance of the action. This still leaves the question open, whether the cognitions that have been reconstructed and validated by agreement suffice to explain the teacher's actions ("Realitätsadäquanz"; "adequacy with respect to reality" – Treiber and Groeben, 1981). In other words: Can the actions of the teacher be explained by these cognitions (validated by agreement)? The answer to this amounts to evaluating the affectivity of those cognitions that one was able to identify.

We have mentioned some of the theoretically and empirically based doubts cast on the adequacy of verbalizations with regard to reality. Of course, there are situations in the daily teaching routine in which the teacher's behaviour is not guided by clearly conscious cognitions ("impulsive actions" or "in which the agents do not have access to the cognitions guiding their actions any more (habits, automated actions)). Even in such cases, though, it is possible to reach an agreement on the cognitions that could be inferred from the teachers' reports. In fact, however, this amounts to the teacher agreeing that his/her reconstructions of plausible cognitions were reconstructed on the epistemological and heuristic idiosyncrasies of the teacher's cognitive structure, but they do not reflect what has been going on in his/her head during the earlier action.

This point of criticism can be immediately refuted, as it relies on the assumption that there is either insufficient or no correspondence between the postactional verbalizations of action-regulating cognitions and the "real" determinants of the action. As regards this argument, we postulated that the perspective defined by this approach toward validating the relation between (verbalized) cognitions and actions is inadequate in many cases (Huber and Mandl, 1982). Validation according to the criterion of adequacy with respect to reality is obviously based on the concept of statistical relations between cognitions and actions. There are many indications, however, that the relations between cognitions – including verbally reconstructed ones – and actions are complex and dynamic. Situative conditions, the process of acting itself and its outcomes not only elicit subjective perceptions, expectations, and evaluations. These very cognitions may change the basis for subsequent orienting or regulative processes surrounding the action. When this is the case, the criterion of "Prognoseadäquanz" (predictive adequacy) seems to be more appropriate for validating verbalized cognitions.

The instructional researcher tends to consider reports on cognitions relating to actions from the past as being retrospective. Accordingly, the verbalized cognitions of the teacher are seen as perceptions, justifications, or explanations (given in a communicative situation with a researcher). The teacher, on the other hand, is possibly structuring his/her experience while reporting his/her earlier action (c.f. Jaeggi, 1979; on the area of therapy). This means that the teacher is actually expressing expectations concerning the relationship between possible actions and possible outcomes, and subjective evaluations of these outcomes. All these are cognitions, which may influence later actions.

In using an epistemological subject model, communicative validation in the form of a dialogue-consensus (Groeben and Scheele, 1977) constitutes the first step in the process of validation. In this process, the researcher and the teacher come to an agreement as to which of the cognitions reported by the teacher should be considered as action regulating. The consensus reached, however, does not necessarily support the assumption that the reconstructed subjective theories really did influence the earlier actions of the teacher about which s/he reported specific cognitions.

Communicative validation has to be followed by action validation, i.e. the truth criterion used in the dialogue-consensus validation has to be supplemented by the falsification-theoretic truth criterion. The procedures for action validation presently discussed in German research on teacher cognitions are the following three (Groeben and Treiber, 1981; Wahl, 1982; Huber and Mandl, 1982):

- The correlation between the teachers' reconstructed subjective theories and his/her overt behaviour.
- Prediction of teaching behaviour based on the reconstructed subjective theories.
- Modification of reconstructed subjective theories through intervention (training procedures). Subsequently, one checks whether this leads to specific changes in overt behaviour.

Correlational studies do not have as much empirical value as testing methods (Treiber and Groeben, 1980). Procedures permitting a precise modification of subjective theories are not yet sufficiently known and tried. In the meantime, however, an elaborated approach exists which we would like to cover briefly (Diff: "Student problems – Teacher problems"). Current research has been focusing on action validation of subjective theories through the prediction of future teacher behaviour (c.f. Wahl, 1981). The work being done here will be discussed in the following sections.

Modification and Progress of Teacher Cognitions

Modification approach

A television study course of the Deutsches Institut für Fernstudien (DIFF) on "Schülerprobleme – Lehrerproleme" (student problems – teacher problems) based its training concept for the management of difficult situations in the classroom on the subjective theories of teachers. The interactional behaviour of the teachers should change as an effect of changes in these cognitive stores (Huber and Handl, 1977; Weinert, 1977; Becker, Huber, Wahl and Weinert, 1978).

Subjective theories have many advantages. They allow the teachers to structure the situations they are confronted with daily in such a way that there is little danger of incapacitation due to too much and too differentiated thinking. However, these systems contain many un-investigated or hardly tested assumptions. Therefore, the actions initiated on the grounds of subjective theories often enhance the conflicts rather than resolving them. However, this experience rarely has an effect on the cognitions of the teacher. The fast succession of imperative decisions to be made makes it hard for the teacher to register all the effects of her/his actions. Just like any other person in everyday situations, teachers are thus in danger of never learning from their "experience", and this is due to the shortcomings in analytical and interpretative processes. However, this is precisely the reason why they remain convinced of their "hold on the situation." On top of the interactive actions, which are mediated by cognitive processes, we must also take into consideration the many impulsive actions and behaviours used by habit. Their process characteristics and antecedent conditions are determined by a long learning history.

The main goal of teacher training is to remove this inability to learn from experience. Situation-outcome expectations and action-outcome expectations have to be corrected by a valid perception of the actual situation in the classroom and the actual interaction effects. This can only be achieved if the teachers improve the rationality of their routine actions. They must learn to see themselves as persons who generate hypotheses and test them, at least in critical situations. In detail, the following results are envisaged by the training:

- Elaborations of techniques for analyzing the classroom situation.
- Modification of dysfunctional elements in the subjective theories.
- Assimilation of scientific information on school problems insofar as they are relevant for the practitioner.
- Extension of the behavioural repertoire and reduction of automatized inadequate behaviours.

First results with this pilot version of the training are available (c.f. Rotering-Steinberg, 1981; Weinert and Rotering-Steinberg, 1981). The material consisted of three volumes, each with 100 pages, concerning people, mostly teachers. Feedback sheets, telephone interviews, content analyses of the phone calls of viewers after the broadcasts, and the statements of two groups of teacher consultants all showed that this approach met with the favour of the teachers. Three surveys during the course revealed that 60% to 80% of the participants had received important new perspectives and theoretical as well as practical impulses from the course; that 25% - 65% considered the suggestions to be of immediate practical use; that 15% - 25% were of the opinion that the suggestions could be used in other classes, situations, etc. than their own. Apparently this first attempt already reached its goal, namely to provide suggestions and aids for school routine. Some problems did appear in the transfer of the training content, and some gaps showed up which are meanwhile taken into consideration in the main phase of the training with an improved set of materials.

Studies on the prognosis of actions

Wahl (1981a; see also 1979, 1981b, c, 1982; Wahl, Schlee, Lutz and Reinhard, 1977; Wahl, Schlee, Krauth and Mureck, 1979a/b) represented the subjective theories of teachers as connections between situations and behaviour possibilities. This representation relied on several dialogues with the teacher and was termed "field of hypotheses." Wahl (1981) maintained that the "field of hypotheses" for "remarkable performances" or "remarkable disruptions" allows some statements about the interpretations a teacher will have for future events in an instructional setting and about which behaviours s/he will show. These assumptions were empirically tested by Wahl by visiting each of the teachers a second time, videotaping new, unusual situations, and computing how well the "new" teaching behaviour could be predicted from the "field of hypotheses."

In investigating this issue, Wahl had a methodical problem: how could one extract a prediction for a new, historically not yet realized situation from the "field of hypotheses"? To realize this procedure, he conferred the task to a "human simulator" or "double" of the teacher, who adapted the "field of hypotheses" of the teacher s/he was simulating. The double was given information on the new teaching situation s/he was to make his/her prediction about. S/he looked at the "field of hypotheses" to identify the type of situations to which the given information was most likely to correspond, decided whether s/he would react or not, looked up the corresponding goals and analyses of causes and effects, and decided on the individual behaviour. All these steps were taken solely based on the information contained in the "field of hypotheses" in front of him/her.

The activity of the "human simulator" consisted of comparing new data with old ones. The behaviour predicted by the double was then compared with the actual behaviour of the teacher in the "new" situation. Both behaviours could be compared either in text form or in picture presentation. The text on the predicted behaviour was transposed to observable behaviour through role-playing on the part of the double. A team of experts including the teacher was shown both videotapes. They rated the similarity, the better the predictive validity of the field of hypotheses. As a control, predictions were made by a double with no field of hypotheses, or with a different one, to judge by its first applications. This procedure, which is rather complex from the methodological point of view, seems to be very promising.

We may thus be quite optimistic about the possibility of validating teacher cognitions and, hence, the subjective theories that regulate their behaviour. First results seem to indicate that the cognitions represented and organized in the "field of hypotheses" bear a strong relationship to over teaching behaviour. Further analyses are concerned with the qualitative aspects of the cognitions in the reconstructed subjective theories that are relevant for action.

On the other hand, it must be remembered, as Rheinberg (1983) pointed out, that scientific research on teacher cognitions has numerous "ignorant competitors." Students are also present in the situations in which the teacher generates her/his subjective theories. How do student predictions compare with the scientific forecasts of teacher behaviour?

Author Reflection Twenty Years On

At the first moment I was astonished about the idea to republish this article. When I read it again carefully, I became even more astonished that a) the theoretical analysis has held up over the years since its publication, and b) the suggestions for modification and prognosis of teacher behaviours have proven to be a solid basis for later programs. The study course described in the paper was broadened and elaborated in the early eighties and summarized in a book (Wahl, Weinert and Huber), which is still widely used in teacher education.

What I would consider today in more detail than just using it as a training technique is the social dimensions of implicit theories, their development, and change. From the point from where I read it today, the analysis is too person-centred, not taking into account the social environment and processes mediating teacher thinking. In other words, the "social cognition" aspects of the approach at large should be considered at least in a preface of an introductory article.

Chapter 28

On the Limitations of the Theory Metaphor for the Study of Teachers' Expert Knowledge

Rainer Bromme

> There is, of course, utility in using root metaphors for the analysis of teaching and education generally. Fresh metaphors, borrowed from other fields, often provide new perspectives on old problems, thereby making it possible to treat those problems in new ways. But metaphors - all metaphors, like all theories - also have a cost. That cost resides in the ways in which they shape our conception of the problems we study. And, because metaphors do not wear their values on their sleeves, we often unknowingly accept the values and assumptions embedded within them when we use them to study or explain educational practice.
>
> (E. Eisner 1979, p. 6)

Summary

In research on teacher cognitions, the metaphor 'theory' is often used to conceptualize the teacher's action-relevant knowledge. Many researchers consider 'theory' as the research guiding metaphor for the object under study i.e. teacher cognitions. This metaphor is specified as 'personal', or 'subjective' theory. The reasons for and against the idea of an analogy between teacher knowledge and scientific theory are discussed. The main objection against using the theory metaphor is that the criteria of scientific rationality and the criteria of effective action are different. Two possibilities of overcoming the limitations of the theory metaphor are discussed: To extend the meaning of 'theory' or to dismiss the theory metaphor in favour of *psychological* concepts about teacher knowledge.

We gratefully acknowledge the assistance of James Calderhead and Günter Seib in the English translation and Bob Halkes and Niels Jahnke for comments on a draft of this article.

Introduction

Any research will require theoretical propositions about the structure of its object. At the outset of a research program, such theoretical propositions are rather more metaphorical and remain so until they are elaborated as theoretical constructs in the course of time. They designate that which is considered the object of research, and the phenomena of the real world that can be interpreted as data. In addition, the theoretical constructs establish the connection between research results concerning a specific problem and the general theory development. To develop metaphorical propositions into elaborated constructs thus is one objective of any research work.

Research on teacher cognitions – which is a quite young field of research – has adhered to this general pattern of scientific research as well; that is, some metaphors have been selected to structure the object of research. For many approaches to teacher cognitions, the concept selected is that of 'theory'.

This concept of theory is supplemented by specifications like 'subjective', 'private', 'personal', 'naive' theory. The term of theory is used to designate the structure of meanings that are mentally represented in the teacher and are relevant for his professional activity. The concept of 'theory' hence fulfills the above functions of theoretical propositions for the course taken by research: it names the object of study and also marks the connection between the special problems and theory development in psychology and educational research. The ideas about 'implicit' theories have not been specially developed for the study of teacher cognitions, rather, they stem from general approaches within personality psychology and social psychology. Thus, many studies on teacher cognitions explicitly refer to Kelly's Personal Construct Psychology, and to Heider's work on the reconstruction of the psychology of the everyday man (see, for example, Hunt's (1976) programmatic paper: Teachers Are Psychologists, Too: On The Application of Psychology To Education).

The sections below will discuss reasons for and against using the theory metaphor for the study of teacher cognitions. Does it make a difference at all which term is used for teacher knowledge?

Of course, this is not a matter of a term that is arbitrary and subject only to an agreement between researchers. The issue of the theory metaphor is not a matter of the term of 'theory', but of the connotative and denotative meaning of this concept. A metaphor's 'aura of meaning' sheds light on certain areas of reality which are still to be studied, and leaves other areas in darkness.

If the methods and theoretical approaches used in a field of research are rather heterogeneous – as is the case in research on teacher cognitions - an explication of the guiding metaphors may be useful. As the reviews by Clark and Peterson (1984), Shavelson and Stern (1981) or Eggleston's volume of 1979 show, there are both interpretative and quantitative studies. There are psychological and sociological approaches. There are experimental designs besides action research. The reason for this variety is that it has been mostly practical problems that have enticed researchers in various disciplines, with different approaches, to become involved in this field. Among these practical problems, there is the intention to facilitate improvement of teacher education and curriculum dissemination by a better understanding of the teacher's cognitive processes. Explication of guiding metaphors does not have the objective of proving that one approach is superior to another. It has the objective to support mutual understanding between different approaches.

Is it possible at all to discuss the usefulness of a metaphor for the study of teacher cognitions without treating the question whether the entire theory it stems from is empirically proven? Or, to use an example: If the concept of 'theory' is taken from Kelly's Personal Construct Psychology, is it not necessary to ask whether Kelly's theory has been empirically proven in order to assess the possible uses of the theory metaphor for the study of teacher cognitions? The objection would be justified if teachers and teacher cognitions had been selected just by accident from the many possible empirical applications of a theory. In such a case, objections raised against a concept at the same time require objections against the theory it belongs to. Our topic here, however, is not to discuss Kelly's or Heider's theories.

Rather, the question here is whether the theory metaphor is useful for the psychological analysis of the teacher's expert knowledge, the basic idea being the following: research into effective teacher behaviour has shown that there are not just a few excellent teacher skills. "In short, effective teaching involves the orchestration of a large number of factors, continually shifting teaching behaviour to respond to similarly shifting needs" (Bennet, 1979, 228; Brophy and Good 1984). It is plausible to assume that this complex demand on the teacher will lead, in the course of professional growth, to developing a corresponding professional knowledge (Doyle 1977, 1979). This knowledge is not merely the result of applying scientific theories and is thus also referred to as craft knowledge (McNamara and Desforges 1978).

We term this knowledge 'expert knowledge'. This concept goes back to the psychological analysis of the thinking of novices and expert in complex task domains such as studies analyzing the thinking of chess players, doctors, architects, physics students etc. (Chi et al. 1983, Shulman and Elstein 1975). In our opinion, studying teacher cognitions within the frame of this approach can contribute to learning more about the conditions of successful teaching (NIE 1975).

Which Analogies Between Scientific Theories and Teacher Knowledge Support Using the Theory Metaphor?

Metaphors are tools to explicate analogies. The reason given for using the theory metaphor is that there are analogies between the knowledge of the everyday man that is relevant for his actions on the one hand, and scientific theories on the other. In the following we shall sketch some of these analogies in order to explain which phenomena of the real world are in favour of the usefulness of the theory metaphor. These are analogies between the *function* of theories and the function knowledge has for (teacher) activity, and they are analogies of the *structure of theories* and the structure of (teacher) knowledge.

In order to stress the analogies between everyday man's knowledge and scientific theories, several studies published in the Federal Republic of Germany use the specification of '*subjective*' theory for everyday knowledge and for teachers' expert knowledge. (An extensive presentation of the analogies is given in Groeben and Scheele 1977, Laucken 1974, on overview about the analysis of everyday knowledge in Laucken 1982). As far as the reasons in favour of the theory metaphor are listed here, we are using this term. Nevertheless we prefer the term 'expert knowledge' because of its origin from theories about problem solving.

Analogies between the function of 'subjective' and scientific theories

G. Kelly's approach of 'man as scientist' emphasizes the analogy between the scientist's observing, forming hypotheses, and experimenting, and the corresponding activities of everyday man (Fransella 1980). Theories are needed for perceiving, explaining, expecting, and for control of activity.

An analogy between 'subjective theories' and scientific theories, for instance, lies in their function for perception. Just as theoretical constructs will influence observation in scientific research, the concepts of 'subjective theories' will structure perception (Treiber 1980).

Thus, teacher perception is structured by characteristics of their pupils such as 'giftedness', 'participation' etc., that is teachers perceive facts of such a kind that can be assigned to these concepts (Hofer 1981). The teacher's perception of the pupils, again, covariates with teacher behaviour in which the differently perceived pupils are differently treated as well. Both the concepts with reference to which the pupils are perceived and the corresponding individualized teacher behaviour is functionally linked to the demands the teacher must meet in order to attain the instructional goals. Such demands, for instance, are control over the subject matter, the uses made of the time within a lesson with regard to the subject matter, control over initiating activities (Cooper et al. 1979). The phenomenon of 'theory dependency' of observations has not only been shown for the teachers' perception of pupils. Rather, it holds for the perception of persons in general. Accordingly, the research into teachers' perception has been influenced by studies of implicit personality theory.

In the 'subjective theories', we do not only find concepts with regard to *perception*, but also to *explanation* of relations and causes - just as scientific theories contain explanatory concepts. Attribution research in particular has subjective explanatory constructs as its object of research. The concept of 'attribution' designates the process of explanation that is the application of a general, stable expectation across a cause-event connection to an empirical phenomenon. For instance, application of the concept of giftedness in order to explain a certain fact observed, i.e. a certain pupil's performance.

A further analogy between the function of scientific and 'subjective theories' lies in establishing both hypotheses and expectations. The development and effect of teacher expectations about pupil performance has been extensively studied in educational psychology (Pygmalion effect, Brophy and Good 1974). Teacher expectations have not only been analyzed with regard to pupils' performance, but for example also with regard to the usefulness of instructional material (Yinger and Clark 1983) or the usefulness of curriculum innovations (Doyle and Ponder 1977).

In the case of observations, explanations and expectations, problems of veridicality, of the appropriateness of the logical conclusions, and of internal consistency arise (Groeben and Scheele 1977). Just as scientific theories can be questioned as to whether they are empirically well founded, the question may be raised in the research on teacher cognitions whether the constructs the teacher uses to explain pupils' performance are empirically proven. Hence, the theory metaphor can be used to conceptualize questions and problems raised with regard to the adequacy and the veridicality of 'subjective theories'.

Analogies between the structure of 'subjective' and scientific theories

In the philosophy of science, theories are conceptualized as systems of statements. Their structure can be imagined to be like that of a tree diagram. The higher one gets in the hierarchy of theorems, the more comprehensive the explanatory power of hypotheses gets, and the more 'data' or 'cases' are *explained* by one hypothesis. The lower one gets in the hierarchy, the easier it is to find statements *descriptive of* the realm of reality to which the empirical theory refers. These statements constitute the model of the realm of reality, the theoretical statements explain the observations described in the model. For our context, it is important that the components of a theory, i.e. its individual concepts and hypotheses, obtain meaning from the context of the theory's other components (Bunge 1967 1, 380ff). The meaning of a theoretical concept is not an isolated property of the concept itself, it is constituted by the other concepts of the theory as well as by the data it refers to.

This image of theories – in this case a description of the *statement view of theories* – can also be found in the explications concerning the structural assumptions of 'subjective theories' in studies on teacher cognitions. Subjective theories are considered to be general propositions (Clark and Peterson 1984, Dann 1983, Laucken 1974).

The analogy to scientific theories resides in the fact that knowledge has been formulated in the shape of statements, and that a hierarchy of statements of varying degree of abstraction and thus of explanatory power is assumed.

For researchers using an ethnomethodological approach, the dependence of the meaning of observations (data) from theoretical concepts is an analogy of special importance to teachers' 'subjective theories'. This analogy has already been mentioned insofar as was said that theories are required for perception. In the case of ethnomethodological approaches, however, it is mainly the aspect of internal consistency of the 'subjective theories' that is stressed. 'Subjective theory' produces a 'universe' within which the agent moves – by creating connections between bits of information. These 'subjective theories', and hence the subjectively generated contexts, may differ widely between different persons – despite the fact that they refer, for the outside observer, to the same facts and objects. [That has been interpreted as a source for communication problems in the classroom (Cicourel et al. 1974, Krummheuer 1983)]. That is an analogy to the competition between alternative theories, referring to the same realm of reality.

A further analogy concerns the question as to whether the character of perception is constructive – is it a reflection of reality or a subjective constitution of meaning. This problem is strongly debated in theory of science as well as in philosophy and psychology (Neisser 1976). By introducing the theory metaphor, the question can be conceptually grasped in research on teacher cognition too (see Pope and Scott in this volume).

Our presentation of the analogies was focussed on the question which 'real world phenomena' of teacher cognitions show similarities with the function and structure of scientific theories. Hence, hints for an empirical justification of using the theory metaphor were sought.

The theory metaphor, however, is used by some authors not merely because they presume that it is empirically adequate. They also consider it to be a useful conceptual tool to facilitate everyday man's reflection about his subjective theories. Thus, Kelly developed the explication of personal constructs by means of the grid technique to enhance in individuals an awareness of their own 'subjective theory'.

Groeben and Scheele (1977) have elaborated the concept of 'subjective theory' in order to facilitate the exchange between the scientist and everyday man by means of criteria and concepts of the philosophy of science, hoping to improve both the constructs of scientific psychology and the constructs of everyday psychology in doing so. Here, we shall not treat these intentions, which aim at using and facilitating the everyday man's and the teacher's activity of reflection by means of the theory metaphor. Rather, the question is which are the possibilities and limits of the theory metaphor for an empirical analysis of teacher's expert knowledge.

Which Differences Between Scientific and 'Subjective Theories' Contradict Using the Theory Metaphor?

After the above section has been devoted to analogies between scientific theories and teacher knowledge, this section will deal with the differences. Just as the analogies supply reasons for using metaphors, the differences show the limits of using the theory metaphor for the study of teacher cognitions. In the following, examples for differences will be presented according to function and structure theories. Of course, nobody using the theory metaphor will claim that there are no differences between activity-relevant knowledge, and scientific theories. This is why it is interesting to see how users of the theory metaphor handle the differences.

Differences between the function of 'subjective' and scientific theories

Within the development of scientific theories, the justification of scientific statements, i.e. of theories, should adhere to fixed rules, to the so-called criteria of rationality. Now there are many examples in psychological literature that show that everyday man violates these criteria of rationality in developing and using his personal knowledge, even if we assume that this is done consciously. Some are presented by Groeben and Scheele (1977, 72 ff).

One of the criteria of rationality, for instance, is the separation of descriptive-analytical from normative-prescriptive statements. Both everyday knowledge and teacher knowledge, however, contain many statements in which this separation is not made. A further 'violation' of the criteria of rationality is the 'distortion' of perception by the perceiver's attitudes and personal values.

A further example for the 'violations' of criteria of rational are the so-called fallacies in decision making under uncertainty which have been well studied in cognitive psychology (Kahnemann, Slovic and Tversky 1982). To reduce complexity, decision makers use heuristics, ignoring information, or evaluating certain data 'inappropriately' from the point of view of maximum use of information. Other researchers working within the same experimental paradigm for fallacies, however, have argued that such 'fallacies' can by no means be designated as irrational. Rather, they are quite functional under everyday conditions of deciding under pressure of time, and in case of complex tasks - and are rational in this sense as well (Einhorn and Hogarth, 1981; for an overview of the controversy see Jungerman 1983).

Such phenomena are also known from research into teacher cognitions. Thus for example Lundgren (1972) was able to show that the basis of teachers' pacing decisions does not depend on their perception of the entire class, but only on a certain group of students in the class. He called this group of students the 'steering group'. These are students situated between the tenth and the twenty-fifth percentile of the class' aptitude scale.

'Neglecting' information is, on the one hand, 'irrational' – but it is functional as it permits the reduction of cognitive load (Bromme and Brophy 1984). It makes no sense, for instance, to assimilate more 'aptitude' information than there are 'treatment' options to choose from. The other 'violations' of the criteria of rationality may well be useful, too – for the purposes of practical action.

The mixing of prescriptive and descriptive statements is even necessary to move from theories to rules for acting (Bunge 1967, II, Bromme and Hömberg 1976). As actions are necessarily intentional, they require decisions about desirable goals. Hence, Clark and Peterson (1984), in their review of the research into teacher cognitions, even combine 'beliefs' and 'theories', i.e. they use the theory metaphor in connection with the description of personal value systems from the very outset. Kelly's characterization of 'constructs' as 'personal' takes this into consideration as well.

It is obvious that the question as to how the 'violations' of the scientific criteria of rationality should be evaluated depends on what is meant by 'rational'. As there is a difference between the criteria of scientific rationality and the functionality of knowledge for activities, this however represents a limitation of the theory metaphor. It implies a concept of rationality that may be appropriate for producing knowledge about reality or for confirming theoretical ideas about reality. It offers, however, no opportunity to analyze the functional appropriateness of action-relevant knowledge. Hence, the use of the theory metaphor will also lead to descriptions of teacher knowledge that show it to be deficient compared to scientific knowledge.

True, this contradicts the explicit intentions of the proponents of the theory metaphor (e.g. Treiber and Groeben 1981), and research into teacher cognitions is not concerned with researchers' evaluations of teacher cognitions. It is, however, a matter of looking for the characteristics of expert-knowledge, and this requires conceptual tools. If we take along nothing but the 'tools' of the theory metaphor, i.e. the criteria for evaluating scientific knowledge, we shall only be able to find results that relate to these criteria. This, however, will easily lead to underestimating expert knowledge, despite all positive intentions.

Differences between the structure of 'subjective' and scientific theories

Subjective theories are described as systems of propositions (Clark and Peterson 1984, Dann 1983, Treiber and Groeben 1981, Wahl et al 1983) that can, in principle, be verbalized by the teacher. This is the crucial structural assumption of the users of the theory metaphor, which is in fact problematic.

The results of research into interactive judgement and interactive decision making in the classroom, however, show that the action-relevant knowledge in actual behaviour is not necessarily consciously present and verbalizable (Shavelson and Stern 1981) - apart from the technical difficulties which would arise from obtaining verbal data during normal action in the classroom (for this problem, see Huber and Mandl 1982).

Reflection will occur only in case of certain 'problematical' situations which require a deviation from the routine of classroom activity or when dealing with events which are unpredictable in principle. If, for instance, expectations with regard to activities and their consequences (what will happen if I ignore Fred's misunderstanding of my question?) would be rehearsed mentally before or during the action, these considerations could be described as use of propositional knowledge during action (in this example: use of an if-then statement concerning teachers' neglect of students misunderstanding).

In which sense does 'knowledge' exist, if it is not used during the activity in a propositional form, or if it is not represented in a propositional form at all (Argyris and Schön, 1976, 9)? There are areas of knowledge that cannot be easily rendered in a propositional form even if the query is made explicitly and independent of an activity, i.e. visual knowledge, which is being studied through psychological research on images (Kosslyn and Pomerantz 1977). It may, however, be expected that such knowledge is immediately action-relevant for the teacher as well. Thus, it is for example part of curricula in the subjects of music, arts, (Eisner 1979) and in mathematics education (Otte 1983), and is influencing instruction decisions in these respects.

The users of the theory metaphor have found various solutions to this problem that shall be briefly discussed here.

Rather frequently the distinction between tacit knowledge vs. explicit knowledge is made referring to Polanyi (1967) or Ryle (1949) who is also mentioned for distinguishing between knowing how and knowing what. These conceptual distinctions are useful to designate the problem, but they do not, of course, provide an answer to the psychological question as to what the structure of the internal representation of 'knowing how' is like.

One possibility to handle this problem in studying teacher cognitions is to use the theory metaphor only for the long term, general knowledge. The theory metaphor, in this context, designates the background of current information processing going on during action. Clark and Peterson (1984), Shavelson and Stern (1981) among others have opted for this solution. In order to characterize the knowledge that is directly relevant for action, they use the concepts of 'script', 'schema', 'plan', and 'image'. This solution has the advantage that the question concerning the structure of action relevant knowledge is framed within psychological theories of information processing, where it can be treated as an empirical problem

This solution, however, has the drawback that it leaves unclear where the difference between 'scripts', 'schemas' etc. and 'theories' lies. Clark and Peterson (1984) characterize 'theory' as propositional knowledge, but this is not distinctive of theories, as schemas and scripts may also be represented in a propositional mode.

Another possibility is to quit using explicitly the theory metaphor for the structure of the action-relevant knowledge – but use it merely as a conceptual frame in order to reconstruct this knowledge. This solution is mainly advocated in several studies in the Federal Republic of Germany (see the overview by Huber and Mandl in this volume). It has the advantage of elucidating that the concept of theory refers to knowledge that can be communicated and made explicit.

Attached to this solution, however, is the disadvantage that nothing more is said about the internal, cognitive representation knowledge in the instant of action. Besides, it suggests again the danger of underestimating the teacher's expert knowledge. Indeed, this solution excludes all knowledge that is not instantly created while action is going on and for which the agents find no verbal representation, or only an unsatisfactory one. This is true, for instance for routines (Bromme 1983, Peters in this volume) or intuitive problem solving.

Hence, results of research into problem solving suggest that there are two different modes of working on a problem used by experts, a serial-analytical mode, and an holistic-intuitive mode (Kuhl 1983). The knowledge used for the serial-analytical mode of problem-solving thinking may be expected to be reconstructable in such a serial way as well, that is in a propositional structure.

How could we reconstruct knowledge used for holistic-intuitive problem solving? Only some first speculations about this question are possible at present. Elbaz' (1983) case study on the structure of teachers' practical knowledge offers some indications. It distinguishes between practical rules, practical principles, and images. While the former two can be conceptualized very well as propositions, the content of an 'image' is better grasped by a metaphor. "The image is a brief, descriptive, and sometimes metaphoric statement which seems to capture some essential aspect of Sarah's (the teacher studied in the case study, R.B.) perception of herself, her teaching, her situation in the classroom or her subject matter, and which serves to organize her knowledge in the relevant area" (1983, 137).

Of course, this case study cannot be taken as empirical proof of the necessity of considering such teachers' metaphors. But the matter here is that the researcher must be aware of the possibility that the experienced teacher's knowledge may be organized in such a holistic way, e.g. by metaphors which are of comprehensive meaning for the teacher, a meaning, however, which cannot be established from the metaphor alone. This leads to the question about what teacher's really mean when answering questions of a researcher.

How to Handle the Limitations of the Theory Metaphor

This section is intended to suggest ways to handle the limitations of the theory metaphor sketched. It enlarges on the possible solutions mentioned above.

To extend the meaning of 'theory'

Our listed objections against the theory metaphor here always refer to a normative concept of 'theory'. Thus, the normative criteria of rationality for the assessment of scientific theories were compared to the problem of the functionality of expert knowledge. It may now be objected that scientists use implicit knowledge as well. Further, it may be objected that these normative criteria do not fit the real course scientific activity takes. They do not apply to the theory metaphor if this means the process of theorizing. Or, more generally: there is a difference between the logic of research and the psychology of scientific work or the actual procedures of scientific research.

The approach chosen by Groeben and Scheele (1977) attempts to solve the problem in this way. The authors stress that the criteria of scientific-ness are also more normative representations that are at present being treated in a more liberal way in the theory of science as well. Consequently, application of these criteria would not characterize the everyday man as substantially more irrational than the scientist.

To extend the meaning of the theory metaphor from theory as a product of scientific activity to the process of theorizing would, however, make the meaning of this metaphor even more indistinct. This would give rise to quite another objection against the theory metaphor: that the production, presentation and use of scientific knowledge is so complex that a correspondence can be found for any aspect of teachers' expert-knowledge and teachers' behavior. In this case, however, the theory metaphor would no longer be precise enough to conceptualize the object of research.

Of course, other interpretations of the theory metaphor are possible. The arguments here always referred to the so-called statement view of theory - a conception of theory that, among other things, does not satisfactorily conceptualize the dynamics of theory development and change. Within the statement view of theories, the gap between normative criteria and actual process of theorizing is quite large. The question would be whether a theory metaphor based on another conception of theory, for example on that of the so-called non-statement view (Sneed 1971, 1977), does not offer better opportunities of conceptualizing the structure, function, and development of expert knowledge? The non-statement view of theories is developed for theories in physics and its application on scientific theories of other disciplines is metaphorically (see e.g. Herrmann 1976, 42) but the use in question here is a metaphorical one anyway. We cannot go into this matter here.

The obvious conclusion is that the users of the theory metaphor should always make explicit the concept of theory on which their considerations are based.

To dismiss the theory metaphor

For a psychological study of teachers' expert knowledge, there is the possibility of using concepts like schema or image as theoretical constructs for teachers' knowledge. In doing so, it must be taken into account, however, that the concept of schema is used to stress another aspect of knowledge than the concept of theory. The concept of schema is used to conceptualize, among other features of knowledge, the fact that action-relevant knowledge is organized with regard to certain situations (as shown by the well-known restaurant example). The strength of the schema concept thus lies in its conceptualization of domain specific knowledge. As opposed to that, the theory concept mainly stresses the generality and relatively domain-independent character of knowledge (a consistent statement within a theory is proven just when it is valid relatively independent of specific situational conditions, a fact which can be easily demonstrated in the case of physical laws). This distinction, however, is not absolute – as the validity of theoretical laws will depend on certain conditions, and as the situation-specific knowledge that is organized, as a schema will also hold for types of situations.

To dismiss the theory metaphor is especially recommendable when it is 'only' intended to explain delimited hypotheses on teachers' knowledge for which precise psychological constructs have been developed. If the intention is, for instance, to analyze the systems of 'rules' which guide action in case of certain problems, one could use the representation mode of knowledge as hierarchies of if-then rules (production-systems) which have been developed for modeling procedural knowledge in problem-solving (Newell 1973).

In section 3, we argued that the theory metaphor might overemphasize the rationality deficits of the structure and function of expert knowledge, because the criteria of rationality for the justification of theoretical statements do not cover the functionality of knowledge for activity. Thus, for example the important feature of expert knowledge is not whether it can be verified, but whether it is functional for a certain activity (Huber and Mandl in this volume).

The functionality of knowledge, however, can be empirically analyzed in the context of the activities and of the demands that the actor has to cope with. It can be analyzed with reference to theories of action. The (implicit or explicit) embeddedness of research on teacher cognitions in theories about teacher actions is emphasized in several contributions to this volume.

The basic theoretical concepts (or metaphors) of action theories refer to goals or intentions, to action and activity and to the object and the external conditions of activity i.e. the task demands the teacher has to cope with. In different approaches, it depends on the type of action under study how the knowledge of the actor is conceptualized. Within the research tradition we were referring to - the analysis of expert problem solving - knowledge is conceptualized as schemata or as production system. That is not to say that this specific conceptualization for teachers' knowledge is appropriate for all research questions. But we want to emphasize here that the usefulness of a metaphor for teachers knowledge will depend of whether it is (at least) compatible to the theory about (teachers) goal-directed activity.

Considering psychological theories about human action, it may well be doubted whether the theory-metaphor is appropriate. At least if one does not want to make the special case of scientists' activities a frame of reference for theories about human activity in general, it is recommendable to dismiss the theory metaphor and to look for more *psychological* concepts about knowledge representation and knowledge use.

Author Reflection Twenty Years On

I think that metaphorical assumptions about the structure of teachers' professional knowledge are still an important issue.

Hugh Munby, Tom Russell and Andrea Martin in their chapter on 'Teachers' knowledge and how it develops' (fourth editions of the Handbook of Research on Teaching) discern between propositional accounts and non-propositional accounts. In my view the 'theory metaphor' is the underlying idea of all propositional accounts. Especially when it comes to the **development** of teachers' professional knowledge propositional accounts have their limitations. Because this issue of development is an important question which needs a lot of further research this research is still of interest for readers.

Chapter 29

Post-Interactive Reflections of Teachers: A Critical Appraisal

Joost Lowyck

Summary

Many researchers in the field of teacher thinking restrict their work to one specific phase of teaching: the pre-interactive, interactive or post-interactive one. Most attention has been paid to the planning phase, less to the interactive thought processes and very few to post-interactive thinking.

This research has been set up as a contribution to a better understanding of the post-interactive reflection of teachers. Both the methodological refinement and the content of after-lesson thoughts are emphasized. The results reveal the problem of reducing the study of teacher thinking to the isolated phases as chronological dissections of the teaching activity. From the teachers' standpoint, planning, interaction and post-interactive reflection are strongly connected, because they think rather in terms of a meaningful content than in chronological terms. The study of the interrelations between the phases needs more attention, to contribute to a more integrative concept of teaching.

Introduction

In most of the publications on teacher thinking the distinction between the three phases of teaching has been taken for granted. It seems to be a workable scheme for several reviews of research (Clark and Peterson, 1985; Shavelson and Stern, 1981).

Whereas process-product studies focused almost exclusively on interactive classroom behaviours (Dunkin and Biddle, 1974), research on teaching mainly stressed the planning phase (Bromme, 1980; Jackson, 1964; Morine, 1973; Stolurow, 1965; Yinger, 1977; Zahorik, 1970). However, some outcomes of these studies reveal problems, due to the isolation of the planning phase from the interactive one (Lowyck, 1980).

Besides the question of the compatibility between the pre-interactive and interactive phase of teaching, there is few insight in what happens after the lesson is over, and how possible reflections during the post-interactive phase influence both future planning and teaching.

In order to gain insight into the relationship between the different phases of teaching, a descriptive-empirical study has been set up (Fets, 1984; Lowyck, 1985).

Post-Interactive Reflection: Conceptualization

It always seems difficult to agree upon definitions, mainly because of the lack of both conceptualization and empirical research. What teachers do after their lessons often has been qualified as 'evaluation' (Shavelson and Stern, 1981). This concept "refers to the phase of teaching where teachers assess their plans and accomplishments and so revise them for the future" (Shavelson and Stern, 1981, p. 471).

These authors attribute the lack of studies in the domain of post-interactive reflection to some psychological evidence, more precisely to the limited ability for learning from past experiences. This conclusion may surprise scholars in the field of teacher education, because of the central notion of 'feed-back' in training procedures. It is quite evident to teacher educators to expect behavioral changes in student teachers, using information from the past. The question is, however, whether (student-) teachers use the appropriate information.

McKay and Marland (1978) use the term "reflection" as a category of interactive thoughts, meaning "units in which the teacher is thinking about past aspects of, or events in, the lesson other than what he has done". These authors limit reflection to those cognitive activities of teachers that concern their not realized teaching behaviours. This interpretation seems too narrow for understanding fully what is meant by reflection. The definition of Clark and Peterson (1985) is more neutral and efficient, speaking about "postactive thoughts". Peters (1984) points to the same interpretation when referring to "post-reflection".

We will use the term "post-interactive reflection" as a descriptive category for the information processing activities of the teacher after a lesson or a broader unit of time. If teachers use some explicit criteria for determining the quality of the information we will speak, then, about evaluation.

An important problem in the conceptualization of post-interactive reflection is the unit of analysis. Yinger (1978) distinguishes the following five units of planning: year, term, unit, week and day. Although this time schedule was very useful for the study of pre-interactive teaching, it nevertheless seems not precise enough for our study. We therefore added two smaller units (lesson moment, lesson) as well as a larger one (career).

Each post-interactive reflection can be conceived of as a moment in the teacher's professional life. Eisner (1979) distinguished four dimensions within curriculum decisions, namely the molar and the molecular decisions on the one hand and the time at which (present) and for which (future) curriculum decisions are made on the other. Post-interactive reflection can be seen as the (present) activity of a teacher at a very precise moment in his professional career, with reference to some moment in the past. These reflections can bear on some activity or event in the future.

An Exploratory Study

Due to the lack of empirical studies on the subject at hand, we explored teachers' post-interactive reflections in a descriptive-empirical way. The results will contribute to a better understanding of teachers' complex functioning as well as to some methodological refinement.

Subjects

As in earlier studies (Lowyck, 1980; 1983) we worked with elementary school teachers. Some similarity of the sample guarantees at least a better comparability of the results from the different projects. Twelve teachers participated (nine females and three males). Their teaching experience was spread between 3 and 32 years, with a mean score of 18. 75 years.

Method

In order to focus the attention of the teachers on the study object, a short letter with instructions has been presented to each of them. They were informed that we were interested in the post-interactive phase of teaching. Moreover, some orienting questions were asked, like: do you think about past teaching events, what is the content of the thoughts, when do you reflect, does it often happen. Then, they were invited to illustrate some thoughts with very concrete instances. Last, the question was asked if their after-lesson thoughts had some impact on their future teaching behaviour.

During November 1983, the retrospections of the twelve teachers were organized in the schools. They lasted between thirty and forty-five minutes. The researcher asked at regular intervals for concrete instances of the reported thoughts. The retrospection was recorded on tape and later on transcribed. This formed, together with both the written answers on the questions and the elaborated lessons plans, the material for the data-analysis.

Analysis of the material

The material was scrutinized in a recurrent, cyclical way. Each protocol was examined successively. After the analysis of the information from the first teacher, a provisional classification of the content was tried out, using rough categories. When inspecting the data of the second teacher, the usefulness of the categories was checked and new categories were generated in order to cover the content of the retrospection. This procedure was repeated with each teacher. At the end, the material was grouped into two kinds of data: the results from the broad retrospection on the one hand, and the information from the concrete instances on the other. In a second cycle, the whole material has been controlled again, in order to assure a maximal integration of all the relevant information into the descriptive categories.

Data from the retrospection

Most teachers experienced it as difficult to distinguish clearly between past and future events, because a lot of thoughts indeed are connected and not ordered chronologically. In our opinion, this has to do with the function of the thought process, namely to gather information from past lessons or periods which is functional for the future.

We categorized the data from the broad retrospection into three main categories: (a) the formal aspects of reflection (systematic/spontaneous, frequent/sporadic, way of recording the thoughts), (b) the content of the reflection (individual pupil, group, teaching behaviour, management, lesson content, colleagues, parents and extra-curricular activities) and (c) the consequences of the reflection.

With regard to the more formal aspects of post-interactive reflection, differences between teachers in the frequency and consciousness of their thoughts was apparent. Teachers can be placed in two categories: never looking at the past or almost always reflecting in a systematic way. That teachers are reflecting systematically is not a function of the frequency but of their habitual way of behaving. If there is some problem with the planning of future lessons and/or with experiences during the past, teachers tend to use short reflections and quickly search for some indicators in the past in order to avoid negative experiences in the future. Here again, the planning and post-interactive reflections are tied together. In the case of negative experiences in the past, teachers try to store that information, together with strategies for some improvement next time. The clearer the problems are, the more teachers will focus on it. For example, when correcting exams, they see the effects of their teaching and reflect on possible causes for the failure of some individual pupils or of the class group.

When asked to report how they fix their post-interactive thoughts, only three teachers mentioned one or another way of laying down the ideas in written form to have at least some lasting support for a reflection in the long run. Most of the thoughts of teachers are, thus, very volatile. Maybe, this is one of the reasons why it is difficult to distinguish between planning and post-interactive reflection: the ideas or thoughts are clustered around one topic, without concern about the exact time the ideas were generated.

As to the content of post-interactive reflections, the following results are of interest. Teachers often retrospect on the behaviour and/or emotions of individual pupils, mainly in the case of difficulties or problems, both in the personal and interactional aspects of functioning. They think about the class as a group, mainly when the lesson was disappointing, at least in the perception of the teacher. Problems with the lesson seem to focus the attention of the teacher to their teaching behaviours, to the timing and to the subject matter. Sometimes, teachers refer to contacts with parents or colleagues who influence their thinking about lessons or events.

About the question if and to what extent post-interactive thoughts influence future behaviours, the results indicate a very broad unit of analysis. Teachers tend to generalize the very concrete experiences into a higher level of abstraction and use these thoughts as a guide for future activities. This finding lays in the line of previous research data (Lowyck, 1980), where teachers report the gradual condensation of lesson content as well as the cumulative integration of pupil reactions as orientations for their planning.

As to the retrospection, the following conclusions are noteworthy. Teachers seldom reflect systematically about past events, although they store selectively some information. Next, the clear dissection between planning and thinking afterwards is non-existing. Teachers very often think in terms of problems or tasks and are not able to use a definite chronological segmentation of the reality. Lastly, the interview as a technique seems not so relevant for exploring the broader aspects of post-interactive reflection, because most teachers reported quite precisely their ideas in the written preparation of the interview. On the contrary: sometimes the verbal interaction seemed to disturb the train of thought, given the tendency of the interviewer to create an atmosphere of mutual talk.

Data from concrete instances

In contrast with the retrospection, teachers are very explicit here, so that the data gathered are richer than in the retrospection. We will use the same classification as for the retrospection, in order to allow the comparison of both techniques.

The analysis of the concrete examples delivers no information about the more formal aspects of post-interactive reflection, as was the case during the retrospection. Only results about both the content and some consequences are available. After a recurrent, cyclical analysis of all the instances, twenty-two subcategories seemed to cover the whole content of the reflection. Each statement referring to the content has been put into one of the 22 categories. Table 1 represents the different categories in a reduced form.

The categorization of after-lesson thoughts could suggest some linearity of reflection, as if teachers only reflect upon one aspect at one time. Although they reflect often on one aspect, in some cases, nevertheless, there is a more complex organization of their thinking. We could speak, then, about combined thoughts. In our material, only 5% of the statements contained more than one aspect. This is possibly not a true sample of reality, because in the wording of their thoughts, teachers could make explicit only a limited part of their thought processes.
Some combinations of categories are situated within the same topic, like individual pupil or class group. A stated problem (1a) is combined with some

1. INDIVIDUAL PUPIL	Reaction of one child	Unsatisfactory performance	(1a)
		Need for diagnosis	(1b)
		Unexpected (positive) performance	
			(1c)
		Interesting statement	
			(1d)
		Disciplinary problem	
	Personality characteristics		(1e)
2. CLASSGROUP	Characteristics of the group		(2a)
	Cognitive achievement	Evaluation (general)	(2b)
		Unsatisfactory results	(2c)
		Satisfactory results	(2d)
	Classroom behaviour and attitudes		(2e)

3.TEACHING BEHAVIOUR	Method or procedure	Content-oriented procedures	(3a)
		Affective procedure	(3b)
		Behavioural procedure	(3c)
	Feelings and background		(3d)
4. OTHER PEOPLE		Extra-curricular contacts with pupils	(4a)
		Parents	(4b)
		Colleagues	(4c)
		Inspection	(4d)
		Non-school contacts	(4e)
5. ORGANIZATION		Course of the lesson	(5a)
		Timing and amount of content	(5b)
6. LESSON CONTENT			(6)

Figure 2. Post-interactive reflection: categories of thought content.

Diagnosis (1b). This leads often to the formulation of a procedure for remediation (3a). Negative performances of pupils (2c) lead often to a closer look at the own teaching behaviour (3c), while a positive result of a pupil (1c) discharges into a positive feeling of the teacher (3d).

There is a fixed way of direction from one group of categories to another. Reflection on several topics leads often to thoughts on the own teaching behaviour, while teachers never reflected about other components, starting from their own teaching behaviour thoughts. Nevertheless, a lot of combinations has been found, starting from reflections on management categories.

Another interesting question is whether there is some relationship between the category of reflection and the unit of analysis from the past. In figure 2 the results are showed for a frequency of 361 thoughts.

Because of the variation of the number of statements in the different cells, it seems not possible to draw definite conclusions. We will limit the report to some hypotheses which could be studied more intensively in future research.

	1 IND. PUPIL	2 CLASS-GROUP	3 TEACH. BEH.	4 OTHER PEOPLE	5 ORGAN-IZATION	6 * CON-TENT
Lesson moment	36	10	10	2	2	3
Lesson	43	53	55	2	8	13
Day	3	3	2	--	--	--
Week	--	--	--	--	--	--
Lesson unit	11	6	10	--	1	--
Month	--	--	--	--	--	--
Tri/semester	1	--	3	--	--	--
Year	22	15	8	13	6	1
Career	--	8	4	4	2	1

* = main classes of thoughts; see figure 1

Figure 2. Number of statements per unit of analysis

- Teachers mainly reflect upon the performances of the classgroup and upon the methods they use in relation to the past lesson. In this case, the evaluative aspect of the reflection is dominant.
- Reflection on difficulties from individual pupils refers, more than the thoughts about the classgroup, to restricted lesson moments.
- Thoughts about the lesson content are restricted to a specific lesson. On the contrary, reflections upon classroom management are related to broader time units.
- Remarks from parents, colleagues and inspection are incentives for the teacher to reflect upon the own teaching over a very broad period, mainly the year already past.

- Pure chronological units (day, week, month, tri/semester) are far less used as units of analysis than more professional relevant units (lesson moment, lesson, lesson unit, year).

Concerning the effects of post-interactive thoughts on future teaching, it is salient that teachers formulate very often their intended changes of teaching behaviour, when reporting some examples. Nevertheless, there is no strict relationship between the entry of the thought (for example the individual pupil) and its consequence (i.e., the classgroup). Teachers often generalize the information gathered from one pupil or a teaching event to broader units in the future. One instance, thus, functions as some point of reference for possible changes in the next professional activity, or, in other terms, as a sufficient representative and valid experience.

The future teaching patterns which are intended as a consequence from the post-interactive reflection, have been grouped into ten categories. Because of the brief description, it is impossible to show all the entries that influence future teaching. We report only the different classes and some main relations between reflection and future action.

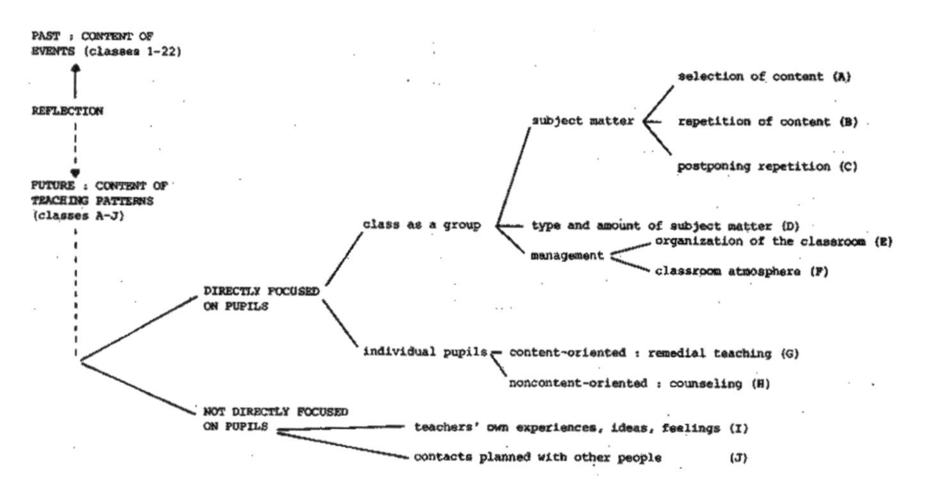

Figure 1. Categories of intended activities as a consequence of post-interactive reflection.

Conclusions

The data gathered both by means of the retrospection and from the concrete instances, reflect a consistent picture of teachers' functioning during the post-interactive phase of teaching. The main conclusions are the following.

The participating teachers reflect upon striking events, behaviours or situations. Whether something is labeled as striking, depends upon the subjective perception of the teacher. Some teachers, for example, are very irritated by pupil talk when not oriented to the task, while others accept this fact as a normal reaction of children who cannot concentrate during the whole day. Pupils' talk is, thus, not always a striking event.

Reflections in the form of post-interactive thoughts will generate in some cases a pattern of problem solving activities. Reflections on an individual pupil lead often to some kind of diagnosis and this by turns to the search for future remedial behaviour. Here we remark the interdependency of the different phases in teaching.

Another observation is that teachers often implicitly work out some naïve teaching heuristic, which would sound as follows: "Repeat in the future what has been perceived in the past as powerful and change only what has appeared amenable to improvement." One could speak rather of rough, sketchy orientations than of precise prescriptions. The adjustment of the teaching activity to coming new situations is not a consequence of the accurate rational planning and reflection. The volatility of thoughts as well as the lack of a systematic control about the reflection content, refer to more intuitive reactions.

Because teaching is a complex activity with different types of behaviours, one cannot restrict the study of teaching to one or another type (Lowyck, 1984). For sure, teachers operate with both explicit thinking and abbreviated behaviours, e.g., routines. The question is now if teachers arrive at the selection of the most important information, in order to direct future behaviour. It is until now unclear how teachers reduce gradually the complexity and how often they use relevant criteria.

The teachers do not seem to be the 'professional, rational decision maker' as has been suggested in the literature on teacher thinking. She is in too many cases dependent upon unclear criteria for making adequate decisions. Moreover, the situation is often dominant, so that sometimes there is more field-steered activity than intentionally.

Discussion

If we evaluate the research reported here, we have to consider some limitations. The small sample, the restricted number of research methods, the broad focus of the research questions and the way of reducing the amount of data, result in the impossibility to be conclusive or exhaustive. The findings only can be used as a platform for further refinement.

In this and previous studies, we intended to describe important variables of teaching as a complex activity. We will now appraise more critically some aspects of (a) the research findings, (b) the methods and (c) the theoretical framework.

Research findings

An analysis of the findings shows that teachers differ strongly in the way their teaching is planned and executed. Moreover, the variables teacher, pupils and subject matter seem to cover most of the teaching variation. It is striking that teachers who have been confronted with complex teaching models, relapse into very simple ones. The insufficiency of teacher training is not a definite argument for explaining this phenomenon. Maybe the lack of a well elaborated 'task environment' of teaching is one of the main causes for the divergent conceptions about teaching, because all a teacher does is called 'teaching.'

The results of the research indicate the great difficulty to discover the essential characteristics of good teaching, only by pure description. In view of effective teaching, the question remains what precisely adequate thoughts and behaviours are and in what situation they are effective. If we cannot find the criteria for good teaching by description only, other ways must be explored. In our opinion, we have to set up more 'constructive' designs: the precise indication of the characteristics of a specific teaching task environment, so that we could control the research input. The integration of the cognitive approach and the process-product studies, enables us to construct a well-defined task, to describe the teaching processes and to measure the learning gains. The elaboration of specific curricular segments could be the platform for such a 'constructive' research. We could call it teaching experiments, if we use this term in a broad sense, including the descriptive power and qualitative analyses developed by the recent cognitive paradigm.

Frequently research on teaching parceled out the complexity of teaching by making the research-entry more dominant than the teaching reality. Such a segmentation leads to a centrifugal tendency: research topics escape from the real functioning of teachers into another reality: that of the researcher.

Methodology

The studies as reported here, mainly use retrospection as a technique, complemented by concrete examples and written materials, such as the lesson plans and the information from diaries. The use of self-reports is a very problematic one, as has been pointed out in recent years (see: Ericsson and Simon, 1980; Calderhead, 1981; Lowyck and Broeckmans, 1985).

When starting our research, the expectation was that a broad approach was the most interesting one to explore the key features of teachers' cognitive processes. Self-reports seemed to be at least necessary to understand what teachers think. In addition, we expected that this broad exploration would indicate the critical cues in expert behaviour. Indeed, the self-reports from experienced teachers could deliver information concerning 'good' teaching. The data from our explorative design however, have shown that the term 'experienced' teachers is too vague a concept, and to some degree irrelevant as an indication for 'good teaching.'

In our study, attention has been paid to the ecological validity of the research situation. Nevertheless, the intended external validity has not been realized in all its aspects. That teachers collaborate on the base of voluntariness is a very specific way of sampling. Moreover, their collaboration with a researcher entails some changes in the teaching situation and in their way of behaving. As such, a teacher is very sensible for suggestions or demands from the researcher. Written instructions, video registrations, interview questions, all reflect some standpoint of the researcher and the teacher. The hidden influences are numerous.

As to the reliability of the retrospection, we observed the difficulty for teachers to situate exactly their cognitive processes on a time line. The fact that retrospection always takes place after the planning or the interactive phase may very well confuse the distinction between different moments of teacher thoughts. If retrospection is replaced by thinking aloud or anticipation of reality, the interactive teaching is affected, then, by the emphasis laid on the objects of thought during the pre-interactive phase. If the retrospection takes place immediately after the lesson, another problem comes to light: teachers are too involved in their activity, so that they do not seem to be able to take some distance from the past lesson, as has been reported in the study of post-interactive thoughts.

An important methodological refinement is realized when confronting the retrospective data with some external information, like lesson protocols or written texts. The use of examples of the thoughts is usable too, because there are possibilities to compare the information both from the broad retrospection and the other materials. In the reported study, we remarked a great difference between the broad retrospection and the concrete illustrations in their quality of reported thoughts. Teachers can better word their teaching in the form of examples, because they use a non-professional language, as Jackson (1968) already observed.

Although the concrete examples are interesting entries to teachers' after lesson thoughts, there nevertheless remains an important restriction. The data refer to discrete moments of the teaching activity, with the consequence that it seems rather impossible to reconstruct the whole process of the interactive phase. Whereas it was quite easy to realize a process analysis of the planning phase, because the teacher had a chronological way of reporting the planning, it was not the case for the interactive and post-interactive phase. Using some associations, the teacher organizes his thoughts not on the basis of a chronological principle, so that the data reflect more the content than the exact timing of thoughts.

The use of retrospection as a self-reporting technique is, among others, responsible for the high degree of rationality as suggested by the data. The whole research context expresses the outlook from the researcher and/or the teacher that it should be possible to give some information about the 'why' of the event. Teachers seldom indicate behaviour as 'routine', but formulate negative statements such as 'I don't know exactly why I do this.' The problem with this kind of information is to know whether the behaviour is so far routinized, that it is impossible to introspect about it, or if other factors account for this fact.

The problems, mentioned above, refer to the necessity of a broad-spectrum of research methods. Due to the refinement of retrospective methods, it seems nowadays possible to combine the external observation with the self-reports of teachers. Whereas the former indicate the realized behaviour, which is the point of reference for the pupils, the latter may contribute to a better understanding of why and how this external behaviour originated. By this way, some cyclical research process appears, in which the acquisitions of different traditions could be combined. It is the time now, for example within studies on academic learning time, to complement the quantitative measures of the real time on task, with qualitative data from teachers and pupils.

Theory building

The research projects as reported earlier, had no well-elaborated theoretical bases. The aim was rather to explore some characteristics of the complexity of teaching in order to contribute, in the long run, to some models and theories of teaching.

Teaching has been described in a nearly axiomatic way as the intentional activity of a teacher within a complex environment. The search for cognitive processes as well as the process analysis brought important features to light. The teacher is an information processor: he perceives, selects and interprets bits of reality on the ground of his ideas and naïve models. Observable behaviour is only fully understandable if the researcher gains access to the hidden world of thinking and feeling. It is the teacher only who can deliver the important information about the determinants of his behaviour. When organizing groupwork in the class, the observer cannot know why a teacher chooses this method: are the reasons pragmatic ones, or based on didactical or social considerations? On the other hand, this does not mean that an external observer is unable to add some meaning to the observed behaviour. A lot of unconscious rules, unknown by the teacher, can be revealed by an observer who can offer his interpretation for falsification by the teacher.

Another remarkable fact is the importance of the specific environment in which the teacher functions. Some behaviours are directly conditioned by situational characteristics, without a high degree of rationality and intentionality. These behaviours are, then, field-steered (van Parreren, 1979).

The contribution to an adequate model of teaching will depend upon the expertise of the scholars in the field to analyse the teaching activity and to structure it along the most relevant features or variables. Although the cognitive approach of teaching is a relatively recent one, there is some danger for a splitting force within the research already done. One of the main shortcomings of the process-product studies has been the lack of conceptualization of the used variables. In a critical analysis of the output of the many studies on teacher effectiveness, Heather and Nielson (1974) already pointed out the main problems in this tradition. The lack of a consistent conceptualization as well as of methodological accuracy were the main reasons for the lack of productivity.

When looking at the cognitive paradigm, it is not easy to avoid the same shortcomings. We do not have a well-elaborated conceptual framework, nor a hard and qualitative methodology. We almost exclusively focus on teacher thoughts, with omission of observable behaviours. We divide the complex teaching activity into a chronological dissection without attention for more meaningful categories. We emphasize the isolated variables within the pre-interactive and post-interactive phase without great concern for the interrelations between the phases.

Now it seems to be the time to focus on the integration of the research work already done. The number of studies has grown steadily, but they only can become surveyable if we collaborate for an acceptable conceptual framework as well as the precise analysis of the models. This work will furnish a platform for further research as an entry to the construction of empirical proved theoretical models, in which intentionality, complexity and situational variables are represented in a meaningful way. In the long run, a consistent theory of teaching is the ultimate aim.

Chapter 30

Teachers and Their Educational Situations They are Concerned About: Preliminary Research Findings

Jan R.M. Gerris, Vincent A.M. Peters and

Theo C.M. Bergen

The authors wish to thank Dr. S. Silverentand and Dr. D Kristensen for their helpful assistance in data collection and data analysis.

Summary

This study focuses upon an exploration of developmental trends that can be inferred from spontaneous verbal reports of secondary school teachers on problematic situations experienced during their professional practice.

We explored quantitative differences in problematic situations between groups of secondary school teachers varying in years of experience. The teachers participating in this study were asked to describe problematic situations they experience in their professional work, keeping in mind four general areas: a) instruction, b) helping, tutoring and counseling pupils, c) school organization and administration and d) external relations of the school. 2692 situation descriptions were collected.

In order to process these situation descriptions the 'Problematic Situation Classification Scheme' (PSCS) was developed. From the results of this study and on the basis of the literature on the development of teachers' concerns it is possible to infer a developmental pattern from self-centered parameters to other-centered and organizational parameters.

Introduction

The long-term goal of the research project is to interpret teachers' behavior in terms of the dynamic person x situation interaction model as described by Endler and Magnusson (1976). This paper deals with the situation side of the interactions. Bergen, Peters, and Gerris (1983) reported on spontaneous verbal reports of secondary school teachers on situations in educational practice that they perceived as problematic. This study focuses on an exploration of a developmental trend that can be inferred from these problematic situations.

Though teachers' concerns and problems are extensively documented, analyses of developmental aspects of problematic situations in the professional practice of teachers are relatively scarce. From the 51 studies concerning problems of both elementary and secondary school teachers identified by Veenman (1982) since 1960, the following rank order of the eight most important problem areas was inferred: 1) discipline, 2) motivating students, 3) handling differences between students, 4) evaluating school performances, 5) rapport with parents, 6) inadequacy of learning and teaching facilities, 7) individual problems of students, 8) time pressure (cf. Veenman, 1982). Concerning the question whether certain periods of years of experience in teaching practice can be characterized by certain areas of teacher concerns, two findings are to be mentioned:

1) the indication of a rather dramatic change (the so-called 'Reality Shock') in experiencing problems is documented in the transition from the pre-service period of teacher education to the first in-service period of the beginning teacher (cf. Mηller-Fohrbrodt et al., 1978);
2) the developmental conceptualization of teachers' concerns as concerns with self as characteristic for the early years and concerns with the students as characteristic for the later years (Fuller, 1969).

This two stage-developmental model was elaborated into a four stage sequence of concerns by Fuller and Bown (1975):

1. pre-teaching concerns, that is to say, non-concern with the specifics of teaching and a low involvement in teaching.
2. early concerns about survival, that is concerns with self-protection, self-adequacy, class-control, adequacy of subject matter, finding a place in the power structure of the school and understanding expectations of supervisors, principals and parents. Concerns about class-control are indicated as the most intense. Furthermore it is postulated that pre-service teachers have more concerns of this type than in-service teachers.
3. teaching situation concerns about limitations and frustrations in the teaching situation (about methods, materials and own teaching performance), which are added to self-survival concerns. It is indicated that these concerns are not about pupils and their learning.
4. concerns about pupils, about recognizing the social and emotional needs of pupils, about relating to pupils as individuals. It is indicated that these concerns are experienced both by pre-service and in-service teachers.

In their review of sources of anxiety and concern reported by teachers Coates and Thoresen (1976) indicate the following sources for beginning teachers: a) their ability to maintain discipline in the classroom, b) students' liking of them, c) their knowledge of subject matter, d) what to do in case they make mistakes or run out of material, e) how to relate personally to other faculty members, the school system and parents. For experienced teachers, the chief sources of anxiety and concern center around: a) time demands, b) problems with pupils, c) large class enrollments, d) financial constraints, and e) lack of educational resources (Coates and Thoresen, 1976).

Method

Subjects

A representative sample of secondary schools was asked to participate in a study of the nature of problematic situations that teachers experience in their professional practice. From a complete listing of the teachers' names, provided by the schools that agreed to cooperate, the investigators identified a stratified sample of 803 secondary school teachers. Stratification criteria were: years of experience in the teaching profession, sex, clusters of subject matter taught. The 253 inventories that could be used represented 31.6% of the total number of 803. Of the 253 subjects, 180 were male and 73 female. The number of subjects per category of experience TI-T6 was: TI: 22 teachers in their first year (M 15, F = 7); T2: 22 teachers in their second year (M = 13, F = 9); T3 24 teachers in their third year (M 19, F = 5); T4: 34 teachers in their fourth year (M = 21, F = 13) 1T5 74 teachers in their fifth-to-tenth year (M = 59, F = 15), and T6 77 teachers with more than ten years of experience (M = 53, F 24).

As in this study, we were only interested in the problematic situations that teachers experience as a function of years of experience, we examined the six groups on differences in sex, cluster of subject matter and number of lessons per week. Sex and subject matter were not contaminated, number of lessons, however, varies with years of experience (X^2 = 48.025, df = 15, P = .0001). Teachers with fewer years of experience tend to give fewer lessons per week than teachers with more years of experience. It is not clear how this finding influences the problematic situations that teachers experience in their professional practice.

Procedure

In order to warrant that teachers described problematic situations representing the whole of their professional practice, they were asked to describe situations keeping in mind four general areas of professional work: 1) instruction, 2) helping, tutoring and counseling pupils, 3) school organization and administration, 4) outside relations of the school. It was emphasized in the instruction that problematic situations should be described as concretely as possible, a problematic situation being conceived of as a description of a concrete situation or event occurring during professional practice which is subjectively experienced as problematic. It was made clear that this study was not intended to collect general problems or concerns of teachers. To assist teachers in giving concrete problematic-situation descriptions they were asked to incorporate the following elements in their descriptions: the action, the persons(s) who is (are) undertaking a certain action, the object(s) of action, the person(s) involved, setting and time of action.

Situation descriptions collected totaled 2692. To reduce these 2692 situation descriptions to a manageable list representing the nature of the raw descriptions as much as possible, a classification scheme was developed by the investigators, making use of two randomly selected packages of a hundred situations each. In

Table 1 the scheme for the classification of problematic situations in the professional practice of secondary school teachers is summarized.

Table 1. The Problematic Situation Classification Scheme (PSCS) for classifying problematic situations in the professional practice of secondary school teachers.

Classification Code		Description
Main-category	Sub-category	
100		Education (macro): situations referring to general educational issues (on a macro level).
200		School as an organization (meso): situations concerning the personal and material power structure of the school as an organization.
	210	(in)adequacy of the organizational structure of the school,
	220	(in)adequacy of facilities of the school (personal and material) that can not be directly influenced by the teacher.
300		The profession of a teacher: burdening elements of the
	310	profession, its status (legal and societal),
	320	influences on the functioning of the teacher himself as a
	330	professional, professionalization of teaching.
400		Situations concerning subjects taught in the school curriculum.
500		Pupils' orientation: situations focusing upon the pupil that are not directly related to a particular teaching situation.
	510	Personal characteristics of the pupil(s), interactions and
	520	social relations between teacher and pupil,
	530	Interactions and social relations between pupils.
600		Teaching-learning situations (micro):
	610	(in)ability of the teacher to cope with learning situations in a classroom context,
	620	reactions of pupils to classroom activities, learning tasks,
	630	disturbances of the progress of activities in the classroom.

The 2692 situation descriptions were categorized according to the PSCS scheme by the three investigators. The categorization focused upon the nature of the situation as described by the teachers. To prevent subjective biases of the individual investigator each situation was coded by two raters. Situations that were coded differently were extensively discussed in the expert group consisting of the three investigators, which resulted in an expert-group code. The average percentage of agreement between individual raters in earlier and later coding phases were respectively 74.5 and 79.7. The percentage of intra-expert-group agreement was 88%, while the average percentage for intra-rater agreement was 75.3.

Results

As could be expected both from research on teachers' concerns and developmental conceptualizations of teachers' concerns in particular, a rather strong relationship was found between type of problematic situations (see the PSCS-categories) and years of experience. This relationship was apparent in the cross-tabulation of type of 2 situations by years of experience in terms of relative frequencies (X^2 = 244.35, P

= .0001). Also in line with the general finding that in different years of experience different types of educational situations are experienced as problematic, higher correlation coefficients can be expected between the rank orderings of the PSCS-categories of adjacent years of experience, and lower ones between non-adjacent years. The rank orderings based on relative frequencies are presented in table 2. Although within the rank orderings based on relative frequency high correlations are found both between adjacent and non-adjacent years of experience (see table 2), coefficients between non-adjacent years tend to be lower, e.g. the row-correlation pattern between T1 – T6 (.65) between T2 – T5 (.31) and between T2 – T6 (.38).

Table 2. Rank order correlations (Spearman-Rho) between groups of teaching experience (T1 – T6) for rank ordering of categories of problematic situations based on relative frequencies.

Groups of teaching Experience	T1	T2	T3	T4	T5	T6
T1 – 1st year	1.00	.85 ***	.93 ***	.85 ***	.64 *	.65 *
T2 – 2nd year		1.00	.76 **	.70 **	.31	.38
T3 – 3rd year			1.00	.89 ***	.75 **	.80 ***
T4 – 4th year				1.00	.81 ***	.84 ***
T5 – 5-10th year					1.00	.94 ***
T6 – 10 years						1.00

* P ≤ .05 ** P ≤ .01 *** P ≤ .001

The rank ordering of problematic situations in each category of experience is presented in Table 3.

From Table 3 it appears that the relative frequency of the three situation clusters belonging to the main category of teaching-learning situations (category 600) shows a declining trend with increasing years of experience. It seems that with more years of experience, teachers experience fewer problematic situations in which they have problems in coping with learning situations in the classroom (code 610), in which they have problems with reactions of pupils to classroom activities (code 620), and fewer problems of discipline (code 630).

The first, second and third year (see table 3) is in accordance with the general findings in most of the studies on teacher problems (Wright, 1975; Williams, 1976; Cruickshank, Kennedy and Myers, 1974; Fuller, 1969; Fuller (and Bown, 1975).

With increasing years of experience, on the other hand teachers report more problematic situations concerning the teaching profession (see table 3, codes 310 and 320), and concerning the (in) adequacy of both the organizational structure and facilities of the school (codes 210 and 220 resp.).

Table 3: Rank ordering of problematic situations based on relative frequencies of the PSCS-categories in various years of teaching experience (T1-T6). | | Means the same position in the rank ordering.

Rank order	T1	T2	T3	T4	T5	T6
1	630	630	630	610	320	320
2	610	610	320	320	210	210
3	320	320	210	630	(310)	(310)
4	210	620	620	(310)	220	510
5	620	510	610	210	630	630
6	410	520	510	620	610	610
	ll					
7	510	410	(310)	510	510	220
8	220	530	410	220	410	620
9	520	210	220	410	620	410
10	(310)	220	520	520	520	110
		ll				
11	530	(310)	530	530	110	520
12	330	330	330	110	530	530
13	_*	110	110	330	330	330

*Category 110 was not used by the group of teachers in their first year of experience.

As could be expected from the hypothesis derived from the conceptualization of Fuller (1969) and Fuller and Bown (1975) the problem area of discipline (disturbance of the progress of activities in the classroom, i.e. code 630) is predominant in the first year and in the second year. The corresponding relative frequencies were 18% and 24%.

Both in early and later years within category 630 situations prevail such as: pupils refusing to observe the disciplinary measures taken by the teacher, getting classes to quiet down when starting the lesson, disturbances at the beginning of the lessons and while the teacher is lecturing. Situations concerning (in)abilities of the teacher to cope with learning situations in the classroom (code 610) take a second position in the rankordering of the classification codes according to relative frequencies (15.06% and 14.12% respectively) (see table 3).

The hypothesis inferred from Fuller and Bown's model that problem areas concerning methods and subject matter (code 410), facilities (code 220) and learning situations (code 610) would dominate in intermediate years of experience - approximately in the second and third year - was not supported. There was no clear pattern of dominance for the problem areas represented by codes 220 and 410 relative to other areas, whereas code 610 takes a second position in the second year, a fifth position in the third year, and a first position in the fourth year.

A third implication of Fuller and Bown's model is that in later years teachers' concerns get focused upon the personal and social needs of the pupil. As a consequence one would expect that the pupil codes of the PSCS-scheme would be relatively dominant in later years of experience. In the rankordering of the PSCS-categories of table 3 there appeared to be no focus on either of the pupil categories as operationalized by codes 510, 520 and 530. The relative frequencies for code 510 were 5.50% (1st year), 10.20% (2nd year), 6.16% (3rd year), 7.20% (4th year), 7.43% (5th - 10th year) and 8.75% (more than 10 years). Finally, it appears from the findings represented in table 3 that in all years of experience the relative frequency of codes 110, 330, 520 and 530 is low. That is to say that teachers in their descriptions of problematic situations hardly ever refer to general educational issues (code 110), the professionalization of their teaching (code 330), their interactions and social relations with pupils (code 520), and relations between their pupils (code 530).

Besides these more quantitative analyses we also noticed qualitative differences within each of the 13 categories, which will be thoroughly analysed elsewhere. In this context we restrict ourselves to a short description of categories 300 and 600.

In early years in category 310 situations are prevalent in which teachers report both physical and psychological tiredness, and pressure of time because of an accumulation of duties. In later years (T3 - T6) situations dominate in which teachers focus on four kinds of complaints:

a. about physical aging processes
b. about spending most of their time on routine matters,
c. about the deadly regularity of the schoolday or –year;
d. about negative evaluations of the teaching profession from the public outside.

For the 320 situations there is a strikingly high position in early years of experience (table 3). Complaints about a lack of positive contacts with colleagues for discussing problems, and about a bad atmosphere between colleagues, seem to dominate in the situation descriptions of category 320 in all years of experience (T1 - T6).

In the situation descriptions belonging to category 610 (teaching-learning situations) for all groups of teaching experience (TI -T6) it is perceived as a problem to meet demands posed by individual differences between pupils, in particular differences in level. It is characteristic for teachers in early years of experience that they mention their inability to divide their attention, to prevent noise while lecturing, to keep track of their lesson plan, and to keep up with the subject curriculum of the school. It is characteristic for later years of experience in category 610 that these teachers report getting 'tired to death' by having to repeatedly explain the same subject matter and by having to motivate pupils again and again (in particular at T5 and T6).

Discussion

The preliminary results reported in this paper are restricted to relative frequencies of thirteen categories of problematic situations teachers are concerned about. A full report and discussion of all the results of the exploration phase is presented in Bergen, Peters and Gerris (1983) and Gerris et al. (1984). The reliability of the Problematic Situations Category Scheme (PSCS), which was developed to reduce the 2692 situation descriptions to a manageable list, was found to be sufficient to warrant future investigations concerning the situation side of our approach to the developments teachers go through.

The main implication of Fuller and Bown's model of development from self to task to pupil with increasing years of experience was not fully replicated. That is, there appeared to be no clear pattern of dominance of the pupil categories in later years of experience. A relative dominance of discipline problems, which are assumed to be characteristic for self-concerns, however, was clearly established in the early years. These results of course should be interpreted with caution, because they are to be checked in further qualitative analyses of the contents of the 2692 situation descriptions.

In order to derive a data-based, developmental hypothesis about changes teachers apparently go through during several years of teaching experience it seems insufficient to label different clusters of teachers' concerns as developmental stages. According to a more structural view, as in the Piagetian and neo-Piagetian tradition, it is essential for the researcher to focus on one or more dimensions resulting in qualitative differences between different stages of development. Observed differences can thus be understood in terms of qualitative changes in the underlying dimension(s).

From the results reported in this study and from the findings in the literature on the development of teachers' concerns, it is possible to infer that with increasing years of experience there is a development from self-centered situations, via focusing on the teaching situation and the functioning as a member of a team, to a focus on the organizational structure of the school. Indications for this hypothesis are found in the rank ordering of PSCS-categories in table 3: dominance of discipline problems (code 630) in early years, in intermediate years a dominance of teaching situations (code 610) and situations referring to colleagues as main influences on the functioning of the teacher (code 320), and in later years of experience a dominance of situations relating to the (in)adequacy of the organizational structure of the school (code 210). That is a development from self-centeredness, via an intermediate orientation on relevant other persons in the school, to an organizational orientation. This would imply a developmental dimension from self and others to the school as an organization. Possibly, teachers are capable of a progressively more self- and person-detached point of view with increasing years of experience.

Although in this hypothetical developmental interpretation the <u>what</u> of teacher development is indicated, it must face several alternative explanations:

1. The feature of self-centeredness (concern with one's own survival and one's own teaching performance) can also be present in the situation descriptions that are categorized on the level of the teaching-learning situation, and the personal and structural relations inside the school. A detailed qualitative analysis, as mentioned before, could offer some more evidence.
2. The findings of relatively more organization-centered parameters in the situation descriptions we have discovered so far could simply be a consequence of the fact that senior teachers have more tasks in the school organization. If this is the case the developmental trend should be explained in terms of distribution of organizational tasks rather than in terms of the development of certain personal characteristics (e.g. a growing ability to handle more complex and more variable situations in the school from a progressively more self-detached point of vicw).

This second possibility could imply a general alternative explanation for findings about changes in teachers' concerns with increasing teaching experience. Is it sufficiently established in research on teacher development to date, that changes in the nature of teachers' concerns and perceived problems are not merely a function of the external features of the school system and the 'rules' the school system applies to beginning or experienced teaches? Or, is there sufficient reason to assume a more personal development of teachers, which implies that development is dependent on interactions between characteristics of both the person and the situation? A discussion of this fundamental question would go beyond the framework of this paper.

Of course, the hypothetical development dimension of a progressively more self- and person-detached point of view with increasing years of experience should be further tested both in cross-sectional and longitudinal studies using years of experience as the independent variable and both quantitative and qualitative features of categories of educational situations as the dependent variables.

A second brief comment concerns the possibility to explain and/or neutralize discrepancies between problems teachers apparently have in paying attention to students as persons on the one hand (cf. In Table 3 pupil categories 510, 520 en 530), and the great need of students for personal support from the teacher on the other hand (cf. Applegate, 1981). In this context further research on characteristics of the teacher that are relevant in interaction with a classification of problematic situations in educational settings, is essential. Further research will concentrate on the following questions:

a. What changes or differences can be observed in the dimensions of meaning underlying the problematic situations between groups of teachers varying in years of experience? That is, by what developmental dimensions can dominant concerns of teachers be characterized (in terms of situation perception and situation reactions);
b. How can the variance in the situation-perception and situation-reaction data be explained in terms of personal characteristics of teachers;
c. What person-situation interactions change qualitatively with increasing years of experience in what direction
d. Which mechanism(s) can account for development through different stages in the progress of becoming a teacher?

Conclusions

Further Explorations of Teachers and Teaching in ISATT Today

ISATT has provided a foundation for forward-looking scholarly thought. The papers included in this volume are not just about events and matters of concern at the time of their presentation. The approach and grounding to thought and research about teachers processes contained herein provided platforms for many continuing important studies and investigations. The further work of these authors and their colleagues, and the considerable effect that each has had on their students in turn, is now producing third and fourth generation ISATT members. All are charged with the responsibility of promoting shared knowledge that develops and creates bonds of inquiry within the learning community. ISATT's legacy encourages creative thought and risk-taking in a supportive, collegial, congenial yet critical environment. Current members seek out challenge and maintain those high standards, while bringing together like-minded scholars to engage in the debate.

The list of books produced from ISATT gatherings and the considerable volume of contributions to the field made through the journal Teachers and Teaching: Theory into Practice has established a respected and prestigious community. While there is much of value in the "what" that is contained in the various example of ISATT research, equal, if not more, attention should be paid to the underlying approaches used to develop, guide and communicate the content. The "how" and "why" questions, that can only be answered by the self of the researcher under the influences of a scholarly environment, feed forward into a spiral of reflective thought that adds eagerness to new projects. "What are you doing next?" and "Where will this lead?" are questions often heard at ISATT gatherings. These are questions well applied to the materials contained in this volume and posed retrospectively to the writers. We ask such questions of current and future thinkers, writers and researcher in education in the spirit of ISATT with an invitation to participate in the sharing of knowledge, perspective and growth.

The purpose of ISATT is to promote, present, discuss and disseminate research on teachers and teaching in order to contribute to theory formation in this field. ISATT members are involved in these activities in order to gain more insight into these aspects of education, add to knowledge, and enhance the quality of education through improved teaching and forms of professional development at all levels of education.

Research on teachers and teaching in schools and higher education encompasses several perspectives. These include but are not limited to: teachers' purposes, beliefs, conceptions, practical activities; theories, narratives, histories, stories, 'voices'; teachers' intentions, thought processes and cognitions, personal practical knowledge; teachers' emotions, thinking and reflection as aspects of professional actions; teachers' thinking and action as influenced by contextual factors in their structural, cultural and social environments; and workplace teaching and learning.

A central intention for this research is to focus on the way teachers themselves understand teaching and their own roles in it. Research is not limited to studying what teachers do but tries also to understand how they think and feel about what they are doing and the cultural contexts in which their work is embedded. Consequently, research is individually as well as socially, psychologically and culturally based. There is an increasing acceptance of the value of research carried out by researchers from complementary research traditions. Researchers from different disciplinary backgrounds at all levels of teaching have come to study such diverse phenomena as teacher planning, decision-making, reflection, teacher understanding of subject matter and of curricula, their judgement of students' work, their beliefs, attitudes, conceptions, implicit theories and thought processes as well as their principles of action, the criteria they apply as professionals and the dilemmas of teaching they encounter. Insights from this wide variety of studies have informed teacher education programmes and curriculum development, contributed to teachers' self-reflection and professional awareness, and provided a growing data bank from which educational policy makers may draw. The growing interest in research from Teacher Thinking and Action Perspectives is an international trend. As an organisation, ISATT is at the forefront. It draws its membership from teacher-researchers worldwide at every academic level and from a range of disciplines publishing in many countries and in their respective languages.

Activities and publications

ISATT offers a biennial Conference with leading researchers as keynote speakers presenting state of the art, cutting edge contributions in the field. Parallel groups for paper presentations provide participants with fora for stimulating discourses of their work, while symposia, workshops, round tables and poster sessions offer possibilities for interactive work on methodological innovations or theory application.

ISATT conferences are renown for their combination of an atmosphere of intellectual stimulation with an amicable collegial climate. ISATT publishes its own international journal in cooperation with CARFAX - Teachers and Teaching- theory and practice. There are four issues per year providing an update on research activities in the field. This refereed journal offers an interesting selection of new research reports and theoretical contributions. ISATT's Publication Series consists of a volume of selected papers from each of the conferences including the keynote speeches. They include:

Halkes, R. and Olson, J.K. (eds). Teacher thinking: a new perspective on persisting problems in education. ISATT, Lisse: Swets and Zeitlinger 1984.

Ben Peretz, M., Bromme , R. and Halkes, R. (eds). Advances of research on teacher thinking. ISATT. Lisse/Berwyn: Swets and Zeitlinger/Swets North America Inc., 1986.

Lowyck, J. and Clark, C.M. (eds). Teacher thinking and professional action. Leuven University Press, 1989.

Day, C., Pope , M. and Denicolo, P. (eds). Insights into teachers' thinking and practice. Falmer Press, 1990.

Day, C., Calderhead, J. and Denicolo, P. (eds). Research on teacher thinking: Towards understanding professional development. Falmer Press, 1993.

Carlgren, I., Handal, G. and Vaag , S. (eds). Teachers Minds and Actions: Research on teachers' thinking and practice. Falmer Press, 1994.

Kompf, M., Boak, R.T., Bond, W.R., & Dworet, D.H. (eds). Changing Research and Practice: Teachers Professionalism, Identities and Knowledge. Falmer Press, 1996.

Lang, M., Olson, J., Hansen, H., Bunder, W. (eds). Changing Schools/Changing Practices: Perspectives on Educational Reform and Teacher Professionalism. Garant, 1999.

Sugrue, C. and Day, C. (eds). Developing Teachers and Teaching Practice: International Research Perspectives, London. Routledge/Falmer Press, 2001.

How to Participate in ISATT

The website www.isatt.org contains information about the organization, the publications, meetings and the executive. National representatives are spread throughout the world in 28 countries and will provide assistance and information on request.

References*

(*Please note original citations unless updated by the author.)

Adams-Webber, J.R. (1979). *Personal Construct Theory, Concepts and Applications*. Chichester, New York, John Wiley & Sons.

Alexandersson, C., Dahlgren, L.O. & Larsson, S. (1979). Om DELTA-projektets nytta. *Bakgrundsmaterial till Skolöverstyrelsens vuxenpedagogiska seminarium i Sigtuna, 1979*. On the use of the DELTA-project.

Alexandersson, C. & Larsson, S. (1980). Betydelsen av erfarenhet? *Bidrag till UHÄseminariet Arbetslivserfarenhetens roll i högskoleutbildningen, Taljöviken, 1980*. The importance of experience.

Alexandersson, C. (1980:03). Amedeo Giorgi's empirical phenomenology. *Reports from the Institute of Education, University of Göteborg.*

Alexandersson, C. (1981:05). Utbildningseffekter - ett exempel pa innehallsrelaterad kvalitativ beskrivning. *Rapporter fran Pedagogiska institutionen, Göteborgs universitet*. Educational impact - an example of content-related qualitative description.

Alexandersson, C. (1982:08). Deltagarnas erfarenhet som en resurs i undervisningvuxenstuderandes uppfattning av ett pedagogiskt begrepp. *Rapporter från Pedagogiska institutionen, Göteborgs uni versitet.* Students' experience as a resource in teaching.

Amarel, M. (1983). Classrooms and Computers as Instructional Settings. *Theory into Practice*, *22*, 260-266.

Anderson, J.R. (1982). Acquisition of cognitive skill. *Psychological Review*, *89*, 369-406.

Anderson, J.R. (1985). *Cognitive Psychology and its implications*. (2nd Ed.), New York, Wiley.

Anderson, N.H. (1970). Functional Measurement and Psychological Judgement. *Psychological Review*, *77*, 53-170.

Anderson, N.H. (1974). Cognitive Algebra. In: Berkowitz, L. (Ed.). *Advances in Experimental Social Psychology*, Vol. 7, New York: Academic Press.

Anderson, N.H. (1976). How Functional Measurement can Yield Validated Interval Scales of Mental Quantities. *Journal of Applied Psychology*, *61*, 677-692.

Anderson, N.H. (1981). *Foundations of Information Integration Theory*. New York: Academic Press.

Anderson, N.H. & Butzin, C.A. (1974). Performance = Motivation x Ability: An Integration Theoretical Analysis. *Journal of Personality and Social Psychology*, *30*, 598-604.

Andersson, E & Lawenius, M. (1983). *Larares uppfattning av undervisning.* Göteborg: Acta Universitatis Gothoburgensis. Teachers' conception of teaching.

Andersson, E. & Lawenius, M. (1983). *Larares uppfattning av undervisning.* Göteborg: Acta Universitatis Gothoburgensis. Teachers' conception of teaching.

Aoki, T.T. (1983). Experiencing ethnicity as a Japanese Canadian: Reflections on a Personal Curriculum. *Curriculum Inquiry*, *13:3*, pp. 321-335.

Apelgren, B. M. (2001). *Foreign language teachers' voices: Personal theories and experiences of change in teaching English as a foreign language in Sweden.* Gothenburg: University of Gothenburg Press.

Applegate, J.H. (1981). Perceived problems of secondary school students. *The Journal of Educational Research , 75*, 49-55.

Aoki, T.T. (1983). Experiencing ethnicity as a Japanese Canadian: Reflections on a Personal Curriculum. *Curriculum Inquiry, 13:3*. pp. 321-335.

Argyris, C. & Schön, D. (1974). *Theory in practice: increasing professional effectiveness*. London.

Argyris, C. & Schön, D.A. (1976). *Theory in practice: increasing professional effectiveness*, Jossey-Bass.

Ausubel, D.P., Novak, J. & Hanesian, H. (1978). *Educational Psychology: A Cognitive View*. New York, Holt, Rinehart & Winston.

Bannister, D. (1970). *Perspectives in personal construct theory*, London Academic Press.

Bannister, D. & Fransella, F. (1977). *A manual for repertory grid technique.* London: Academic Press.

Bartlett, F. (1964). *Thinking. An experimental and social study*. London, Unwin University Books.

Barz, M., Uttendorfer-Marek, I., Wagner, A.C., Weidle, R. & Maier-Stormer, S. (1981). Denkknoten. *In:Psychologie Heute, 5.*

Batcher, E. (1981). *Emotion in the classroom.* New York: Praeger.

Bateson, G. (1973). *Steps to an Ecology of mind.* London: Paladin Books.

Beck, A.T. (1979). *Wahrnehmung der Wirklichkeit und Neurose.* Munchen

Becker, G.E., Huber, G.L., Mandl, H., Wahl, D. & Weinert, F.E. (1978). Konzeptionsrahmen für einen Fernstudien-Lehrgang "Lehrertraining: Probleme entwicklungs -, verhaltens- und lerngestörter Schüler". Tübingen: DIFF.

Becker, G..E., Huber, G.L., Mandl, H., Wahl, D. & Weinert, F.E. (1981). Konzeptionsrahmen für das Fernsehkolleg Schülerprobleme – Lehrerprobleme. In S. Rotering-Steinberg (Ed.). Fersehkolleg Schülerprobleme – Lehrerprobleme (pp. 5-28). Tübingen: DIFF.

Bennet, S. (1978). Recent research on teaching: a dream, a belief, and a model. *British Journal of Education Psychology, 48*, 127-147.

Ben-Peretz, M. (1981). *The form and substance of teacher's lesson planning.* Paper presented at the AERA meeting, Los Angeles.

Ben-Peretz, M. (1984). Kelly's theory of personal constructs as a paradigm for investigating teacher thinking. In: Halkes, R. & Olson, J.K. *Teacher thinking: a new perspective on persisting problems in education.* Lisse: Swets & Zeitlinger.

Ben-Peretz, M. & Katz, S. (1980, April). *Curriculum perception profile of language teachers*. Paper presented at the annual meeting of the American Educational Research Association, Boston.

Ben-Peretz, M. Katz, S. & Silberstein, M. (1982). Curriculum interpretation and its place in teacher education programs, *Interchange, 13:4*, 47-55.

Ben-Peretz, M. & Katz, S. (1983, July). *From simplicity to complexity differences in the construct systems of student-teachers.* Paper presented at the Fifth International Congress on Construct Psychology Boston.

Berg, R. (1983). Resisting Change: What the Literature Says About Computers in the Social Studies Classroom. *Social Education*, 314-316.

Bergen, T.C.M., Peters, V.A.M. & Gerris, J.R.M. (1983). Een exploratieve studie naar de probleemsituaties van ervaren en beginnende docenten tijdens hun beroepsuitoefening. In: S. Veenman en H. Coonen (Red.,), *Onderwijs en opleiding*. Lisse: Swets & Zeitlinger.

Berk, L. (1980). Education in lives: Biographic narrative in the study of educational outcomes. *Journal of Curriculum Theorizing, 2:2*, pp. 88-155.

Berlak, A. & Berlak, H. (1981). *Dilemmas of schooling : Teaching and social change*. London: Methuen.

Berliner, D.C. & Tikunoff, J. (1976). The California Beginning Teacher Evaluation Study; overview of the ethnographic study. *Journal of Teacher Education* XXVII (1). 24-30.

Bieri, J. (1955). Cognitive complexity-simplicity and predictive behaviour. *J. abnormal and social psychology*, *51:2*, 263-268.

Borg, W.R. & Stone, D.R. (1974). Protocol materials as a tool for changing teacher behaviour. *The Journal of Experimental Education, 43*, 34-39.

Bowers, K.S. (1979). Knowing more than we can say leads to saying more than we can know: On being implicitly informed. Final draft of a paper presented at a symposium on "The situation in psychological theory and research." Stockholm.

Boyd, E.M. (1981). Reflection in experiential learning: case studies of counsellors. Unpublished Ed.D. thesis, Univ. of Toronto.

Broadbent, D.E. (1964). *Perception and Communication*. Oxford, Pergamon Press.

Bromme, R. (1979). Das Denken von Lehrern bei der Vorbereitung von Uterrricht – eine empirische Untersuchung zu kognitiven Prozessen von Mathematiklehrern. Dissertation. Universität Oldenburg.

Bromme, R. (1980). Die alltaglichen Unterrichtsvorbereitungen von Mathematiklehrern. Zu einigen Methoden und Ergebnissen einer Untersuchung des Denkprozesses. In: *Unterrichtswissenschaft.* nr. 2, S. 142-156.

Bromme, R. (1981). *Das Denken von Lehrern bei der Unterrichtsvorbereitung* Beltz, Weinheim.

Bromme, R. (1982). *How to analyze routines in teacher's thinking processes during lesson planning*. Bielefield, Institut fηr Didaktik der Mathematik der Universitat Bielefield (Occasional Paper 24).

Bromme, R. (1982). *How to analyze routines in teachers' thinking processes during lesson planning*. Paper presented at the annual meeting of the AERA, New York. (ERIC Document Reproduction Service No. ED 223546).

Bromme, R. (1986). Der Leherer also Experte – Skizze eines Forschungsansatzes. To be published in: Neber, H. (Ed.). *Angewandte Problem-lösepsychologie*. Münster: Aschendorff.

Bromme, R. & Brophy, J. (1986). Teachers' cognitive activities. In (Eds.). *Basic Components of Mathematics Education for Teachers* (in press).

Bromme, R. & Hoemberg, E. (1976). *Einfuehrende Bemerkungen zum Problem der Anwendung Psychologischen Wissens (Technologieproblem*. (Materialien und Studien Vol. 4). Bielefeld: Institut fuer Didaktik der Mathematik der Universitaet Bielefeld.

Bromme, R. & Hömberg, E. (1980). Methodische Probleme und Möglichkeiten der Untersuchung sprachlich gefaßter handlungstregulierender Kognitionen. In W. Volpert (Ed.), Beiträge zur psychologischen Handlungstheorie (pp. 105-120). Bern: Huber.

Bromme, R. & Juhl, K. (1984). *Students' understanding of tasks in the view of mathematics teachers* (Occasional Paper No. 58). Bielefeld: Institut für Didaktik der Mathematik. (ERIC Document Reproduction Service No. ED 254418). [German version published in: *Zeitschrift für Empirische Pädagogik und pädagogik und pädagogische Psychologie*, *9*, 1-14.]

Brophy, J.E. (1982). How teachers influence what is taught and learned in classrooms. *The Elementary School Journal*, *83(1)*, 1-14.

Brophy, J.E. & Good, T.L. (1976). *Teacher-student relationship.* New York: Holt, Rinehart & Winston.

Brophy, J. & Good, T. (1984). Teacher behavior and student achievement. In M.C. Wittrock (Ed.). *Handbook for research on teaching, third edition* (chap 11, in press). New York: Mcmillan.

Brown, A.F. (1964). Exploring personal judgements with discriminant perception analysis. *Proceedings of 3rd Canadian conference on educational research.* Ottawa: CCRE, 229-238.

Brown, A.F. (1974). A perceptual taxonomy of the effective-rated teacher, in *The employment of teachers* (D. Gerwin, ed.) Berkeley: McCutchan.

Brown, A.F. (1978). Teacher selection: attaining congruence between posted and operative criteria, to Can. Soc. for Study of Ed.

Brown, A.F. (1982). Interpersonal administration: overcoming the Pygmalion effect. *Interchange*, 13:4, 15-26.

Brown, A.F. (1984). How to change what teachers think about teachers: affirmative action in promotion decisions. In R. Halkes and J.K. Olson, eds. *Teacher thinking: a new perspective on persisting problems in education.* Lisse, Holland: Swets & Zeitlinger, 197-209.

Brown, A.F. & Ritchie, T.J., Eds. (1982). Personal Constructs in Ed. Theme number of *Interchange*, 13:4, pp. 82.

Brown, A.F., Rix, E.A. & Cholvat, J. (1983 Fall). Changing promotion criteria: cognitive effects on administrators' decisions. *J. Experimental Ed.*, 52:1, 4-10.

Buchmann, M. (1984). The use of research knowledge in teacher education and teaching. *American Journal of Education*, 92(4), 421-439.

Buchamann, M. & Schwille, J. (1983). Education: The overcoming of experience. *American Journal of Education*, *92(1)*, 30-51.

Buchmann, M. (1983). *Role over person: Justifying teacher action and decisions* (Research Series No. 135). East Lansing: Michigan State University, Institute for Research on Teaching.

Bunge, M. (1967). *Scientific Research T & 11.* Heidelberg: Springer.

Burton, F. (1985, January). *Off the bench and into the game: A classroom teacher's conception of doing action research.* Paper presented at the Meadow Brook Research Symposium on Collaborative Action Research in Education. Oakland University, Rochester, MI.

Bussis, A.M., Chittenden, E. A. & Amarel, M. (1976). *Beyond surface curriculum.* Boulder, Colorado Westview Press.

Butt, R.L., Olson, J. & Daignault, J. (Eds.) (1983). *Insiders realities, outsiders dreams: Prospects for_curriculum (Curriculum Canada IV).* Vancouver: Center for the Study of Curriculum and Instruction, University of British Columbia and the Canadian Association for Curriculum Studies.

Butt, R.L. & Raymond, D. (1989). Studying the nature and development of teachers' knowledge using collaborative autobiography. *International Journal of Educational Research, 13(4)*, 403-419.

Cacioppo, J.T. & Petty, R.E. (1980). Social psychological procedures for cognitive response assessment: The thought-listing technique. In T.V. Merluzzi, C.R. Glass & M.Genest (Eds.), Cognitive assessment. (Prepublication manuscript). New York: Guilford Press.

Calderhead, J. (1981). Stimulated recall: A method for research on teaching. *British Journal of Educational Psychology, 51*, 180-190.

Calderhead, J. (1984). Teachers' Classroom Decision-Making. London: Holt, Rinehart and Winston.

Cawthron, E.R. & Rowell, J.A. (1978). Epistemology and Science Education, *Studies in science_Education, 5*, pp 31-59.

Champagne, A.B. & Klopfer, L.E. (1981). *Using the ConSAT: A Memo to teachers.* (Reports to educators R.T.E. 4). Pittsburgh, Learning Research and Development Center, University of Pittsburgh.

Chi, M., Glaser, R. & Reese, E. (1982). Expertise in problem solving. In R. Sternberg (Ed.). *Advances in the psychology of human intelligence* (Vol 1, pp. 7-76). Hillsdale, NJ: Lawrence Erlbaum.

Cicourel, A.V. (Ed.). (1974). *Language use and school performance.* New York: Academic.

Clandinin, D.J. (1983). *A Conceptualization of image as a component of teacher personal practical knowledge.* Unpublished doctoral dissertation, University of Toronto.

Clandinin, D.J. (1985). Personal practical knowledge: A study of teachers' classroom images. *Curriculum Inquiry, 15(4)*, 361-385.

Clandinin, D.J. & Connelly, F.M. (1995). *Teachers' professional knowledge landscapes, (Advances in Contemporary Educational Thought Series, Volume 15,* Jonas F. Soltis, Editor.). New York: Teachers College Press.

Clandinin, D.J. & Connelly, F.M. (2000). *Narrative inquiry: Experience and story in qualitative research.* New York: Jossey-Bass.

Claparede, E. (1971). Die Entdeckung der Hypothese. In: Carl Friedrich Graumann (Hrsg.), Denken. Koln/Berlin.

Clark, C.M. (1978-79). A new question for research on teaching. *Educational Research Quarterly, 3, 4*, 53-58.

Clark, C.M. (1980). Choice of a model for research on teacher thinking. *Journal of curriculum studies (12),* 41-47.

Clark, C.M. (1984, June). *Research in the service of teaching.* Paper presented to the Contexts of Literacy Conference, Snowbird, Utah.

Clark, C.M. & Peterson, P.L. (1984). Teachers' Thought Processes. In M.C. Wittrock (Ed.). *Handbook of research on teaching. third edition.* New York: Mcmillan (in press).

Clark, C.M. (1995). *Thoughtful teaching.* London: Cassell.

Clark, C.M. & Florio, S. (1983). The Written Literacy Forum: Combining research and practice. *Teacher Education Quarterly, 10(3).*

Clark, C.M. & Lampert, M. (1985). *What knowledge is of most worth to teachers? Insights form studies of teacher thinking.* (Occasional Paper No. 86). East Lansing, MI: Michigan State University. Institute for Research on Teaching.

Clark, C. & Peterson, P. (1986). Research on teacher thinking. In M. Wittrock (Ed.), *Handbook of research on teaching* (3rd Ed). New York: Macmillan.

Clark, C.M. & Peterson, P.L. (1986). Teachers' thought processes. In M. Wittrock (Ed.), *Handbook of Research on Teaching*, Third Edition. (pp. 255-295). New York: Mcmillan.

Clark, C.M., Wildfong, S. & Yinger, R.Y. (1978). *Identification of salient features of language and activities*, Mimeographed, East Lansing, Michigan State University.

Clark, C.M. & Yinger, R.J. (1971). Research on Teacher Thinking, *Curriculum Inquiry, 7*, No. 4.

Clark, C.M. & Yinger, R. J. (1977). Research on Teacher Thinking. *Curriculum Inquiry 7, 4*, 279-304.

Clark, C.M. & Yinger, R.J. (1977). Teacher thinking, In: P.L. Peterson & H.J. Walberg, *Research on teaching,* McCutchan, Berkeley.

Clark, C.M. & Yinger, R.J. (1979). *Three studies on teacher planning*. East Lansing, Institute for Research on Teaching, Michigan State University.

Clark, C.M. & Yinger, R.J. (1980). *The hidden world of teaching: implications of research on teacher planning*. East Lansing, Michigan: Institute for Research on Teaching, Michigan State University. (Research Series No. 77).

Cogan, M.L (1973). *Clinical Supervision*, Boston, Mass.

Coates, T.M. & Thoresen, C.E. (1976). Teacher anxiety: A review with recommendations. *Review of Educational Research, 46(2),* 159-184.

Collins, C. (1979). The pragmatic rationale for educational research in its phenomenological horizons. Paper presented at A.E.R.A.

Combs, A.W. (1967). *The professional education of teachers: A case study and a theoretical analysis*. Boston: Allyn & Bacon.

Cone, R. (1978, March). *Teachers' decisions in managing student behavior: A laboratory simulation of interactive decision making by teachers*. Paper presented at the annual meeting of the American Educational Research Association, Toronto, Canada.

Confrey, J. (1982). Content and pedagogy in secondary schools. *Journal of Teacher Education, 33(1)*, 13-16.

Connelly, F.M. & Clandinin, D.J. (1984). Personal practical knowledge at Bay Street School: Ritual, personal philosophy and image. In R. Halkes & J.K. Olson (Eds.), *Teacher thinking: A new perspective on persisting problems in education* (pp. 134-148). Lisse: Swets and Zeitlinger B. V.

Connelly, F.M. & Clandinin, D.J. (1984). Teachers' personal practical knowledge. In R. Halkes & J. Olson (Eds.), *Teacher thinking: A new perspective on persisting problems in education.* Heirewig, Holland: Swets.

Connelly, F.M. & Clandinin, D.J. (In press). On narrative method, personal philosophy and narrative unities in the study of teaching. *Journal of Research in Science Teaching.*

Connelly, F.M. & Clandinin, D.J. (1985). Personal practical knowledge and the modes of knowing: Relevance for teaching and learning. *NSSE Yearbook, 84(2)*, 174-198. In Elliott W. Eisner (Ed) *Learning the Ways of Knowing*, the 1985 Yearbook of the National Society for the Study of Education. Chicago: The University of Chicago Press, (in press).

Connelly, F.M. & Clandinin, D.J. (1988). *Teachers as curriculum planners: Narratives of experience*. New York: Teachers College Press.

Connelly, F.M. & Clandinin, D.J. (1990). Stories of experience and narrative inquiry. *Educational Researcher 19 (5)*:2-14.

Connelly, F.M. & Clandinin, D.J. (1999). *Shaping a professional identity: Stories of educational practice*. New York: Teachers College Press.

Connelly, F.M. & Clandanin, D.J. (1983). Studies in personal practical knowledge: image and its expression in practice. A paper presented to the founding conference of the International Study Association on Teacher Thinking, Tilburg, Holland.

Connelly, F.M. & Elbaz, F. (1980). Conceptual bases for curriculum thought. *A.S.C.D. Yearbook, 1980.*

Cooper, M., Burger, J. & Seymoer, G. (1979). Classroom context and student ability as influences on teacher perceptions of classroom control. *American Educational Research Journal, 16*, 189-196.

Creemers, B.P. & Westerhof, K.J. (1982). Onderzoek naar routines in het leerkrachtengedrag. In: Halkes, R. & Nijof, W.J. (Eds.), *Planning van onderwijzen*. Lisse, Swets & Zeitlinger.

Cruickshank, D.R., Kennedy, J.J. & Meyers, B. (1974). Perceived problems of secondary school teaches. *The Journal of Educational Research, 68*, 154-159.

Cusick, P. (1982). *A study of networks among professional staffs in secondary schools* (Research Series No. 112). East Lansing: Michigan State University, Institute for Research on Teaching.

Cvitkovik, M. (1986). Literacy: The problems in literacy seminar. McLuhan Program in Culture and Technology *Newsletter*, *7*, March 1, 3.

Dahlgren, L.O. (1978). Vuxen att studera. Stencil; Grown-up to study.

Dahlgren, L.O. (1978:65). Effects of university education on the conception of reality. Paper presented at the fourth international conference on improving university teaching, July 26-29, Aachen, FRG. *Reports from the Institute of Education*, University of Göteborg.

Dahlgren, L.O. (1979). Forskning kring undervisning och inlarning i vuxenuthildningen. Stencil; Research on teaching and learning in adult education.

Dahlgren, L.O., Larsson, S., Liden, E. & Saljo, R. (1978:162). Deltagarnas kunskaper och erfarenheter som resurs i vuxenuthildning - en projektpresentation. *Rapporter från Pedagogiska institutionen, Göteborgs universitet.* Students' knowledge and experience as a resource in adult education - a presentation of a project.

Dann, H.-D. (1982). Was man bei der Uberprüfung der Handlungswirksamkiet subjektiver Theorien alles falsch machen kann. In H.-D. Dann, W. Humpert, F. Krause & K.-C. Tennstädt (Eds.), Analyze und Modifikation subjektiver Theorien von Lehrern (pp. 184-193). Forschungsbericht 43. Konstanz: Universität Konstanz, Zentrum I Bildungsforschung SFB 23.

Dann, H.-D. (1983). Subjektive Theorien: Irrweg oder Forschungsprogramm? Zwischenbilanz eines kognitiven Konstrukts. In L. Montada, K. Reusser & G. Steiner (Edo.). *Kognition und Handel* (pp. 77-92). Klett-Cotta.

Dann, H.-D., Humpert, W., Krause, F., Olbrich, Ch. & Tennstädt, K.-Ch. (1979). Bericht des Teilprojekts "Aggression in der Schule". In Zentrum I Bildungsforschung/SFB 23 (Ed.), Wissenschaftlicher Arbeits- und Ergebnisbericht 1976-1978 (pp. 161-200). Universität Konstanz.

Dansereau, D.F., McDonald, B.A., Collins, K.W., Garland, J., Helley, C.D., Diekhoff, G.M. & Evans, S.H. (1979). Evaluation of a learning strategy system. In: O'Neill, H.F. and Spielberger, G.D. (Eds.), *Cognitive and affective learning strategies*. New York, Academic Press.

Darroch-Lozowski, V. (1982). Biographical narrative as the expression of existence. In V. Darroch & R.J. Silvers (Eds.), *Interpretive human studies* (pp. 215-227). Washington, D.C.: University Press of America, Inc.

Day, C. (1981). *Classroom-based in-service teacher education: the development and evaluation of a client centred model.* Occasional Paper 9, University of Sussex Education Area, Brighton, UK.

De Corte, E. (1979). Objecten,doelen en methodolgie van de onderwijspsychologie. *Tijdschrift voor Onderwijsresearch*, 4, 5, 209-218.

De Corte, E. (1982). Het speurwerk over het plangedrag van leerkrachten onder de loep. In: Halkes, R. & Nijhof, W.J. (Eds). *Planning van Onderwijzen.* Lisse: Swets & Zeitlinger, 77-93.

Deforges, C. & McNamara, D. (1977). One man's heuristic is another man's blindfold: some comments on applying social science to educational practice. *British Journal of Teacher Education*, *3*, *1*, 27-39.

Deforges, C. & McNamara, D. (1979). Theory and practice: methodological procedures for the objectification of craft knowledge. *British Journal of Teacher Education*, *5*, *2*, 145-152.

Denicolo, P. (1983). (Personal communication).

Denzin, N.K. (1978). *The research act: a theoretical introduction to sociological methodology*. New York, McGraw-Hill.

Deutsches Institut für Fernstudien. (1979). Fernsehkolleg Schülerprobleme – Lehrerprobleme: 1. Mangelnde Mitarbeit, 2. Schlechte Leistungen, 3. Gestörter Unterricht. Tübingen: DIFF.

Dewey, J. (1910). *How to think.* Boston, Heath.

Dewey, J. (1971). *How we think: A restatement of the relation of reflective thinking to the education process.* Chicago, IL: Henry Regnery. (Original work published 1933).

Diamond, C.T.P. (In Press). Fixed Role Treatment: An invitation to conjecture. In Keen, T and Pope M (Eds). *Kelly in the Classroom.* Cybersystems Ltd., Canada.

Dobslaw, G. (1983). Zum *Aufgabenverständnis von Schülern im Mathematikunterricht – eine Untersuchung su Lehrererklärungen über den Verstehensproseß von Schülern.* Unpublished master's thesis, Bielefeld University, Bielefeld.

Domas, S.J. & Tiedeman, D.V. (1950). Teacher competence: an annotated bibliography. *Journal of Experimental Education*, 19, 101-218.

Donaldson, M. (1978). *Children's Minds*. London: Fontana.

Donaldson, M. (1978). *Children's minds*. Glasgow: W. Ollins.

Douglas, J.D. (Ed.) (1970). *Understanding everyday life*. Chicago: Aldine.

Doyle , W. (1977). *Learning the classroom environment: An ecological analysis of induction into teaching*. Paper presented at the annual meeting of the AERA, New York. (ERIC Document Reproduction Service No. ED 135782).

Doyle, W. (1979). Classroom tasks and students' abilities in Peterson, P. L. & Walberg, H. J. (Eds) Research on Teaching: concepts, findings and implications. Berkeley, California: McCutchan.

Doyle, W. (1979). *The task of teaching and learning in classrooms*. (R&D Rep. No. 4103). Austin: Research and Development Center for Teacher Education, The University of Texas at Austin.

Doyle, W. (1980). Research on teaching in classroom environments. In G.E. Hall, S.M. Hord & G. Brown (Eds.), *Exploring issues in teacher education: questions for future research* (pp. 501-517). Austin, Texas: Research and Development Centre for Teacher Education, University of Texas at Austin.

Doyle, W. & Ponder, C.A. (1976). *The Practicality Ethic in Teacher Decision-Making*, Mimeo, North Texas State University, Denton, Texas, USA.

Doyle, W. & Ponder, G.A. (1977/78). The practicality ethic in teach decision making. *Interchange, 8 (3),* 1-12.

Driver, R. (1973). *The representation of conceptual frameworks in young adolescent science students*. Ph.D. Thesis, Illinois: University of Illinois.

Duffy, G. (1977). *A Study of teachers' conceptions of reading*. Paper presented to the National Reading Conference, New Orleans.

Dunkin, M.J. & Biddle, B.J. (1974). *The study of teaching*. New York, Holt, Rinehart & Winston.

Edwards, R. (1979). *Contested terrain: The transformation of the workplace in the 20th century*. New York: Basic Books.

Eggleston, J. (Ed.). (1979). *Teacher decision making in the class*. London: Routledge.

Einhorn, H.J. & Hogarth, R.M. (1981). Behavioral decision theory: Process of judgment and choice. *Annual Review of Psychology, 32*, 53-88.

Eisner, E. (1979). *Cognition and Curriculum*. New York: Longman.

Eisner, E.W. (1983). The art and craft of teaching. *Educational Leadership*, January.

Elbaz, F. (1981). The Teachers 'practical knowledge': Report of a study. *Curriculum Inquiry, 2*, 1981, pp 43-73.

Elbaz, F. (1983). *Teacher thinking: A study of practical knowledge*. London: Croom Helm and New York: Nichols Publishing Company.

Elliott, J. (1976). Preparing Teachers for Classroom Accountability *Education for Teaching*, Summer 1976, Number 100.

Elliott, J. (1977). *Some Key Concepts Underlying Teachers' Evaluation of Innovation*. Mimeo, Cambridge Institute of Education.

Elliott, J. (1982). Institutionalizing Action Research in Schools, in Elliott J and Whitehead D. *Action Research for Professional Development and Improvement of Schooling, 5*, Cambridge C A R N Publications.

Ellis, A. (1977). *Die rational-emotive Therapie*. Munchen.

Endler, N.S. & Magnusson, D. (1976). Personality and person by situation interactions. In: N.S. Endler en D. Magnusson (Eds.), *International psychology and personality*. Hillsdale, N.J.: Erlbaum.
Eraut, M. (1978). Accountability at school level - some options and their implications. In A. Becher and S. Maclure, *Accountability in Education*, NFER Publishing.
Eraut, M. (1982). What is learned in In-Service Education and How. *British Journal of In-Service_Education*, Vol 9, No. 1, Autumn.
Erickson, F. (1986). Qualitative methods in research on teaching. In M. Wittrock (Ed.), *Handbook of Research on Teaching*, Third Edition. (pp. 119-161). New York: Mcmillan.
Ericsson, K.A. (1978). Verbalizing internal states and process: Retrospective verbal reports. Paper presented at the Annual Meeting of the APA, Toronto.
Ericsson, K.A. & Simon, H.A. (1978). Retrospective verbal reports as data. CIP Working Paper No. 388, Pittsburgh.
Ericsson, K.A. & Simon, H.A. (1980). Verbal reports as data. *Psychological Review*, *87*, 215-251.
Fattu, N.A. (1965). A model of teaching as problem solving. In: Macdonald, J.B. & Leeper, R.R. (Eds.), *Theories of instruction*. Washington, Association for Supervision and Curriculum Development.
Feiman, S. & Floden, R.E. (1980). A consumer's guide to teacher development. *The Journal of Staff Development*, *1(2)*, 126-147.
Feiman-Nemser, S. & Floden, R. (1984). *The cultures of teaching* (Occasional Paper No. 74); East Lansing, Michigan: Michigan State University, The Institute for Research on Teaching.
Fenstermacher, G.D. (1980). What needs to be known about what teachers need to know? In G.E. Hall, S. M. Hord and G. Brown (Eds). *Exploring issues in teacher education: Questions for future_researchers*. Austin, Texas. Research and Development Center for Teacher Education.
Ferguson, M. (1982). Karl Pribram's Changing Reality. In: Wilber, K. (ed.), *The Holographic Paradigm*. Boulder & London (Shambala).
Fets, B. (1984). *De reflectie bij leerkrachten over het eigen onderwijzen*. Leuven, Afdeling Didactiek (niet gepubliceerde licentiaatsverhandeling).
Fishbein, M. & Aijzen, I. (1972). Attitudes and Opinions. *Annual Review of Psychology*, *23*, 487-544.
Fitts, P.M. (1965). Factors in complex skill training. In: Glaser, R.M. (Ed.), *Training research and education*. New York, Wiley.
Flanders, N. (1979). *Analyzing Teaching Behaviour*, Mass: Addison-Wesley.
Flanders, T. (1983). Teachers realities, needs, and professional development. In Butt, R.L., Olson, J., and Daignault, J. (Eds.). *Insiders realities, outsiders dreams: Prospects for curriculum change*. Vancouver: Centre for the Study of Curriculum and Instruction and the Canadian Association of Curriculum Studies.
Floden, R.E. (1983, February). *Actively learning to be expert: A new view of learning*. Paper presented at the Annual Meeting of the American Association of Colleges for Teacher Education, Detroit.
Florio, S. & Walsh, M. (1981). The teacher as colleague in classroom research. In H.T. Trueba, G.P.

Fogarty, J., Wang, M. & Creek, R. (1983). A descriptive study of experienced and novice teachers' interactive instructional thoughts and actions. *Journal of Educational Research*, 77, 22-32.

Fransella, F. (1980). Man as scientist. In A. Chapman & D. Jones (Eds.). *Models of Man* (pp. 243-260). Leicester, U.K.: British Psychological Society.

Fraser, R. et al (Undated). Learning Activities and Classroom Roles With and Without the Computer. ITMA Collaboration, University of Nottingham, Xerox.

Fried, C. (1978). *Right and wrong*. Cambridge, MA: Harvard University Press.

Fuller, F.F. (1969). Concerns of teachers: A developmental conceptualization. *American Educational Research Journal, 6(2),* 207-226.

Fuller, F.F. & Bown, O.H. (1975). Becoming a teacher. In K. Ryan (Ed.), *Teacher Education* (pp. 25-52). The 74th yearbook of the N.S.S.E. Chicago: University of Chicago Press.

Fuller, F.F. & Manning, B.A. (1973). Self-Confrontation Reviewed, *Review of Educational Research*, *43*: 469-528.

Fuchs. (1968). How Teachers Learn to Help Children Fail. In N. Keddie (1973). *Tinker .. Tailor .. The Myth of Cultural Deprivation.*

Fuller, F.F. (1969). Concerns of teachers: A developmental characterization. *American Educational Research Journal, 2*, 207-226.

Gage, N.L. (Ed.). (1963). *Handbook of research on teaching*. Chicago, Rand McNally.

Gage, N.L. (1963). Paradigms for research on teaching. In N.L. Gage (Ed.), *Handbook of research on teaching* (pp. 91-141). Chicago: Rand McNally.

Gerris, J.R.M., Peters, V.A.M. & Bergen, T.C.M. (1984). *Educational situations teachers are concerned about: theoretical and empirical elaborations on person-and situation characteristics from a developmental perspective.* Nijmegen: Institute for Teacher Education, University of Nijmegen, (submitted for publication).

Getzels, J.W. & Jackson, P.W. (1963). The teacher's personality and characteristics. In: Gage, N.L. (Ed.), *Handbook of research on teaching.* Chicago, Rand McNally.

Giacquinta, J. (1975). Status risk-taking: a central issue in the initiation and implementation of public school innovations. *Journal of Research and Development in Education*, 8, 1, 102-113.

Glaser, B.G. & Strauss, A.L. (1967). *The Discovery of Grounded Theory*. Chicago.

Glass, G.V. (1976). Primary, secondary and meta-analysis of research. *Educational researcher*, 5, 3-8.

Glaser, R. (1976). Components of a psychology of instruction, toward a science of design. *Review of educational research, (46).* 1-24.

Goffman, E. (1969). *The presentation of self in everyday life*. London.

Good, C.V. (Ed.). (1973). *Dictionary of Education*. New York, McGraw-Hill.

Goodland, J.I. (1983). *A place called school*. New York: McGraw-Hill.

Goodman, P. (1972). *Compulsory Miseducation*. Penguin Books, London.

Goodman, J. (1984). Reflection and teacher education: A case study and a theoretical analysis. *Interchange*, *15(3)*, 9-26.

Gouldner, A.W. (1968). The sociologist as partisan: Sociology and the welfare state. *The American Sociologist*, *3(2)*, 103-116.

Greeno, J.G. (1978). Understanding and procedural knowledge in mathematics instruction. *Educational Psychologist*, *12*, 262-282.

Greeno, J., Glaser, R. & Newell, A. (1983, April). *Summary: Research on cognition and behavior relevant to education in mathematics, science, and technology*. Research paper submitted to the National Science Board Commission on Pre-college Education in Mathematics, Science and Technology by the Federation of Behavioral Psychological and Cognitive Sciences.

Groeben, N. & Scheele, B. (1977). *Argumente fuer eine Psychologie des reflexiven Subjekts,* Damrstadt: Steinkopff.

Grumet, M. (1978). Supervision and situation: A methodology of self-report for teacher education. *Journal of Curriculum Theorizing, 1(1)*, 191-257.

Grumet, M. (1980). Autobiography and reconceptualisation. *Journal of Curriculum Theorizing, 2:2*, pp. 155-158.

Guba, E. & Lincoln, Y. (1981). *Effective evaluation.* San Francisco: Jossey-Bass.

Habermas, J (1972). *Knowledge and Human Interest.* London.

Haley, J. (1963). *Strategies of Psychotherapy*. Grune & Stratton.

Halkes, R. & Deijkers, R. (1984). Teachers' teaching criteria. In R. Halkes & J.K. Olson (Eds.), *Teacher thinking: A new perspective on persisting problems in education* (pp. 149-162). Lisse: Swets & Zeitlinger B.V.

Halkes, R. & Olson, J.K. (Eds.) (1984). *Teacher thinking: a new perspective on persisting problems in education.* Lisse: Swets & Zeitlinger.

Hall, G.E. & George, A.A. (1978). *Stages of concern about the innovation: The concept, verification, and implications*. Austin: University of Texas, Research and Development Center for Teachers Education.

Hall, G.E. & Loucks, S.F. (1978). Teacher concerns as a basis for facilitating and personalizing staff development. *Teachers College Record, 80(1)*, 36-53.

Hall, G. & Rutherford, W. (1976). Concerns of teachers about implementing team teaching. *Educational Leadership*. 227-234.

Hall, G.E., Loucks, S.F., Rutherford, W.L. & Newlove, B.W. (1975). Levels of use of the innovation: A framework for analyzing innovation adoption. *The Journal of Teacher Education, 26(1)*, 52-56.

Hanke, B., Lohmuller, J.-B. & Mandl, H. (1980). Schulerbeurteilung in der Grundschule. Munchen: Oldenburg.

Hargreaves, A. (1982). The rhetoric of school-centered innovation. *Journal of Curriculum Studies, 14(3)*, 251-266.

Harre, R. (1979). *Social Being.* Oxford: University Press.

Hasselgren, B. (1981). *Ways of apprehending children at play.* Göteborg: Acta Universitatis Gothoburgensis.

Hasselgren, B. (1984). "Teacher thinking" - mot en kognitiv undervisningsteknologi" *Inlagg till NFPF:s kongress i Oslo, mars 1984*. Uppsala universitet.

Hawkins, D. (1974). I, thou, and it. In D. Hawkins (Ed.), *The informed vision: Essays on learning and human nature* (pp. 48-62). New York: Agathon Press.

Hayes-Roth, B. & Hayes-Roth, R. (1978). *Cognitive processes in planning*. R-2366-ONR. A report prepared for the Office of Naval Research.

Heath, R.W. & Nielson, M.A. (1974). The research basis for performance-based teacher education. *Review of Educational Research*, 1, 4-16.

Heckhausen, H. & Rheinberg, F. (1980). Lernmotivation im Untericht, erneut betrachtet. Unterrichtswissenschaft 8, 7-47.

Heisenberg, W. (1971). *Physics and Beyond.* New York, Harper & Row.
Herrmann, T. (1976). *Die Psychologie un ihre Forschungsprogramme.* Goettingen: Hogrefe.
Heymann, H.W. (1980). *Probleme der Erforschung subjectiver Unterrichtsmethoden von Mathematiklehrern.* Paper presented at the 29th Conference of the AEPF, Munchen.
Highet, G. (1969). *The art of teaching.* New York: Alfred A. Knopf.
Hills, G. (1983, June). *A philosophical look at contemporary conceptions of teaching.* Paper presented at the meeting of the Canadian Society for the Study of Education, Vancouver, B.C.
Hill (Rix), E.A. (1988). Understanding the disoriented senior as a personal scientist. Chapter D3 in F. Fransella and L. Thomas, (Eds.) *Experimenting with Personal construct Psychology.* London: Rutledge & Kegan Paul.
Hofer, M. (1975). Die Validitat der impliziten Personlichkeitstheorie von Lehrern. Unterrichtswissenschaft 3, 5-18.
Hofer, M. (1977). *Entwurf einer Heuristik fur eine theoretishc geleitete Lehrer- und Erzieherbildung.* Diskussingspapier Nr. 10. Bericht aus dem Psychologischen Istitut der Universitat Heidelberg.
Hofer, M. (1981). *Informationsverarbeitung und Entscheidungsverhalten von Lehrern.* München: Urban und Schwarzenberg.
Hofer, M. (1981). Schuelergruppierungen in Urteil und Verhalten des Lehrers. In M. Hofer (Ed.). *Informationsverarbeitung und Entscheidungsverhalten yon Lehrern* (pp. 192-222). Muenchen: Urban u. Schwarzenberg.
Hofer, M. (1986). *Sozialpsychologie erzieherischen Handelns.* Göttingen: Verlag für Psychologie, Dr. C.J. Hogrefe.
Hofer, M. & Dobrick, M. (1978). Die Rolle der Fremdattribution von Ursachen bei der Handlungssteuerung des Lehrers. In D. Gorlitz, W.-U. Meyer & B. Weiner (Eds.), Bielefelder Symposium uber Attributionen (pp. 51-69). Stuttgart: Klett-Cotta.
Hofer, M. & Dobrick, M. (1981). Naïve Ursachenzuschreibung und Lehrerverhalten. In M. Hofer (Ed.), Informationsverarbeitung und Entscheidungsverhalten von Lehrern (pp. 11-158). Munchen: Urban & Schwarzenberg.
Hofer, M., Dobrick, M., Tacke, G., Pursian, R., Grobe, R., Preuss, W. & Rosner, H. (1982). Bedingungen und Konsequenzen individualisierenden Lehrerverhaltens. Abschlubericht an die Deutsche Forschungsgemeinschaft (Ho 649/3). Braunschweig.
Hofman, J. & Kremer, L. (1980). Attitudes towards higher education and course evaluation, *Journal of Educational Psychology 72*, 5: 610-617.
Hoz, R., Mahler, S., Yeheskel, N., Tomer, Y. & Elbaz, F. (1984). *Project for the Evaluation of Teacher Education.* PETE annual report 1984. Beer-Sheva, Ben-Gurion University of the Negev.
Huber, G.L. & Mandl, H. (1977). Konzeptionsrahmen fur einen Fernstudien-Lehrgang "Lehrertraining: Probleme entwicklungs-, verhaltens- und lerngestorter Schuler." Papier im Auftrag des Deutschen Istituts fur Ferstudien. Tubingen.

Huber, G.L. & Mandl, H. (1979). Spiegeln Lehrerurteile uber Schuler die implizite Personlichkeitsstruktur der Beurteiler oder der Beurteilungsbogen? Zeitschrift fur Entwicklungspsychologie und Padagogische Psychologie 11, 218-231.

Huber, G.L. & Mandl, H. (1982). *Methodological questions in describing teacher cognitions.* Paper presented at the annual meeting of the American Educational Research Association, New York.

Huber, G. & Mandl, H. (Eds.). (1982). *Verbale Daten.* Weinheim: Beltz.

Huber, G.L. & Mandl, H. (1982). Verbalisationsmethoden zur Erfasung von Kognitionen im Handlungszusammenhang. In G.L. Huber & H. Mandl (Eds.), Verbale Daten: Erhebung und Auswertung (pp. 11-42). Weinheim: Beltz.

Huber, G. & Mandl, H. (1984). Access to teacher cognitions: problems of assessment and analysis. In: Halkes, R. & Olson, J.K. (Eds.). *Teacher thinking: a new perspective on persisting problems in education.* Lisse: Swets & Zeitlinger.

Hughes, M.M. Utah study of the assessment of teaching. In: Bellack, A.A. (Ed.). (1970). *Theory and research in teaching.* New York, Teachers College Press.

Hundeide, K. (1977). *Piaget i kritisk lys.* Oslo: Cappelens Forlag. Piaget in critical light.

Hunt, D.E. (1976). Teachers are psychologists, too: on the application of psychology to education. *Canadian Psychological Review, 17*, 210-218.

Hunt, D.E. (1982). How to be your own best theorist.. *Theory into Practice, 19, 4*, 287-293.

Hunter, M. (1980). Six types of supervisory conferences, *Educational Leadership*, *5*: 408-413.

Hutchins, R.M. (1936). *The Higher Learning in America.* Yale University Press, New Haven, Conn.

Illich, I. (1971). *De-schooling Society.* Harper and Row, New York.

Jackson, P.W. (1964). The conceptualization of teaching. *Psychology in the schools*, 232-243.

Jackson, P.W. (1968). *A process analysis of teaching.* Leuven, K.U. Leuven, Departement Pedagogische Wetenschappen, Afdeling Didactiek en Psychopedagogiek, Report No. 21 (EDRS-ED 190513).

Jackson, P.W. (1968). *Life in classrooms.* New York: Holt, Rinehart and Winston.

Jaeggi, E. (1979). Kognitive Verhaltenstherapie. Weinheim: Beltz.

Jae-on K. & Muller, C.W. (1978). *Factor analysis, statistical methods and practical issues.* London: Sage Publications.

James, W. (1969). The moral philosopher and the moral life. In J.K. Roth (Ed.), *The moral philosophy of William James.* New York: Thomas Y. Crowell.

Janesick, V.J. (1981). Developing grounded theory: Reflections on a case study of an architectural design curriculum. In Schubert, W.H. and Schubert, A.L. (Eds.). *Conceptions of curriculum knowledge: Focus on students and teachers.* Publication of A.E.R.A.-SIG group Creation and Utilization of Curriculum Knowledge.

Janesick, V. (1982). Of snakes and circles: Making sense of classroom group processes through a case study. *Curriculum Inquiry, 12(2)*, 161-189.

Jennings, T.W. (1982). On ritual knowledge. *Journal of Religion, 62, 2*, 111-127.

Jungermann, H. (1983). The two camps on rationality. In R. Scholz (Ed.). *Decision_making under uncertainty* (pp. 63-86). Amsterdam: North Holland.
Kahneman, D., Slovic, P. & Tversky, A. (Eds.). (1982). *Judgment under uncertainty:_Heuristics and biases.* New York: Cambridge University Press.
Katz, L.G. (1972). Developmental stages of preschool teachers. *Elementary School Journal, 73*, 50-54.
Keddie, N. (1971). Classroom Knowledge, Young M F D (Ed). *Knowledge and Control*, Collier-Macmillan.
Kelly, G.A. (1955). *The Psychology of Personal Constructs_*(Vols. 1 and 2). New York, W.W. Norton and Co. Inc.
Kelly, G.A. (1955). *The psychology of personal constructs (2 vols).* New York: Norton.
Kelman, H. (1961). Three Processes of Social Influence, *Public Opinion Quarterly*, Vol 25, Princeton, USA.
Kerlinger, F. (1967). The first- and second-order structures of attitudes toward education. *American Educational Research Journal*, 4, 191-205.
Kerr, D.H. (1983). Teaching competence and teacher education in the United States. In L. Shulman & G. Sykes (Eds.), *Handbook on teaching and policy* (pp. 126-149). New York: Longman.
Kirk, R. (1969). *Experimental Designs: Procedures for the Behavioural Sciences.* Belmont, CA: Brooks/Cole.
Kosslyn, S.M. & Pomerantz, J.R. (1977). Imagery, propositions, and the form of internal representations. *Cognitive Psychology, .9*, 52-76.
Kounin, J. (1970). *Discipline and group management in classrooms.* New York: Holt-Rinehart & Winston.
Krampen, G. (1979). Erziehungsleitende Vorstellungen von Lehrern. *Zeitschrift für experimentelle und angewandte Psychologie.* Band XXVI, 1, 94-112.
Kremer, L & Ben-Peretz, M. (1984). Teachers' self-evaluation – concerns and practices, *Journal of Education for Teaching, 10, 1*, pp. 53-60.
Kroma, S. (1983). *Personal practical knowledge of language in teaching: An ethnographic study.* Unpublished doctoral dissertation University of Toronto.
Krummheuer, G. (1983). Das Arbeitsinterim im Mathematikunterricht. In H. Bauersfeld et al. (Eds.). *Lernen und Lehren von Mathemat1k* (pp. 57-106). Koeln: Aulis.
Kuhl, J. (1983). Emotion, Kognition und Motivation II. Die funktionale Bedeutung der Emotionen fuer das problemloesende Denken und fuer das konkrete Handeln. *Sprache und Kognition, 1*, 228-253.
Kun, A., Parsons, J. & Ruble, D.N. (1974). Development of Intergration Processes Using Ability- and Effort-Information to Predict Outcome. *Developmental Psychology*, *10*, 721-732.
Lacey, C. (1977). *The Socialisation of Teachers*, London, Methuen.
Lakatos, I. (1970). Falsification and the Methodology of Scientific research programmes. In Lakatos, I and Musgrave, A. (Eds) *Criticism and the Growth of Knowledge.* Cambridge University Press.
Lakoff, G. & Johnson, M. (1980). *Metaphors we live by.* Chicago: University of Chicago Press.

Lampert, M. (1985). How do teachers manage to teach? Perspectives on problems in practice. *Harvard Educational Review*, 55, 178-194.

Larkin, J.H., McDermott, J., Simon, D.P. & Simon, H.A. (1980). Models of competence in solving physics problems. *Cognitive Science*, 4, 317-345.

Larsson, S. (1979). Studier i larares omvarldsuppfattning: Betygs inverkan på undervisning eller Pontius Pilatus i klassrummet. *In: Gandrup, P, Kvale, S &_ark, P (Eds): Karakterer - alternativer till karakterer. Rapport från* NFPF-symposium, 23-25 oktober,_1979. Studies in teachers' conceptions of their professional world: Grades impact on the instructional process or Pontius Pilatus in the classroom.

Larsson, S. (1980:06). Studier i larares omvdrldsuppfattning: Deltagarnas erfarenheter som en resurs i vuxenutbildningen. *Rapporter från Pedagogiska institutionen,_Göteborqs universitet.* Studies in teachers' conceptions of their professional world: Students' experiences as a resource in adult education.

Larsson, S. (1982:01). Teachers' interpretation of the concept "experience". In: Abrahamsson, K (ed): Cooperative education, experiential learning and personal knowledge. *UHÄ-rapport.* (a).

Larsson, S. (1981:16). Studier i larares omvarldsuppfattning: Lararskicklighet. *Rapporter fran Pedagogisk institutionen, Göteborgs universitet.* Studies in teachers' conceptions of their professional world: Teaching skill (a).

Larsson, S. (1981:17). Studier i larares omvärldsuppfattning: Särdrag hos vuxna. *Rapporter från Pedagogiska institutionen, Göteborqs universitet.* Studies in teachers' conceptions of their professional world: Characteristics of adult students (b).

Larsson, S. (1982:07). Studier i larares omvärldsuppfattning: Intentioner och restriktioner i undervisningen. *Rapporter från Pedagogi ka institutionen, Göteborgs universitet.* Studies in teachers, conceptions of their professional world: Intentions and restrictions in the instructional process.

Larsson, S. (1982:14). Studier i lärares omvärldsuppfattning: Kunskaps intresse och kunskapssyn i läroämnen. *Rapporter från Pedagogisk institutionen, Göteborgs_universitet.* Studies in teachers' conceptions of their professional world: Cognitive interest and conception of knowledge in subjects taught (c).

Larsson, S. (1983). Teachers' interpretation of the concept of "experience". In: *Boot,_R. & Reynolds, M. (eds): Learning and Experience in Formal Education.* Manchester Monographs (a).

Larsson, S. (1983). Paradoxes in teaching. *Instructional Science, 12*, 355-365. (b).

Larsson, S. (1984). Describing teachers' conceptions of their professional world. In R. Halkes & J.K. Olson (Eds.) *Teacher thinking: A new perspective on persisting problems in education* (pp. 123-133). Lisse: Swets & Zeitlinger B.V.

Laucken, U. (1974). *Naive Verhaltenstheorie.* Stuttgart: Klett.

Laucken, U. (1982). Aspekte der Auffassung und Untersuchung von Umgangswissen. *Schweizer Zeitschrift: fuer Psychologie und ihre Anwendungen, 41*, 07-113.

Lazarus, R.S., Averill, U.R. & Opton, Jr., E.M. (1974). The psychology of coping: Issues of research and assessment. In G.U. Coelho, D.A. Hamburg & J.E. Adams (Eds.), Coping and adaption (pp. 249-315). New York: Basic Books.

Leinhardt, G. (1983, April). *Overview of a program of research on teachers' and students' routines, thoughts, and execution of plans.* Paper presented at the meeting of the American Educational Research Association (AERA).
Leinhardt, G. (1983, April). *Routines in expert math teachers' thought and actions.* Paper presented at the annual meeting of the American Educational Research Association, Montreal.
Leinhardt, G. & Smith, D. (1984). *Expertise in mathematics instruction: Subject matter knowledge.* Paper presented at the annual meeting of AERA, New Orleans.
Leinhardt, G., Weidman, C. & Hammond, K.M. (1984, April). *Introduction and integration of classroom routines by expert teachers.* Paper presented at the annual meeting of the American Educational Research Association, New Orleans.
Leithwood, K. & MacDonald, R. (1981). Reasons given by teachers for their curriculum choices. *Canadian Journal of Education*, 6, 2, 103-116.
Lesgold, A.M. (1983). *Acquiring expertise.* Technical Report PD-5. Pittsburgh, University of Pittsburgh, Learning Research and Development Center.
Lindblad, S. & Hasselgren, B. (1983). Teachers changing teaching. A study of conceptions. Paper presented at the first ISATT-symposium, October 26-28, 1983, Tilbury University, The Netherlands. *Pedagogiska institutionen, Göteborgs universitet.*
Lipsky, M. (1980). *Street level bureaucracy.* New York: Russell Sage.
Little, J.W. (1981). *School success and staff development: The role of staff development in urban desegrated schools* (NIE No. 400-79-0049). Boulder: University of Colorado, Center for Action Research.
Lortie, D. (1975). *School-teacher.* Chicago: University of Chicago Press.
Lowyck, J. (1979). *Prozessanalyse d. Unterrichtsverhaltens.* Unveroffentlichtes Manuskript, presented at the symposium The cognitive approach on teaching" in Leuven.
Lowyck, J. (1980). *A process analysis of teaching.* Leuven, Belgium: Katholieke Universiteit Leuven, Departement Pedagogische Wetenschappen, Afdeling Didactiek en Psychopedagogiek. (Report No. 21).
Lowyck, J. (1982). Het plannen van onderwijzen: een handelingstheoretische benadering. In R. Halkes & W.J. Nijhof (Eds.), *Planning van onderwijzen* (pp. 49-63). Lisse, the Netherlands: Swets & Zeitlinger.
Lowyck, J. (1983). De leerkracht als probleemoplosser: een kritische analyse. In E. De Coret & P. Span, *Studies over anderwijsleerprocessen.* Leuven, Helicon.
Lowyck, J. (1984). Reflectie van leerkrachten na afloop van de les. Negatieve leservaringen stemmen tot nadenken. *Didacktief*, 14, 20-24.
Lowyck, J. (1984). Teacher thinking and teacher routines: a fifurcation? In: Halkes, R. & Olson, J.K. (Eds.). *Teacher thinking: a new perspective on persisting problems in education.* Lisse: Swets & Zeitlinger.
Lowyck, J. & Broeckmans, J. (1985). Technieken voor zelfrapportering in het onderzoek van onderwijzen. In R. Halkes & R.G.M. Wolbert (Eds.), *Docent en methode.* Lisse, Swets & Zeitlinger.
Luer G. (1973). *Gesetzmissige Denkablufe beim Problemlosen.* Weinheim.

Lundgren, U.P. (1972). *Frame factors and the teaching process*. Stockholm: Almquist & Wiksell.
Lydecker, A.F. (1981). *Teacher planninq of social studies instructional units*, Paper presented at the AERA meeting, Los Angeles.
Macdonald, B. & Ruddock, J. (1971). Curriculum research and development projects: barriers to success. *British Journal of Educational Psychology, 41.*
MacIntyre, A. (1982). *After Virtue*. Notre Dame: University of Notre Dame Press.
Mackay, D.A. & Marland, P. (1978). Thought processes of teachers. *Eric reports*, ED 151 328. Washington.
Mackay, D. A. & Marland, P. W. (1978). *Thought processes of teachers.* Paper presented at the American Educational Research Association conference, Toronto.
Magoon, A.J. (1977). Constructivist approaches in Educational Research *Review of Educational Research*_Vol. 47, No. 4, pp 651-693.
Mandl, H. & Huber, G.L. (1983). Subjektive Theorien von Lehrern. Psychologie in Erziehung und Unterricht 30, 98-112.
Marland, P. (1977). *A study of teachers' interactive thoughts*. Unpublished doctoral dissertation, The University of Alberta, Edmonton.
Marland, P.W. (1983). *Models of teachers' interactive thinking*. Unpublished paper, James Cook University, Townsville, Australia.
Marton, F. & Svensson, L. (1978:158). Att studera omvärldsuppfattning. Två bidrag till metodologin. *Rapport från Pedagogiska institutinen, Göteborgs universitet.* To study conceptions of the world around us.
Marton, F. (1981). Phenomenography - describing conceptions of the world around us. *Instructional Science, 10*,_177-200.
Marx, R.W. (1978). *Teacher judgements of students' cognitive and affective outcomes*. Unpublished doctoral dissertation, Stanford University.
Marx, R. & Peterson, P. (1980). The nature of teacher decision making, in: B.R. Joyce & C.C. Brown, *Flexibility in teaching*, Longman, New York.
McCutcheon, G. (1980). How do elementary school teachers plan? The nature of planning and influences on it *The Elementary School Journal, 81*, 4-23.
McKeon, R. (1952). Philosophy and action. *Ethics, 62, 2*, 79-100.
McKloskey, M., Caramazza, A. & Green, B. (1980). Curvilinear motion in the absence of external forces: Naïve beliefs about the motion of objects. *Science*, 210, 1139-1141.
McLaughlin, M.W. & Marsh, D.D. (1978). Staff development and social change. *Teachers College Record*, 80(1), 69-94.
McNair, K. & Joyce, B. (1979). *Teachers' thoughts while teaching: the South Bay study, part 2.* Research report of the Institute of Research on Teaching, University of Michigan, East Lansing.
McNamara, D. & Desforges, C. (1978). The social sciences, teacher education and the objectification of craft knowledge. *British Journal of Teacher Education, 4(1),* 1-20.
Mead, G.H. (1938). *The philosophy of the act*. Chicago: University of Chicago Press.
Meichenbaum, D. (1977). *Kognitive Verhaltensmodifikation.* Munchen.

Meichenbaum, D.W. & Butler, L. (1979). Cognitive ethology: The streams of cognition and emotion. In K. Blankstein, P. Pliner & J. Polivy (Eds.), Advances in the study of communication and affect: Assessment and modification of emotional behaviours. (Vorpublikation) Vol. 6. New York: Plenum Press.

Merton, R. (1957). The role-set: Problems in sociological theory. *British Journal of Sociology*, *8*, 106-120.

Michener, E.R. (1978). Understanding mathematics. *Cognitive Science*, 2 361-383.

Miles, M.B. (1981). Mapping the Common Properties of Schools, in Lehming R; Kane M; *Improving Schools: Using what we know*. Sage Publications, Beverley Hills.

Miles, M.B. & Huberman, A.M. (1984). *Qualitative Data Analysis*, Beverly Hills, Sage Publications.

Miller, G.A., Galanter, E. & Pribram, K.H. (1960). *Plans and the structure of behaviour*. New York: Holt, Rinehart and Winston.

Miller, G.A., Galanter, E. & Pribram, K.H. (1973). Strategien des Handelns. Stuttgart: Klett.

Mintz, S. & Yarger, S.J. (1980). *Conceptual level and teachers' written plans*, Paper presented at the AERA meeting, Boston.

Mischel, T. (1964). Personal constructs, rules and the logic of clinical activity. *Psychological Review*, *71*, 180-192.

Mischel, W. (1973). Toward a cognitive social learning reconceptualization of personality. *Psychological Review, 80*, 252-283.

Morine, G. (1973). Planning skills: paradox and parodies. *Journal of teacher education*, 2, 135-143.

Morine, G. (1976). *A Study of teacher planning*, BTES special study C, Far West Laboratorium, San Francisco.

Moustakas, C. (1972). *Teaching as learning*, New York, Ballantine Books.

Müller-Fohrbrodt, G., Dann, H.D. & Cloetta, B. (1978). *Der Praxisschock bei jungen Lehrern*. Stuttgart: Klett Verlag.

Munby, H. (1982). *A qualitative study of teachers' beliefs and principles*. Austin.

Munby, H. (1983, April). *A qualitative study of teachers' beliefs and principles*. Paper presented at the annual meeting of the American Educational Research Association, Montreal, Canada.

Munchen Russell, B. (1956). *A Logic and Knowledge*. London.

National Institute of Education (NIE) (1975). *Teaching as clinical information processing*. Washington D.C.: NIE Conference on studies in teaching, Panel 6.

Neisser, U. (1976). *Cognition and reality*. San Francisco: Freeman.

Newell, A. (1973). Production systems: Models of control structures. In W.C. Chase (Ed.). *Visual information processing* (pp. 463-526). New York: Academic.

Newell, A. & Simon, H.A. (1972). *Human problem solving*. Englewood Cliffs, Prentice-Hall.

NIE (1975). *NIE Conference on studies in teaching. Panel 6: teaching as clinical information processing*. Washington, U.S. Department of Health, Education and Welfare.

Nisbett, R.E. & Wilson, T.C. (1977). Telling more than we can know: Verbal reports on mental processes. Psychological Review 84, 231-259.

Novak, J.D. (1980). *Handbook for the learning how to learn program.* Ithaka, Cornell University.
Novak, J.D. (1982). Psychological and Epistemological Alternatives, In: Modgil, S. and Modgil, C. (Eds.) *Jean Piaget: Consensus and controversy.* London, Holt, Rinehart and Winston, 331-349.
Nuthall, G. & Snook, I. (1973). Contemporary models in teaching. In : Travers, R.M. (Ed.), *Second handbook of research on teaching.* Chicago, Rand McNally.
Olson, J. (in press). Microcomputers and the Classroom Order. *New Education.* ERIC 245671.
Olson, D. (1977). From utterance to text: The bias of language in speech and writing, *Harvard Education Review*, 47, 257-281.
Olson, J.K. (1980). Teacher constructs and curriculum change. *Journal of Curriculum Studies, 12, 1*, 1-11.
Olson, J.K. (1980 April). *Teacher constructs and curriculum change: Innovative doctrines and practical dilemmas.* Paper presented to annual meeting of American Education Research Association, Boston, Massachussets.
Olson, J.K. (1981). Teacher influence in the classroom: A context for understanding curriculum translation. *Instructional Science, 10*, 259-275.
Olson, J.K. (1982). Classroom knowledge and curriculum change. In: Olson, J.K. (Ed.). *Innovation in the Science Curriculum.* London: Croom Helm.
Olson, J.K. (1983). A reflexive conception of change and its consequences for innovation activity. Paper given to the OTG-Onderwijsvernieuwingen. Tilburg University, The Netherlands.
Olson, J. & Russell, T. (Eds.). (1983). *Science education in Canadian schools Vol. III: Case studies.* Ottawa: Science Council of Canada.
Olson, J. & Reid, W. (1982). “Studying Innovations in Science Teaching: The Use of Repertory and Techniques in Developing a Research Strategy.” *European Journal of Science Education*, 4, 193-201.
Olson, D., Torrence, N. & Hildyard, A., (Eds.) (1985). *Literacy, language and learning: The nature and consequences of reading and writing.* Cambridge U.K.: Cambridge University Press.
O’Shea, T. (1984). “The Open University Micros in Schools Project.” CAL Research Group Technical Report No. 47.
Otte, M. (1983). Textual Strategies. *For the Learning of Mathematics, 3* (3), 15-28.
Parry, R. (1978). Elementary school principal effectiveness: perceptions of principals and superintendents. Unpublished Ed.D. thesis, University of Toronto.
Peters, J. et al. (1982). *Onderzoek naar voorbereiding en uitvoering van leerkrachten.* Groningen, the Netherlands: Rijksuniversiteit Groningen, Vakgroep Onderwijskunde.
Peters, J.J. (1984). Teaching: intentionality, reflection and routines. In R. Halkes & J.K. Olson (Eds.), *Teacher thinking: a new perspective on persisting problems in education.* Lisse, Swets & Zeitlinger, 19-34.
Peters, T.J. & Waterman, R.H. (1982). *In search of excellence.* New York: Harper & Rowe.
Peterson, P. L. & Clark, C. M. (1978). Teachers’ reports of their cognitive processes during teaching *American Educational Research Journal, 15*, 555-565.

Peterson, P.L., Marx, W. & Clark, C.M (1978). Teacher planning, teaching behavior and student achievement, *American Educational Research Journal, (3),* 417-432.

Philips, S.U. (1972). *The invisible culture: Communication in classroom and community on the Warm Springs Indian Reservation.* New York: Longman.

Piaget, J. (1967 (1927)). *The child's conception of the world.* London: Routledge & Kegan Paul Ltd.

Pinar, W. (1980). Life history and educational experience. *Journal of Curriculum Theorizing, 2:2*, pp. 59-212.

Pinar, W. (1981). Life history and educational experience. *Journal of Curriculum Theorizing, 3:1*, pp. 259-286.

Pinar, W. F., Reynolds, W. M., Slattery, P., & Taubman, P. (1995). *Understanding curriculum.* N.Y: Peter Lang.

Polanyi, M. (1958). *Personal knowledge: Towards a post-critical philosophy.* Chicago: The University of Chicago Press.

Polanyi, M. (1958). *The study of man.* Chicago: University Press.

Polyani, M. (1967). *The Tacit Dimension*, Garden City, New York: Doubleday.

Polya, G. (1966). *Vom Lösen mathematischer Aufgaben* (2 Bände). Basel: Birkhäuser.

Ponder, G. & Doyle, W. (1977). Teacher practicality and curriculum change: an ecological analysis. Paper presented at the AERA annual meeting. New York.

Pope, M.L. (1978). *Constructive Alternatives in Education*, Ph.D. Thesis, Brunei University.

Pope, M.L. (1981 August). *In True Spirit: Constructive Alternativism in Educational Research.* Paper presented at the 4th Int. Congress on Personal Construct Psychology. Brock University, St. Catharines, Canada.

Pope, M.L. & Denicolo, P.M. (2001). *Transformative education: Personal construct approaches to practice and research.* London: Whurr.

Pope, M.L. & Gilbert, J K (1983). Personal experience and the construction of knowledge in science. *Science Education, 67 (2),* pp 193-203.

Pope, M.L. & Keen, T R (1981). *Personal Construct Psychology and Education.* Academic Press, London.

Pope, M.L. & Scott, E. (1984). Teachers' epistemology and practice. In R. Halkes & J.K. Olson (Eds.), *Teacher thinking: A new perspective on persisting problems in education* (pp. 112-122).

Popkewitz, T.S., Tabachnik, B.R. & Zeichner, K.M. (1979). Dulling the senses: research in teacher education. *Journal of Teacher Education*, *30, 5*, 52-61.

Posner, G., Strike, K., Hewson, P. & Gertzog, W. (1982). Accommodation of a scientific conception: Toward a theory of conceptual change. *Science Education*, 66(2), 211-227.

Postman, N. & Weingartner, L. (1971). *Teaching as a Subversive Activity.* Penguin Books, London.

Potter, D.A. (Ed.) (1973). *A critical review of the literature : teacher performance and pupil growth.* Princeton, Educational Testing Service.

Preece, P.F.W. (1976). Mapping cognitive structure: A comparison of methods. *Journal of Educational Psychology*, 68, 1-8.

Preece, P.F.W. (1978). Exploration of semantic space: Review of research on the organization of scientific concepts in semantic memory. *Science Education*, 62, 547-562.

Purkey, S.C. & Smith, M.S. (1983). Effective schools: A review. *The Elementary School Journal*, 83(4), 427-452.

Ragsdale, R. (1982). *Computers in the Schools: A Guide for Planning*. Toronto: OISE.

Rathje, H. (1982). *Untersuchungen zur Integration von schülerbezogenen Leistungsfaktoren im prognostischen Urteil von Lehrern.* Unveröffentlichte Diplomarbeit. Technische Universität Braunschweig.

Resnick, L.B. & Ford, W.W. (1981). *The psychology of mathematics for instruction*. Hillsdale, N.J.: Lawrence Erlbaum.

Rheinberg, F. (1983). Kognitionspsychologische Ansatze zur Analyse des Lehrerverhaltens. Habilitationsvortrag. Fakultat fur Philosophie, Padagogische Psychologie, Universitat Bochum.

Rheinberg, F. & Elke, G.E. (1978). Wie naiv ist die 'naïve' Psychologie von Lehrern? In Eckensberger, L.H. (Ed.). *Bericht über den 31. Kongreß der Deutschen Gesellschaft für Psychologie* (pp. 45-48). Göttingen: Hogrefe.

Rheinberg, F. & Elke, G.E. (1979). Wie naiv ist die "naïve" Psychologie von Lehrern. In L. Eckensberger (Ed.), Bericht uber den 31. Kongreb der Deutschen Gesellschaft fur Psychologie in Mannheim, Vol. 2 (pp. 45-47). Gottingen: Hogrefe.

Rheinberg, F. & Hoss, J. (1979). Störungen and Mitarbeit im Unterricht. Eine Erkundungsstudie zu Kounins Kategorisierung des Lehrerverhaltens. *Zeitschrift für Entwicklungspsychologie und Pädagogische Psychologie*, 11, 244-249.

Rix, E.A. (1980). Understanding promotion criteria through understanding personal constructs. Unpublished Ph.D. thesis, University of Toronto.

Rix, E.A. (1981). Dynamics of promotion decisions. *Comment on education, 12*: *1*, 4-12.

Rix, E.A. (1982). Sift and shift effect: a cognitive process in understanding the psychological dynamics of administrators' personnel decisions. *Interchange*, 13:4, 27-30.

Rix, E.A.H. (1983 July). *Towards a reflective Epistemology of Educational practice: Canadian_Educational applications of PCP* Paper presented at the 5th International Congress on Personal Construct Psychology, Boston, Mass.

Rix, E.A. (1985). Towards a reflective epistemology of educational practice: Canadian applications of PCP. In *Anticipations of Personal Construct Psychology* (A.W. Landfield and F. Epting, Eds.). Lincoln, Nebraska: Univ. of Nebraska Press, (in press).

Rix, E.A. & Parry, R. (1981). Principal candidates' changing perceptions of promotability, paper to Can. Soc. for Study of Ed.

Rogers, C. (1973). Die *Klient-bezogene Gesprachstherapie.*

Rosenshine, B. (1971). *Teaching behaviours and student achievement*. London: National Foundation for Educational Research.

Rosenshine, B. (1979). Content, time and direct instruction. In: Peterson, P. & Walberg, H. (Eds.). *Research on teaching*. Berkeley: McCutchan.

Rosenshine, B. & Furst, N. (1971). Research on teacher performance criteria. In: Smith, B.O. (Ed.), *Research in teacher education.* Englewood Cliffs, Prentice-Hall.

Rosenshine, B. & Furst, N. (1973). The use of direct observation to study teaching. In: Travers, R.M. (Ed.), *Second handbook of research on teaching.* Chicago, Rand McNally.

Ross, C. (1970). *Ross Educational Philosophical Inventory.* ERIC. Document Reproduction Service No. ED 053995.

Rotering-Steinberg, S. (Ed.). (1981). Fersehkolleg Schulerprobleme – Lehrerprobleme. Ein Training fur schwierige Situationen in der Klasse. Bericht uber die Pilotphase. Tubingen: DIFF.

Roth, K. (1984). Using classroom observations to improve science teaching and curriculum materials. In C.W. Anderson (Ed.), *Observing science classrooms: Perspectives from research and practice.* 1984 Yearbook of the Association for the Education of Teachers in Science. Columbus, OH: ERIC Center for Science, Mathematics, and Environmental Education.

Ryans, D.C. (1960). *Characteristics of Teachers, Their Description, Comparison and Appraisal*, Washington, D.C., American Council on Education.

Ryle, G. (1949). *The concept of mind.* London: Hutchinson.

Säljö, R. (1981). Skattesystemet i medborgarperspektiv. II. Hur betala vi (direkt) skatt? *Bidrag till symposiet "Omvärldsuppfattning", Tylösand, 21-23 oktober*, 1981. The tax-system from the perspective of the citizens.

Säljö R. (1982). *Learning and understanding.* Göteborg: Acta Universitatis Gothoburgensis.

Schafer, R. (1981). Narration in the psychoanalytic dialogue. In W.J.T. Mitchell (Ed.), *On narrative* (pp. 25-49). Chicago: The University of Chicago Press.

Schallert, D.L. (1982). The significance of knowledge: a synthesis of research related to schema theory. In: Otto, W. and White, S. (Eds.) *Reading expository material.* New York: Academic Press.

Schank, R. & Abelson, R. (1977). *Scripts, plans, goals, and understanding.* Hillsdale, N.J., Lawrence Erlbaum Associates.

Scheffler, I. (1965). *Conditions of knowledge.* Glenview, IL: Scott, Foresman.

Scheffler, I. (1977). Justifying curriculum decisions. In A.A. Bellack & H.M. Kliebard (Eds.), *Curriculum and Evaluation* (pp. 497-505). Berkeley, CA: McCutchan.

Schlee, J. (1982). Beobachtung zur Modifikation subjektiver Theorien von Lehrern. In H.-D. Dann, N. Krause & K.C. Tennstadt (Eds.), Analyse und Modifikation subjektiver Theorien von Lehrern. Ergebnisse eines Kolloquims (pp. 213-218). Universitat Konstanz.

Schmidt, W.H. & Buchmann, M. (1983). Six teachers' beliefs and attitudes and their curricular time allocations. *The Elementary School Journal*, 84(2), 162-171.

Schön, D. (1983). *The reflective practitioner.* New York: Basic Books.

Schubert, W.H. & Schubert, A.L. (1982). Teaching curriculum theory. *Journal of Curriculum_Theorizing, 4:2*, pp. 97-111.

Schwab, J.J. (1976a). Education and the state: Learning community. In *The great ideas today* (Annual yearbook of Encyclopedia Britannica, pp. 234-271). Chicago, IL: Encyclopedia Britannica.

Schwab, J.J. (1976b). Teaching and learning. *The Center Magazine*, 9(6), 36-45.
Schwab, J.J. (1978). Education and the structure of the disciplines. In I. Westbury & N.J. Wilkof (Eds.), *Science, curriculum and liberal education* (pp. 229-272). Chicago, IL: University of Chicago Press.
Schwab, J. (1983). The practical 4: Something for curriculum professors to do. *Curriculum Inquiry, 13, 3*, in press.
Schwarzer, R. (1979). Sequentielle Prädiktion des Schulerfolgs. *Zeitschrift für Entwicklungspsychologie und Pädagogishce Psychologie*, 11, 170-180.
Shavelson R.J. (1973). What is the basic teaching skill? *Journal of Teacher Education, 24*, 144-151.
Shavelson, R.J. (1976). Teacher's decision-making. In: Gage, N.L. (Ed.), *the psychology of teaching methods*. Chicago, University of Chicago Press.
Shavelson, R. Cadwell, J. & Izu, T. (1977). Teachers' sensitivity to the reliability of information in making pedagogical decisions. *American Educational Research Journal*, 14, 83-97.
Shavelson, R.J. & Stern, P. (1981). Research on Teachers' Pedagogical Thoughts, Judgements, Decisions and Behaviour, *Review of Educational Research*, Winter 1981, Vol 51, No 4, pp 455-498.
Shenigold, K. (1981). *Issues Related to the Implementation of Computer Technology in Schools: A Cross-Sectional Study*. Bank Street College of Education. Technical Report No. 1.
Shipman, M. (Ed). (1976). *The Organization and Impact of Social Research*, R K P Ltd.
Shklar, J.N. (1984). *Ordinary vices*. Cambridge, MA: Belknap Press of Harvard University Press.
Shulman, L.S. (1981). Recent developments in the study of teaching. In: Tabachnick, B.R., Popkewitz, T.S., Szekely, B.B. (Eds.). *Studying teaching and learning, trends in Soviet and American Research*. New York: Praeger, 87-101.
Shulman, L.S. (1986). Those who understand: Knowledge growth in teaching. *Educational Researcher*, 15(2), 4-14.
Shulman, L.S. & Carey, N.B. (1984). Psychology and the limitations of individual rationality: Implications for the study of reasoning and civility. *Review of Educational Research*, 54, 501-524.
Shulman, L.S. & Elstein, A.S. (1975). Studies of problem solving, judgment, and decision making: implications for educational research. In F. N. Kerlinger (Ed.). Review of Research in Educations (vol. 3, pp. 3-42). Itaska, Ill.: Peacock.
Simon, A. & Boyer, E.G. (Eds.) (1970). *Mirrors for behavior.* Philadelphia, Research for Better Schools.
Simon, A. & Boyer, E.G. (Eds.). (1970). *Mirrors for behavior.* Philadelphia, Research for Better Schools.
Simons, H., Weinert, F.E. & Ahrens, H.J. (1975). Untersuchungen zur differential-psycholgischen Analyse von Rechenleistungen. *Zeitschrift für Entwicklungspsychologie und Pädagogishce Psychologie*, 7, 153-169.
Sizer, T.R. (1984). *Horace's compromise: The dilemma of the American high school*. Boston, MA: Houghton Mifflin.

Smith, E.L. & Anderson, C.W. (1984). *The planning and teaching intermediate science study: Final report* (Research Series No. 147). East Lansing: Michigan State University, Institute for Research on Teaching.
Smith, L.M. & Geoffrey, W. (1968). *The complexities of an urban classroom.* New York: Holt, Rinehart and Winston.
Smith, E.R. & Miller, F.D. (1978). Limits on perception of cognitive processes: A reply to Nisbett and Wilson. Psychological Review, 85, 355-362.
Sneed, J. (1971). *The logical structure of mathematical physics.* Dordrecht: Reidel.
Sneed, J. (1977). The structural approach to descriptive philosophy of science. In M. de Mey, R. Pinxten, M. Poriau & F. Vandamme (Eds.). *International workshop on the cognitive viewpoint* (pp. 359-366).
Stenhouse, L.A. (1975). *An Introduction to Curriculum Research and Development*, Heinemann Educational Books Ltd.
Stogdill, R.M. (1965). *Managers, employees, organizations: A study of 27 organizations.* Columbus: Ohio State University.
Stolurow, L.M. (1965). Model the master teacher or master the teaching model. In: Krumboltz, J.D. (Ed.), *Learning and the educational process.* Chicago, Rand McNally.
Strawson, P.F. (1974). *Freedom and resentment: And other essays.* New York: Metheun.
Svensson, L. (1976). *Study skill and learning.* Göteborg: Acta Universitatis Gothoburgensis.
Swift, D. (1983). *Images of Science in Secondary Schools. Interim Research report,* Mimeograph IED University of Surrey.
Sykes, G. (1983). Public policy and the problem of teacher quality: The need for screens and magnets. In L. Shulman & G. Sykes (Eds.), *Handbook on teaching and policy* (pp. 97-125). New York: Longman.
Tabachnick, B.R. & Zeichner, K.M. (1985). *The teacher perspectives project: Final Report.* Madison: Wisconsin Center for Education Research.
Taylor, R. (1970). *Good and evil: A new direction.* London: Macmillan.
Taylor, P.H. (1970). *How teachers plan their courses*, Slough, Bucks, National foundation for educational research in England and Wales.
Tennstadt, K.-Ch. (1982). Veranderbarkeit subjektiver Theorien. In H.-D. Dann, W. Humpert, F. Krause & K.-Ch. Tennstadt (Eds.), Analyse und Modifikation subjektiver Theorien von Lehrern. Forschungsbericht 43 (pp. 228-239). Konstanz: Universitat Konstanz, Zentrum I Bildunsforschung, SFB 23.
Thelen, H.A. (1973). Profession anyone? In D.J. McCarty (Ed.), *New perspectives on teacher education* (pp. 194-213). San Francisco, CA: Jossey-Bass.
Thiele, H. (1978). *Steuerung der verbalen Interaktion durch didaktische Intervention. Eine empirische Untersuchung zum Effekt von drei Methoden zum Lehrverhaltenstraining.* Dissertation Technische Universitat Braunschweig.
Thiele, H. (1981). Zur Beeinflussung des Entscheidungsverhaltens im Unterricht: Eine empirische Untersuchung zu einem theoriegeleiteten Lehrertraining. In: Hofer, M. (Hrsg.): *Informationsverarbeitung und Entscheidungsverhalten von Lehrern.* Muchen, 273-311.
Thiele, H. (1983). *Trainingsprogram Gesprachsfuhrung im Unterricht. Kognitives Lehrtraining zum Selbststudium.* Bad Heilbrunn.

Tillema, H.H. (1983). *Leerkrachten als ontwerpers*. (Teachers as instructional designers). Utrecht, Rijksuniversiteit.

Tomlinson, P. (1982). Personal Construct Psychology and Education Book review. *Journal of Further and Higher Education, 6 (3),* pp 95-99.

Treiber, B. (1980) . *Erklaerung von Foerderungeffekten* in *Schulklassen durch Merkmale subjektiver Unterrichtstheorien ihrer Lehrer*. (Diskussionspapier No. 22). Heidelberg: Psychologisches Institut Universitaet Heidelberg.

Treiber, B. (1980). Erklarung von Forderungseffekten in Schulklassen durch Merkmale subjektiver Unterrichtstheorien ihrer Lehrer. In W. Michaelis (Ed.), Bericht uber den 32. Kongreb der Deutschen Gesellschaft fur Psychologie in Zurich, Vol. 2 (pp. 631-634). Gottingen: Hogrefe.

Treiber, B. (1980). *Erklarung von Forderungseffecten von Schulklassen* urcr ale subjektiver Unterrichtsideen ihrer Lehrer. Report of the Psychologisches Institut der Universitat Heidelberg.

Treiber, B. & Groeben, N. (1980). Handlungsforschung in epistemologischer Rekonstruktion. In H. Moser & P. Zedler (Eds.), Theorie und Aktion. Leverkusen: Leske & Budrich.

Treiber, B. & Groeben, N. (1981). Handlungsforschung und epistemologisches Subjektmodell. *Zeitschrift fuer Sozialisationsforschung und Erziehungssoziologie, 1*, 117-138.

Trieber, B. & Groeben, N. (1981). Handlunsforschung und epistemologisches Subjektmodell. Zeitschrift fur Sozialisationsforschung und Erziehungssoziologie, 1, 117-138.

Turner, R.L. (1964). Teaching as problem-solving behavior: a strategy. In: Biddle, B.J. & Ellena, W.J. (Eds.), *Contemporary research on teacher effectiveness*. New York, Holt, Rinehart and Winston.

Tyler, R.W. (1950). *Basic Principles of curriculum and instruction*, Chicago University Press, Chicago.

Ulich, D., Hauber, K., Mayring, P., Alt, B., Strehmel, P. & Grunwald, H. (1981). Kognitive Kontrolle in Krisensituationen. In W. Michaelis (Ed.), Bericht uber den 32. Kongreb der Deutschen Gesellschaft fur Psychologie in Zurich, Vol. 2 (pp. 634-635). Gottingen: Hogrefe.

Ulich, D. & Hausser, K. et al. (1981). *Kognitive Kontrolle in Krisensituationen: Arbeitslosigkeit bei Lehrern.* Fortsetzungsantrag an die Deutsche Forschungsgemeinschaft, Munchen.

Unwin, D. & McAleese, R. (Eds.). (1978). *Encyclopaedia of educational media, communication and technology*. London, Macmillan.

Van Manen, M.J. (1975). An exploration of alternative research orientations in social education. *Theory and Research in Social Education, 3:1.* pp. 1-28.

Van Parreren, C.F. (1979). *Het handelingsmodel in de leerpsycholgie.* Rede ter opening van de lessen in het kader van de buitenlandse Francqui-leerstoel aan de Brije Universitcit tc Brussel, Brussel.

Van Parreren, C.F. (1981). Activiteit als object van de pscyologie: noch 'inwendig', noch 'uitwendig'. *Netherlands Tijdschrift voor de Psychologie, 36*, 185-196.

Veenman, S.A.M. (1982). Problemen van beginnende leraren: uitkomsten van een literatuur recherche. *Pedagogische Studiën. 59 (11),* 458-470.

Von Wright, G.H. (1972). On so-called Practical Inference, *Acta Sociologica, 15 (1).*

Wagner, A.C. (1973). Changing teaching behaviour: A comparison of microteaching and cognitive discrimination training. *Journal of Educational Psychology, 64*, 299-305.

Wagner, A.C. (1984). Conflicts in consciousness: imperiative cognitions can lead to knots in thinking. In: Halkes, R. & Olson, J.K. (Eds.). *Teacher thinking: a new perspective on persisting problems in education.* Lisse: Swets & Zeitlinger.

Wagner, A.C., Barz, M., Maier-Stormer, S., Uttendorfer-Marek, I. & Weidle, R. (1984). Bewusstseinskonflikte im Schulalltag. Denkknoten bei Lehrern und Schulern erkennen und losen Weinheim.

Wagner, A.C., Maier, S., Uttendorfer-Marek, I. & Weidle, R. (1980). Die Analyse von Knoten und Handlungsstrategien von Lehrern und Schulern. Unterrichtswissenschaft, 8, 382-392.

Wagner, A.C., Maier-Stormer, S., Uttendorfer-Marek, I., Weidle, R. (1981). *UnterrichtspsVchogramma. Was in den Kopfen von Schulern und Lehreri vorgeht.* Reinbek.

Wagner, A.C., Maier, S., Uttendorfer-Marek, I. & Weidle, R. (1981). Unterrichtsprogramme. Was in den Kopfen von Lehrern und Schulern vorgeht. Reinbek bei Hamburg: Rowohlt.

Wagner, A.C., Uttendorfer-Marek, I. & Weidle, R. (1977). Die Analyse von Unterrichtsstrategien mit der Methode des "Nachtraglichen Lauten Denkens" von Lehrern und Schulern zu ihrem unterrichtlichen Handel. In: *Unterrichtswissenschaft 5*, 244-250.

Wagner, A.C. & Weidle, R. (1982). Knots in Cognitive Processes - some Empirical Results and Implications for Cognitive Psychotherapy. In: *Proceedings of the_Ist European Conference on Psychotherapy* Research Vol. 11, Frankfurt: Lang.

Wahl, D. (1976). Naïve Verhaltenstheorie von Lehrern. Projektbericht Nr. 1 (unpublished paper). Weingarten: Padagogische Hochschule.

Wahl, D. (1979). Methodische Probleme bei der Erfassung handlungsleitender und handlungspsychologischer Theorien von Lehrern. Zeitschrift fur Entwicklungspsychologie und Padagogische Psychologie, 11, 208-217.

Wahl, D. (1981). Methoden zur Erfassung handlungssteuernder Kognitionen von Lehrern, in: M. Hofer, *Informationsverarbeitung und Entscheidungs verhalten_von Lehrern,* Manchen.

Wahl, D. (1981a). Subjektive psychologische Theorien: Moglichkeiten zur Rekonstruktion und Validierung, am Beispiel der handlungssteuernden Kognitionen von Lehrern. In W. Michaelis (Ed.), Bericht uber den 32. Kongreb der Deutschen Gesellschaft fur Psychologie, Vol. 2, 625-631. Gottingen: Hogrefe.

Wahl, D. (1981b). Methoden zur Erfassung handlungssteuernder Kognitionen von Lehrern. In M. Hofer (Ed.), Informationsverarbeitung und Entscheidungsverhalten von Lehrern (pp. 49-77). Munchen: Urban & Schwarzenberg.

Wahl, D. (1981c). Psychologisches Alltagswissen im Unterricht. In H.-J. Fietkau & D. Gorlicht (Eds.), Umwelt und Alltag in der Psychologie (pp. 67-90). Weinheim: Beltz.

Wahl, D. (1982). Handlungsvalidierung. In G.L. Huber & H. Mandl (Eds.), Verbale Daten: Erhebung und Auswertung (pp. 259-274). Weinheim: Beltz.

Wahl, D., Schlee, J., Lutz, M. & Reinhard, W. (1977). Naïve Verhaltenstheorie von Lehrern. Antrag an die Deutsche Forschungsgemeinschaft. Weingarten: Padagogische Hochschule.

Wahl, D., Schlee, J., Krauth, J. & Mureck, J. (1979a). Naïve Verhaltenstheorie von Lehrern. Zwischenbericht uber die Projektarbeit. Weingarten: Padagogische Hochschule.

Wahl, D., Schlee, J., Krauth, J. & Mureck, J. (1979b). Naïve Verhaltenstheorie von Lehrern. Fortsetzungsanstrag an die Deutsche Forschungsgemeinschaft. Weingarten: Padagogische Hochschule.

Wahl, D., Schlee, J., Bischofsberger, U., Krauth, J. & Mureck, J. (1981). Naïve Verhaltenstheorie von Lehrern. Zwischenbericht uber die Projektrabeit an die Deutsche Forschungsgemeinschaft. Weingarten: Padagogische Hochschule.

Wahl, D., Schlee, J., Krauth, J. & Mureck, J. (1983). Naïve Verhaltenstheorie von Lehrern. Abschulbbericht eines Forschungsvorhabens zur Rekonstruktion und Validierung subjektiver psychologischer Theorien. Oldenburg: Univesitate Oldenburg, Zentrum fur padagogische Berufspraxis.

Waller, W. (1961). *The sociology of teaching*. New York: Russell & Russell. (Original work published 1932).

Watzlawick, O., Beavin, J.J. & Jackson, D.D. (1969). *Menschliche Kommunikation -Formen, St6rungen,* Paradoxien. Bern.

Watzlawick, P., Wakland, J.H. & Fisch, R. (1974). *L6sungen. Zur Theorie und Praxis_menschlichen Handelns.* Bern.

Wehling, L. & Charters, W. (1969). Dimensions of teacher beliefs about the teaching process. *American Educational Research Journal*, 6, 7, 29.

Weick, K. (1976). Educational organizations as loosely coupled Systems. *Admin. Sci. Quart., 21*, 1-19.

Weidle, R. & Wagner, A.C. (1982). Die methode des Lauten Denkens, In: Huber, G.L. & Mandl, H., *Verbale Daten.* Weinheim: Beltz.

Weinert, F.E. (1977). Padagogisch-psychologische Beratung als Vermittlung zwischen subjektiven und wissenschaftlichen Verhaltenstheorien. In W. Arnold (Ed.), Texte zur Schulpsychologie und Bildungsberatung, Vol. 2 (pp. 7-34). Braunschweig: Westermann.

Weinert, F.E. (1978). Kommentar zum Beitrag von Hofer und Dobrick. In D. Gorlitz, W.U. Meyer & B. Weiner (Eds.), Bielefelder Symposium uber Attribution (pp. 65-69). Stuttgart: Klett.

Weinert, F.E. (1978a). Expose zum Fernstudienlehrgang "Lehrertraining: Probleme entwicklungs-, verhaltens- und lerngestorter Schuler". Papier in Auftrag des Deutschen Instituts fur Fernstudien. Tubingen.

Weinert, F.E. & Rotering-Steinberg, S. (1981). Schulerprobleme – Lehrerprobleme. Ein Lehrertraining fur schwierige Situationen in der Klasse. Bericht uber erste Erfahrungen mit einem Fernstudienprogramm. Zur Lehrerweiterbildung. Unterrichtswissenschaft, 9, 64-69.

Welford, A.T. (1968). *Fundamentals of skill.* London, Methuen.

Whiting, H.T.A. (1975). *Concepts in skill learning*. London. Lepus Books.

Wilkening, F. (1979). Combining of stimulus dimensions in childrens' and adults' judgements of area: An information integration analysis. *Developmental Psychology*, 15, 25-33.
Williams, L.E. (1976). *Perceptions of the problems of beginning teachers and the relationship these problems to selected variables*. University of Georgia.
Wilson, J. (1979). *Fantasy and commonsense in education*. New York: John Wiley.
Wilson, S. (1977). The use of ethnographic techniques in educational research. *Review of Educational Research*, *47*, 245-266.
Wilson, T.C. & Nisbett, R.F. (1978). The accuracy of verbal reports about the effects of stimuli on evaluations and behaviour. Social Psychology, 41, 118-131.
Wise, A.E. (1979). *Legislated learning: The bureaucratization of the American classroom*. Berkeley: University of California Press.
Wittrock, M.C. (1986). Students' thought processes. In M.C. Wittrock (Ed.), *Handbook of Research on Teaching*, Third Edition, (pp. 297-314). New York: Mcmillan.
Woodruff, A.D. (1967). Cognitive models of learning and instruction. In: Siegel, L. (Ed.): *Instruction – Some contemporary viewpoints*. Scranton, 55-98.
Wright, A.W. (1975). *The problems of beginning elementary teachers in Newfoundland schools*. University of Northern Colorado, 1975.
Yinger, R.J. (1977). *A study of teacher planning,* description and theory development using ethnographic and information processing methods, doctoral dissertation, Michigan State University.
Yinger, R.J. (1978). *A study of teacher planning: description and a model of preactive decision making*. East Lansing, Michigan State University, (Research Series no. 18).
Yinger, R.J. (1979). Routines in Teacher Planning, *Theory in Practice*, Vol XVIII, No. 3.
Yinger, R. J. (1980). A study of teacher planning. *The Elementary School Journal, 80*, 107-127.
Yinger, R.J. (1985, December). *Learning the language of practice*. Paper presented at the Symposium on Classroom Studies of Teachers' Personal Knowledge. Ontario Institute for Studies in Education, Toronto.
Yinger, R.J. & Clark, C.M. (1982). *Understanding teachers' judgements about instruction: The task, the method, and the meaning*, (Research Series No. 121). East Lansing, MI: Michigan State University, Institute for Research on Teaching.
Yinger, R. & Clark, C. (1983). *Self reports of teacher judgement*. Paper presented at the annual meeting of the AERA, Montreal.
Young, R.E.A. (1981). Study of Teacher Epistemologies. *Australian Journal of Education, 25 (2),* pp 194-208.
Zahorik, J.A. (1970). The effect of planning on teaching, *Elementary School Journal*, *(3)* 143-156.
Zeichner, K.M. & Teitelbaum, K. (1982). Personalized and inquiry-oriented teacher education: An analysis of two approaches to the development of curriculum for field-based experiences. *Journal of Education for Teaching*, 8(2), 95-117.

Zeichner, K. & Tabachnick, B.R. (1985). The development of teacher perspectives: Social strategies and institutional control in the socialization of beginning teachers. *Journal of Education for Teaching*, 11, 1-25.
Zeiher, H. (1973). *Gymnasiallehrer und Schulreformen*. Stuttgart.

Index